THE MEASURE AND MEANING OF TIME IN MESOAMERICA AND THE ANDES

Dumbarton Oaks Pre-Columbian Symposia and Colloquia

THE MEASURE AND MEANING OF TIME IN MESOAMERICA AND THE ANDES

ANTHONY F. AVENI
Editor

DUMBARTON OAKS RESEARCH LIBRARY AND COLLECTION
WASHINGTON, D.C.

Printed in the United States of America
by Sheridan Books, Inc.

LIBRARY OF CONGRESS CATALOGING-IN-PUBLICATION DATA

The measure and meaning of time in Mesoamerica and the Andes / Anthony F. Aveni, editor.
pages cm. — (Dumbarton Oaks Pre-Columbian Symposia and Colloquia)
Includes bibliographical references and index.
ISBN 978-0-88402-403-3

1. Time—Social aspects—Central America—History—Congresses.
2. Time—Social aspects—Andes Region—History—Congresses.
3. Indians of Central America—Social life and customs—Congresses.
4. Mayas—Social life and customs—Congresses.
5. Indians of South America—Andes Region—Social life and customs—Congresses.
6. Indian chronology—Congresses.
7. Maya chronology—Congresses.
8. Indian calendar—Central America—History—Congresses.
9. Indian calendar—Andes Region—History—Congresses.
10. Maya calendar—History—Congresses.

I. Aveni, Anthony F.

F1434.2.S63M43 2015
304.237—dc23
2014017475

General editor: Colin McEwan
Art director: Kathleen Sparkes
Design and composition: Melissa Tandysh
Managing editor: Sara Taylor

Volume based on papers presented at the Pre-Columbian Studies symposium "The Measure and Meaning of Time in the Americas," held at Dumbarton Oaks Research Library and Collection, Washington, D.C., on October 5–6, 2012.

www.doaks.org/publications

CONTENTS

PART II REGISTERING PRE-COLUMBIAN TIME

PREFACE

As Tony Aveni explains in his introduction, the chosen theme for this volume had its genesis in chance remarks shared in the course of a conversation some years earlier with Joanne Pillsbury, my predecessor as director of Pre-Columbian Studies. Joanne had already laid the groundwork for the symposium and preparations were well in hand when I arrived at Dumbarton Oaks in the fall of 2012. Truth be told, I couldn't have wished for a more exciting, challenging, and thought-provoking theme to begin my own tenure. In my opening remarks at the symposium, I recalled what a privilege it had been to participate many years before in the third conference on World Archaeo-astronomy that convened in 1990 at St. Andrews University, Scotland. This was a genuinely innovative meeting of minds riding the wave of interest in this relatively new "inter-discipline" at which, as I recall, Tony had informed and animated the gathered company at every turn with his infectious enthusiasm. Nearly twenty-five years later, it is clear that Tony's intellectual reach and restless curiosity, ranging across cultural astronomies past and present, combined with the fieldwork he has undertaken in both Mexico and Peru, equip him uniquely to lead the broader evaluation of timely themes in the Pre-Columbian Americas that are explored here.

Beginning in the 1970s with "The Cult of the Feline," "Death and the Afterlife in Pre-Columbian America,'" and "The Sea in the Pre-Columbian World," many Dumbarton Oaks symposia have consciously sought to foster dialogue between Mesoamerica and the Andes. "The Measure and Meaning of Time in the Ancient Americas" extends and deepens this practice of choosing a topic around which to organize thought-provoking comparisons between the two main regional foci of Pre-Columbian civilization. The essays included here venture far beyond the standard comparisons of the contrasting calendrical systems found in these two major areas of the ancient Americas to consider underlying strata of perception, cognition, and recording practices applied to the measure and meaning of time. To choose just one illustration of the potential such comparisons offer for generating new insights, I refer the reader to the concluding overview and reflections on the outcomes of the symposium in which Tony comments on the parallels between the storied Popol Vuh and Huarochirí manuscripts. Both derive from mythic traditions that guided and informed ritual practices in "deep time" and speak to the interrelated nature of space-time relations. Tony observes that these native ontologies subvert conventional "western" ways of experiencing and categorizing time and space. Ritual movements served to construct and inscribe collective memory in the landscape and in the later additions of the built environment, both of which were also deployed as aide-mémoire. In his 2010 publication *Circling the Square: How the Conquest Altered the Shape of Time in Mesoamerica*, Tony contrasts indigenous American and European approaches to the perception, manipulation, and representation of time in space. This calls to mind Tom Zuidema's earlier exploration of *Squaring the Circle* (1973) in the Andes. Beyond the notable differences, some intriguing similarities can be observed in how space-time relations were organized in the Andes and Mesoamerica, respectively, and these remind us of the rich avenues of comparative enquiry that still lie ahead.

Many individuals and much teamwork are required to organize and execute a Dumbarton Oaks symposium and the subsequent publication of a volume. In the first place, I wish to thank Jan Ziolkowski, the director of Dumbarton Oaks, for his encouragement and support of our work and activities in Pre-Columbian Studies. Mary Pye, the interim director of Pre-Columbian Studies, offered advice and insights that helped effect a seamless transition in the program at the time of my arrival; she generously helped to oversee the latter stages of preparations for the symposium. The input and suggestions of Elizabeth Boone, chair of the board of senior fellows at the time, as well as the senior fellows themselves (Barbara Arroyo, Tom Cummins, Charles Stanish, Gary Urton, and Leonardo López Luján) contributed directly and indirectly to the success of this enterprise. The superb organizational skills of Pre-Columbian Studies program coordinator Emily Gulick Jacobs were brought to bear. Emily and I were ably supported in the running of the symposium by Susannah Italiano, Jane Padelford, and the Dumbarton Oaks facilities, event, and security staff. Our Pre-Columbian Studies librarian, Bridget Gazzo, created a concurrent temporary on-site and online exhibition, *The Ancient Future: Mesoamerican and Andean Timekeeping*, that coincided with the symposium.

We counted on the talented publications department, led by director Kathleen Sparkes, as well as the skills and dedication of art and archaeology editor Sara Taylor. I thank them, as well as the copyeditor Sarah Soliz and the designer Melissa Tandysh for their skills and goodwill, which, as always, underpin the quality of the final outcome. Together we are indebted to the two reviewers who offered constructive and insightful critiques for strengthening the volume. With the knowledge, experience, and hard work of all the colleagues acknowledged above, and the vision and guidance of our symposiarch, Anthony Aveni, we are pleased to place before the reader the volume *The Measure and Meaning of Time in Mesoamerica and the Andes*.

CMcE 7.14.14

ACKNOWLEDGMENTS

On behalf of the participants, I gratefully acknowledge comments on the manuscript by Catherine Allen and Norman Hammond, and I commend the discussants at the conference for submitting written questions, from which the presenters benefited in drafting their final papers. I thank the fellows and three directors of study in the Pre-Columbian Studies program—Joanne Pillsbury, Mary Pye, and Colin McEwan—who have worked diligently with me to bring this work to completion. And I offer special thanks to the meticulous work-a-day efforts of Sara Taylor, Emily Gulick Jacobs, Katie Caruso, Kathy Sparkes, Sarah Soliz, and the rest of the Dumbarton Oaks staff and to Diane Janney, my assistant at home base.

Anthony F. Aveni
Russell Colgate Distinguished University Professor of Astronomy, Anthropology, and Native American Studies
Colgate University

1

Timely Themes

An Introduction to the Measure and Meaning of Time in Mesoamerica and the Andes

ANTHONY F. AVENI

> The whole world is just one big clock, but it is one which different people can read very differently because what we can see, out there in the objective world, is only so to speak, the hands of the clock, but not the clock-face in relation to which, and to which alone, the configuration of the hands assumes its particular temporal meaning.
>
> —Alfred Gell, *The Anthropology of Time*

THE CONFERENCE FROM WHICH THIS volume derives was conceived in an elevator in the Library of Congress at the Dumbarton Oaks symposium in the fall of 2006, when David Webster jokingly remarked to then-director of Pre-Columbian Studies Joanne Pillsbury: "Why don't we host the 'last Maya meeting' in 2012?" Turning to me, Joanne said, also jokingly, "Tony, you should be the organizer." Fast-forward six years, the end of the world a mere seventy-seven days away, and the symposium began.

That American popular culture is as obsessed with "time and the Maya" as the Maya themselves were—and still are—obsessed with timekeeping in general is perplexing enough. Victoria Bricker tells me that of all the Western technological inventions encountered by the present-day Maya, none seems to have been given more attention, at least by her Zinacantecan informants, than the wristwatch. At festivals held in the homes of cargo holders, she was constantly being asked what time it was so that the rituals would be timed correctly. I have argued that the preoccupation with what I call the "Y12 phenomenon" has had much to do with a revitalized utopianism born out of America's Puritan heritage come upon hard times. But a share of the hype should be allotted to the widespread dissemination of the seeds of falsely interpreted Maya knowledge via the Internet and at lay conferences and workshops.

Seizing the day, and with the assistance of the Dumbarton Oaks senior fellows, I tried to put together a dialogue that would transcend the issue of why the Maya mania of 2012 whipped contemporary culture into a frothy fatalistic frenzy. Together, we assembled a group of international scholars from diverse, cross-cutting disciplines and charged them with dealing with the subject of the measure and meaning of time across Mesoamerican and Andean cultures.

Few words in the dictionary have managed to acquire as many definitions as the word *time* (see Aveni 1989 for some of them). On the one hand, time is the idea of successive states of existence or the recognition of an order or sequence in our states of consciousness. To use a spatial metaphor, one can think of time as a continuously running line, straight or curved, on which recognizable events are pegged as a series of moments. On the other hand, time is a measure of duration, a metric quantity that is continuous, homogeneous, unchangeable, and never ending. In both these senses, at least in the Western view, time becomes a kind of reality that lies outside of human existence. Like other tangible things, it is already there to waste, spend, kill, lose, or get caught up in. Time is, as Newton put it, "of itself, and flows equably without relation to anything external" (Cajori 1947:6).

We sense time by taking the measure of its duration: once we used our pulses, then instruments, from sundials to atomic clocks. Watching natural phenomena for generations—such as the stars turning through seasons, the moon waxing and waning over months, and flower petals opening at dawn and closing at dusk—and applying the pre-Socratic and Aristotelian conviction that time is related to motion have led us to measure time through mechanical models. Western chronology began with the invention of models, or simulacra as the Greeks called them, that further concretized time via sand falling through an overturned hourglass, a pendulum bobbing back and forth, or an encased, vibrating crystal wrapped about one's wrist.

Historian David Landes (1983) regards the mechanical clock as the result of an interest in measuring time, not the origin of that interest. The one-handed version of medieval times led to a model with two hands, and in the age of expansion of trade and commerce that followed, the hours and their parts chimed, gonged, and clattered their way into our cycle of daily activities. Today, we have managed to divide time into epochs, eras, ages, centuries, years, months, weeks, days, hours, minutes, and seconds; atomic clocks cleave it further into milli-, micro-, nano-, pico-, and femtoseconds (if the distance to the moon equaled one second, a femtosecond could be represented by the width of a human hair).

High-speed modernity has caused us to rethink the meaning of time, to relativize and further subjectivize it. The discovery in quantum physics that time is discontinuous at the subatomic level was another revolutionary development that took place in the twentieth century. While all of this may sound like pure physics, such considerations have had a decided impact on our interpretations of causality and free will.

History, too, was regarded as a chain of events, a process whereby every happening contributed to the cause of future events. Like the scientists, historians of the nineteenth century seemed confident in their ability to give definite, objective answers to the questions they entertained. One supposedly studied archival documents without preconceptions then wrote a story about them—a story, as historian Hayden White (1973:141) puts it, "attested to by the documents in such a way as to make the story itself the explanation of 'what happened' in the past."

Unlike indigenous American understandings of time, our Western way of temporal thinking has had the effect of separating human history from events in the history of nature, which appear to be strung out on a different, far lengthier processional time line. By creating a history of nature, scientific empiricism effectively robbed theology of any hold it once had on the notion of one-way development through linear time. At the start of the twentieth century, all the sciences, with the exception of physics, which still primarily concerned itself with the discovery of immutable laws, dealt with temporal processes that could be studied, in the already well-established empirical tradition, by the observation and evaluation of physical evidence—what science historians Stephen Toulmin and June Goodfield (1965) aptly call the "testimony of things" as opposed to human history, or the "testimony of people." Thus, the fossil record along with observations of the variation of extant biological species led to the notion of life as an extraordinarily lengthy developmental chain of being.

In the nineteenth century, geological evidence from studies on stratigraphy and sedimentation pushed the traditional linear time barrier back even further, and in the century to follow, the cosmological discovery of the expansion of the universe, revealed through the red shifts measured in the spectra of distant galaxies, wrenched away any remaining theological grip on the word *time*. Like the new interstate highway running alongside the old state road, connected by neither exit nor entry ramp, the development of nature, like that of humanity, came to be viewed as a one-way process. The discovery of the New World, the assassination of Lincoln, the fall of the Berlin Wall—all blips on one time line—could never be related causally to the eruption of Mount Vesuvius, the appearance of Halley's Comet, or the blizzard of 1888, which are event markers on time's alternate route.

By the beginning of the twentieth century, the apprehension of time in human history lay on no more solid ground than the positivist's comprehension of time in natural history. Nor were the two thought to be connected. Historians came to realize the difference between a chronology, or the simple arrangement of facts on a timeline, and a narrative, or the story one writes about them. Like the physicist who perturbs an electron while attempting to observe and describe its position by illuminating it with a single photon of light, each storyteller affects the meaning of a set of facts by "emplotting" the events from his or her own "viewpoint." It is in the plot, writes philosopher Paul Ricoeur (1984:xi), that we "reconfigure our confused, informed, and at the limit mute temporal experience."

Rooted in the late nineteenth- and early twentieth-century German and French schools of philosophy, and central to modern philosophical debates concerning temporal sensation and consciousness, is the issue of whether there can be a transcendental phenomenology of time—that is, can one have a direct intuitive access to time's essential structure and meaning? Rejecting scientific objectivism, Edmund Husserl (1965) argued that the essence of time can be discovered by examining its structure and that God is both the end and the motivating force of consciousness and reason. Martin Heidegger (1985) and Maurice Merleau-Ponty (1962) are among those philosophers who sought to subjectivize time. Heidegger, in particular, placed heavy emphasis on the past by viewing all temporal experience as part of the dialectical process of coming into being. Rather than helplessly sliding along a teleological pathway toward death, one participates in the management of time by taking action along with others who share a communal heritage. Accordingly, one does not merely exist as a moment in the world but brings oneself into being in and through the traditions one chooses to carry on (cf. Eberl, this volume for further discussion).

At this point readers may be wondering why I have chosen to structure my introduction to the work of a volume on the measure and meaning of time in Mesoamerica and the Andes around Western perspectives related to understanding and expressing time's meanings. I have done so to remind us that though we may seek to apprehend time in the Other, we remain inescapably imprisoned in our own culture's way of knowing it. Edward Evans-Pritchard's reaction to the Nuer conception of time is appropriate:

> Though I have spoken of time and units of time, the *Nuér* have no expression equivalent to "time" in our language, and they do not speak of time as though it were something real, which passes, can be wasted, can be saved; and, I do not think they ever experience the same feeling of fighting against time or having to coordinate activities with an abstract passage of time. . . . The *Nuér* are fortunate. (Evans-Pritchard 1969:3)

But some Western temporal concepts do resonate with our own Americanist studies, such as stories about time, the idea of time as a dialectical process, schemes for reckoning time, the old linear versus cyclic time debate, and the relationship between human and natural temporality. As readers will learn, these concepts are very much fused in the cultures we study.

Alfred Gell's intriguing clock commentary that I chose as the epigraph to my introduction comes from his commentary on Maurice Bloch's proposed distinction between two kinds of time: cognitively universal time and cyclic ritual time. Based on his ethnographic studies of tribal societies in New Guinea and India, Gell concluded that while ideas of time do arise when nature is the subject of human activity, the contexts in which nature becomes that subject are not uniform enough to give rise to universals when it comes to cognizing time (Bloch 1989:85). What time we feel or think it is, argues Gell, has nothing to do with a bio-psychologically functioning internal clock, but rather is determined by a feedback process imposed on the mind by particular human tasks or activities. How we cognize time, in other words, is related not to how we perceive it, that is, by reading a clock (which need not be a mechanized device), but rather by "beliefs we hold about the world" (Gell 1992:95).

The subjectification of time implied in Gell's analysis resonates with many expressions and understandings of temporality in Mesoamerican and Andean cultures that appear in the following essays. For example, William Barnes shows us how what we could call eternal mythic and ever-changing historic times were conflated by Mexican artists who brought the ancient gods to life in the personage of the current ruler. Likewise, Jalh Dulanto demonstrates how Andean mythic cosmogonic principles became revivified as specific historic events in the early colonial Huarochirí Manuscript. And Linda Brown writes about how Tz'utujil Maya ritual specialists reanimate object-beings from a previous era, as she explores the periodic reawakening of sacred objects during the liminal period between day and night.

Liminality, or the unclearly structured state of temporal transition (e.g., the time between day and night or between the seasons, the Maya Uayeb, as well as the five unlucky days at the end of the year and our twelve days of Christmas), emerged as a prominent theme in the symposium papers. Among them is that of Alfredo López Austin. What else might one expect from a scholar who wrote an entire book on the Mesoamerican mythology of the opossum, a decidedly liminal creature (López Austin 1990)? He deals with what happens in the liminal zone between the space-time of creatures (the ecumene) and that of the supernatural (the anecumene) in and beyond Nahuatl thought.

Much (some say too much) has been written on the age-old argument about linear versus cyclic time. Remember the old saw that linear time exists only in the Western literate tradition, while all other cultures perceive time as cyclical? This idea is based largely on historian of religion Mircea Eliade's (1954, 1959) idea of a "paradise of archetypes" governed by time structures based on the renewal of the regenerative powers inherent in all biorhythms. Time's periodic regeneration presupposes that all personal and cosmic re-creation repeats the first act of genesis or becoming. By reactualizing creation via ritual, the believer relives over and over the eternal oscillation between chaos and order. Past events never become a part of history; consequently, they do not create the burden of time characteristic of linear temporal development. Even though historians (e.g., Farriss 1987) and anthropologists (e.g., Bricker 1981; Goody 1977; Leach 1966) have challenged this dichotomy between preliterate cyclic and literate linear views of time, the idea that "traditional" cultures think about time as if it were frozen in an unalterable pattern of events persists and continues to permeate a good deal of the popular literature (cf. Hallpike 1979; Ricoeur 1975).

The work of this volume explores other culture-specific ways of understanding how the present structures the past and vice versa. One can argue, along the lines of the metastructure to universal history disclosed by modern chaos theory, that if a long enough period of Maya history is observed, one will discover that a pattern exists. One need only devise "a cycle so large that it could in theory encompass any conceivable linear sequence, to account for any contingency" (Farriss 1987:574). In this vein, Mayanist Markus Eberl employs the epigraphic record to introduce us to the linear temporal concept of *tz'ak* (to put in order). Further, Eberl demonstrates the role it plays in preserving the Maya concept of spacetime (one word) by replicating the world of creation in the Maya city. On the other hand, Harvey and Victoria Bricker emphasize the cyclicity of Maya time, especially in the codices. Their essay delves into the fascinating world of number and its role in commensurating, or fitting together, historical, calendrical, and astronomical periodicities.

No treatment of time in the cultures of Mesoamerica and the Andes would be complete without some discussion of the *ceque* system of Cuzco, an ingenious radial, quadripartite mnemonic related to social hierarchy, politics, and religious worship that overlay the Inca capital. My own contribution relies in part on the interpretations of Tom Zuidema, who has devoted an enormous portion of his scholarly life to studying the system. Zuidema's (2010) overview of the Inca calendar and ceque system has recently appeared in print. My essay centers around what sense can be made of the system when it is reflected against what we know about world calendrical models and how his model has been adapted to other situations.

Also related to the Andes, Stella Nair's essay on the landscape of memory explores a different kind of temporal text, one used by the Inca lords who brought time and space together to boldly and visibly inscribe their place in history by altering, carving, shaping, and naming places in the environment they claimed as their own. Reading Nair's essay on architecture as a medium for sensing time, one can understand why chronicler Bernabé Cobo was so meticulous about pinpointing the location of the huacas that make up Cuzco's ceque system.

Being interested in comparative studies, I was delighted to hear John Monaghan's presentation on the study of calendars that build on cycles in which he refers to the Balinese system. I was even more pleased to learn that he would deal with the confrontation between the two worlds of calendar making that clashed during the Hispanic contact period. This is a subject I have engaged as well. Not wishing to take up space in the proceedings with an essay of my own, I shall seize the opportunity for a short discourse that Monaghan's essay invites by giving a brief synthesis of my thoughts about the subject of syncretic processes that change the way one thinks about and manages time (cf. Aveni 2012 for a full exposition).

I believe the embeddedness of the convention of the circle as a way of organizing and conveying knowledge of the world at the time of Hispanic contact cannot be overstated. The persistent imagery of wheels that grind out time and the perfectly round shape of the Platonic and Aristotelian universe remind us of the gears on both the inside and the outside of the mechanical clock. Sonorous tones emanating from the clock's round countenance parsed out both sacred and secular time from every centrally located church or town hall in sixteenth-century Europe. Motion in the cosmos found expression in the forms of circular shells and orbs in Copernicus's heliocentric model of the universe, introduced in the middle of that century (1543). I argue that such representations of time and space descended from the Greek ideal of the circle. So it seems logical that the native intellectuals of the Americas, who served as informants and assistants to the early chroniclers, would have found it necessary to adapt to and make use of the circular form that dominated contemporary Hispanic ways of knowing and expressing the universe of space-time. This is what we find in the calendar wheels of Diego de Landa, Bernardino de Sahagún, Diego Durán, and Toribio Motolinia. But the circular format was not entirely unknown in Mesoamerica, as I have been careful to point out.

Imagine one of these scribes confronting alien ways of knowing time and trying to make sense of two vastly different systems of time reckoning, one inherited, the other forcibly thrust upon him. Imagine, too, how puzzled that scribe must have been when forced by an inquisitor priest to answer questions relating to the meaning of symbols and numbers in the old codices and not being able to offer up the anticipated answers.

Little wonder that following the burning of more than 99 percent of the Maya codices shortly after contact, native calendar keepers felt the need to create clandestine documents, such as the *Books of Chilam Balam* and the *Annals of the Cakchiquel*, documents that, like the Aymara coat of arms discussed by Tristan Platt, constitute an amalgam of conflicting temporal views seeking resolution. How else could these people preserve their sacred heritage?

Faced belatedly with the need to understand the mathematical correlations among the time periods conceived in the native mentality, both for the sake of the conversion process and, for some, out of general interest in the problem, the Spanish chroniclers relied very heavily on seeking likenesses between the native and European systems via the familiar circular forms of expression. Such a course of action further contributed to their failure to grasp many of the nuances of Mesoamerican timekeeping.

When I examined Postclassic-period calendrical documents, I found that both Maya and Central Mexican interpretations of time, especially the *tzolk'in* (*tonalpohualli*), were expressed diagrammatically in square or quadripartite form. This concept was likely derived from the well-documented native perception of a quadripartite universe, with directions conceived not as cardinal points of the compass, but as borders or sides of space. While using circles to express the passage of Maya *k'atuns* (the so-called *k'atun* round) seems to have worked quite well in colonial times, we have little evidence before contact that these time units were conceived in circular form (with a few exceptions). To judge from their comments, one suspects that the chroniclers initially did not bother to study indigenous time-related documents very carefully and only later, after they had destroyed most of them, began to take their content seriously. Then they resorted to a severely biased interview process to acquire information about native time and timekeeping.

Details of the forced overlay of the European circular paradigm on indigenous quadripartite temporality are clearly reflected in several of the examples I studied; for example, the rounded spiral form of Motolinia's calendar wheel likely took its cues from native rectilinear diagrams, such as page twenty-five of the Borgia Codex, that were intended to be read in spiral fashion. Chroniclers made several attempts to express the 365-day festival calendar, or *xihuitl*, in the form of a circle depicting eighteen months of twenty days, plus an additional five days, thus likening the xihuitl to the liturgical seasonal round in which festivals (such as All Saint's Day) were pinned down more precisely to specific dates of the year. So, too, was the *trecena*, or thirteen count, forced to resemble the twelve hours on the face of the ubiquitous European clock or the format of the Western zodiac. Other European concepts that encroached on native forms of expressing space and time included the Greek elements, the wind compass, the synodic month, the olympiad, the jubilee (a close approximation to the indigenous fifty-two-year cycle), and, not surprisingly given the concurrent (late sixteenth-century) European calendar reform, leap years.

The clash of contact between two worlds of time and space further emerged in our symposium papers. Tristan Platt addresses it in relation to the Andes in his essay, which deals with pre-Hispanic symbols of power and beauty that were recast and transformed in a colonial coat of arms in Spanish heraldic format. Platt argues that this artifact served as an instrument to prolong Aymara governance during a delicate period of negotiation with Spain over their relationship with the Inca.

I had often wondered to what extent Felipe Guaman Poma's drawings of the four world ages in *El primer corónica y buen gobierno* (Murra and Adorno 1980) were based on Western intrusion or were copied from Mexican artifacts like the Aztec Sun Stone. Juan Ossio addresses the possibility that the idea of world ages one reads about in Western deep-time history was indeed part of an indigenous Andean messianic tradition.

And regarding world ages, when I first read historian Richard Landes's (e.g., Landes et al. 2003) works on the Western fascination with timing apocalyptic moments since Y1K (he tells us that over one hundred thousand such predictions have been proffered in Western history), I believed he would be the ideal scholar to keynote our symposium. His presentation offers a gateway between America's contemporary fascination with millennialism and the manifold aspects of temporality that occupied indigenous people across the Americas.

For want of time, I cannot mention many other relevant themes, questions, and problems. Believing that the perspectives raised in the discussion period following the oral presentations might aid presenters in revising their papers for publication, we revived an old Dumbarton Oaks custom of recording the discussions and submitting that record to each presenter. I believe this procedure enables readers to engage with a more well-integrated document than a mere collection of papers.

Following Landes's paper on time in the Western perspective, the presentations are offered in two broad overlapping topical sections. Part I is entitled "Sensing Pre-Columbian Time." It consists of presentations by López Austin, Brown, Eberl, Aveni, Nair, and Dulanto, and it brings together works that address the diverse ways in which cultures of Mesoamerica and the Andes experience time. Part II, "Registering Pre-Columbian Time," addresses ways of recording and remembering time. It is made up of works by Bricker and Bricker, Barnes, Ossio, Platt, and Monaghan. My personal overview concludes the volume.

The study of other cultures, a famous anthropologist once remarked, gives us a chance to expose what we think we know to a buffeting by other ideas and customs best able to call our own knowledge into question. This volume articulates the diverse participatory, human action-oriented ways in which Mesoamerican and Andean people live and lived time. Our symposium offers an appropriate antidote to Nobel laureate in physics Steven Weinberg's (1988:154) famous confession about the Big Bang and cosmic time: "The more the universe seems comprehensible, the more it also seems pointless." The people whose stories of time are told by the volume's contributors almost exclusively assume responsibility for continuing the world, believing that only through human action will things stay on course. Such action is not about personal salvation; it is collective, and each individual bears a responsibility to contribute to the effort.

REFERENCES CITED

Aveni, Anthony F.

1989 *Empires of Time: Calendars, Clocks, and Cultures.* Basic, New York.

2012 *Circling the Square: How the Conquest Altered the Shape of Time in Mesoamerica.* Transactions of the American Philosophical Society, vol. 102, pt. 5. American Philosophical Society, Philadelphia.

Bloch, Maurice

1989 *Ritual History and Power.* Athlone Press, London.

Bricker, Victoria

1981 *The Indian Christ, the Indian King: The Historical Substrate of Maya Myth and Ritual.* University of Texas Press, Austin.

Cajori, Florian (translator)

1947 *Sir Isaac Newton's Mathematical Principles of Natural Philosophy and His System of the World.* University of California Press, Berkeley and Los Angeles.

Eliade, Mircea

1954 *Cosmos and History: The Myth of the Eternal Return.* Translated by Willard R. Trask. Princeton University Press, Princeton, N.J.

1959 *The Sacred and the Profane: The Nature of Religion.* Translated by Willard R. Trask. Harcourt Brace, New York.

Evans-Pritchard, Edward

1969 *The Nuer.* Oxford University Press, New York.

Farriss, Nancy

1987 Remembering the Future, Anticipating the Past: History, Time, and Cosmology among the Maya of Yucatan. *Comparative Studies in Society and History* 29(3):566–593.

Gell, Alfred

1992 *The Anthropology of Time: Cultural Constructions of Temporal Maps and Images*. Oxford University Press, Oxford.

Goody, Jack

1977 *The Domestication of the Savage Mind*. Cambridge University Press, Cambridge.

Hallpike, Charles

1979 *The Foundations of Primitive Thought*. Clarendon Press, Oxford.

Heidegger, Martin

1985 *History of the Concept of Time*. Indiana University Press, Bloomington.

Husserl, Edmund

1965 *The Phenomenology of Internal Time Consciousness*. Translated by James S. Churchill. Edited by Martin Heidegger. Harper, New York.

Landes, David S.

1983 *Revolution in Time: Clocks and the Making of the Modern World*. Harvard University Press, Cambridge, Mass.

Landes, Richard, Andrew Gow, and David Van Meter

2003 *The Apocalyptic Year 1000: Religious Expectations and Social Change, 950–1050*. Oxford University Press, Oxford.

Leach, Edmund

1966 Sermons by a Man on a Ladder. *New York Review* 20(October):28–31.

López Austin, Alfredo

1990 *The Myths of the Opossum: Pathways of Mesoamerican Mythology*. University of New Mexico Press, Albuquerque.

Merleau-Ponty, Maurice

1962 *Phenomenology of Perception*. Translated by Colin Smith. Routledge and Kegan Paul, London.

Murra, John, and Rolena Adorno

1980 *El primer nueva corónica y buen gobierno por Felipe Guamán Poma de Ayala*. Siglo Veintiuno, Madrid.

Ricoeur, Paul

1975 *Les cultures et le temps: Études préparées pour l'Unesco*. Payot, Paris.

1984 *Time and Narrative*, vol. 1. Translated by Kathleen McLaughlin and David Pellauer. University of Chicago Press, Chicago.

Toulmin, Stephen, and June Goodfield

1965 *The Discovery of Time*. University of Chicago Press, Chicago.

Weinberg, Steven

1988 *The First Three Minutes*. Basic, New York.

White, Hayden

1973 *Metahistory: The Historical Imagination in Nineteenth-Century Europe*. Johns Hopkins University Press, Baltimore, Md.

Zuidema, R. Tom

2010 *El calendario inca, tiempo y espacio en la organización ritual del Cuzco: La idea del pasado*. Fondo Editorial del Congreso del Perú, Fondo Editorial de la Pontificia Universidad Católica del Perú, Lima.

2

From Counting Down to Counting Out

On the Relationship between Apocalyptic and Normal Time in the Western Passion for Precise Time Measurement

RICHARD LANDES

Arguably, no other culture has had the Promethean passion for precise time measurement and its use in organizing purposeful human activity that we find in the West, beginning in the later medieval and continuing to grow in the modern and postmodern periods. While some cultures, like the Chinese and Maya, have developed a remarkable attention to time measurement (Aveni 1980:ch. 4; Bricker and Bricker, this volume; Landes 1983:15–32), none have pursued the matter as far as the West. In the fourteenth century, mechanical clocks became one of the West's greatest technological innovations, one that rapidly diffused, first from town to town (clock towers), then to households (standing clocks), and finally, in the early modern period, to individuals (watches).[1] Nor were these timepieces mere ornaments: they were cultural drivers of both individual and collective behavior. Indeed, one could argue that this cultural passion is one of the few to survive not only undiminished by the advent of modernity but rather intensified by that "Great Transformation"—a driver of modernity.

David Landes, in his seminal work *Revolution in Time* (1983), argues that this uniquely Western drive for precision in timekeeping came from, rather than created, the fascination with time (Dohrn-van Rossum 1992). Already in the tenth century, he argues, one finds in monasteries a need for precise timekeeping instruments to mark prayer times, instruments that, unlike in both Islam and Judaism, which were also known for such concerns, needed to work at night and in cold climes, when one could rely neither on sundials nor on water clocks. The French version of "Frère Jacques" recalls this need to awaken he who will warn the rest—"Are you sleeping? Sound the bells for [predawn] Matins"—which is lost in the English translation that turns the imperative into an intransitive verb as "Morning bells are ringing."

This essay attempts to record the "prehistory" of this fascination with chronological precision by tracing it to an earlier (and continuing) effort to date the apocalyptic moment, the time when, for Christians at least, Jesus would return in power and glory and judge the quick and the dead. In particular, it will focus on a chronological tradition that dated the Parousia to the year 6000 after creation, a tradition of exceptional vigor in (Latin) Christianity from the second to the ninth Christian centuries. The year 1000 (and at the latest, the year 1033) was the last and most powerful of these "millennial" dates, which, in addition to the failure of their carefully calculated apocalyptic expectations, produced a remarkably vigorous tradition of both chronological and calendrical studies, what the historian of Western historiography Bernard Guenée (1980:152–153) called "la fièvre computistique" of the tenth through twelfth centuries.

This "fièvre computistique" that Guenée identified had a predecessor in the eighth and ninth centuries, which we will examine more carefully. For the moment, let me just mention that at both these periods of computistical overheating, a distinct relationship existed between time measurement and another fevered activity—liturgy. This close correlation offers the link between my computists and those time-disciplined monks who drove the need for mechanical clocks.

Before beginning, however, let me provide a few terminological remarks for those who wish to follow my discussion (Landes 2011:3–61):

Apocalyptic: Gk. *apocalypsis*, "revelation"
sense of imminence (changes behavior radically)
scenario of expectations about apocalyptic transformation.

Eschatological: Gk. *eschaton*, "last things"
the end of the saeculum
Last Judgment
the good go to heaven, the evil go to hell
divinely driven (passive), least dangerous.

Millennial (Chiliastic): L. *mille* + *anni*, Gk. *chiliastos*
perfect society *on* this *earth* (i.e., in the flesh, in history)
necessarily political, strong active component of transforming the world, most dangerous.

Roosters: apocalyptic prophets who crow to the believers, "Awaken! The Day of the Lord dawns!"

Owls: antiapocalyptic opponents who insist that believers should "go back to sleep. It's the middle of the night, the foxes are out, and the Master sleeps."[2]

Annus Mundi (AM): a chronology based on the age of the world, a feature of both Jewish and Christian historiography. The first Christian AM (I) placed the incarnation (AD) in the year 5500 from creation of the world (and therefore the year 6000 in 500 CE), the second (AM II) in 5199 (6000 in 801), and the third (AM III) in 3952 (6000 in 2048).

Apocalyptic Time (Moment): the normally suppressed and rejected claims of the roosters breach the public sphere and carry the day, inducing (brief) moments of widespread expectations that sweep aside the owls and produce public acts inspired by a sense of the imminence of the End (Figure 2.1).

Saeculum: a Latin term that was used by Christians, especially Augustine, for the time-space continuum and that refers both to this world and to time. The French use of the term (*siècle*) refers both to the profane world and to a hundred-year period (century).

Most historians who describe the apocalyptic passions that mark the origins of Christianity generally assume that they passed away, to be replaced with more sober, ecclesiastical thought. An alternative approach argues that apocalyptic episodes repeat frequently in Christianity for centuries (even millennia) and that only the impression of the documentation, which systematically favors the owls writing in what one might call the "retrospective

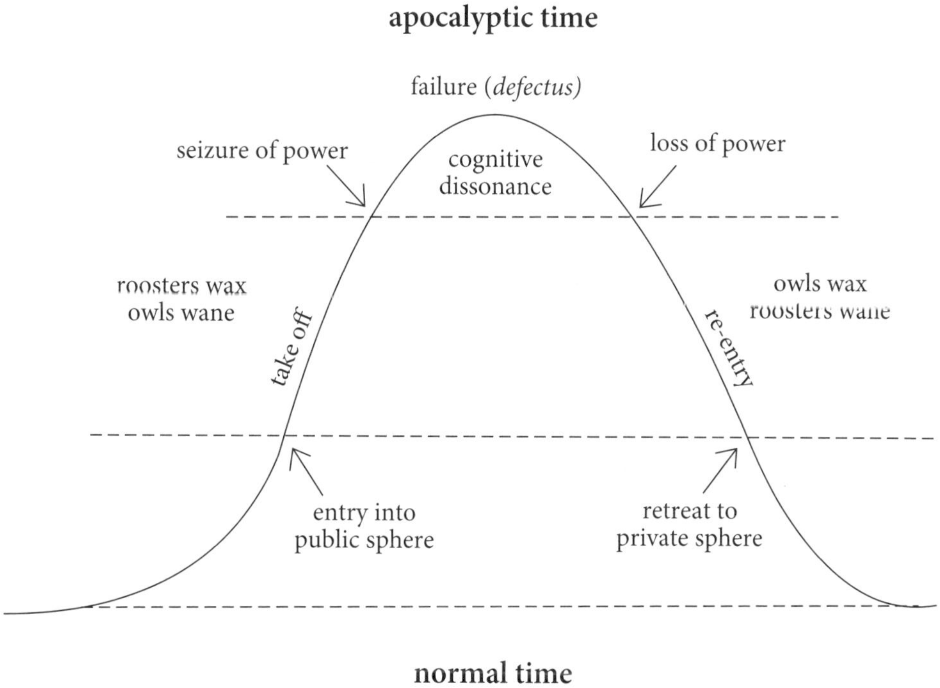

figure 2.1
Apocalyptic curve. (Illustration by Richard Landes.)

perfect," permits one to dismiss as minor the role of roosters in every Christian generation (Landes 2011:62–88).

The tale as I understand it begins with the need for a countdown. As anyone who has tried to meditate for a long and set period of time knows, when your knees start to hurt, counting breaths can help immeasurably in surviving to the end of the hour. Of course, that is not what you are supposed to do, and it does not necessarily enhance the experience, but sometimes it seems like the only way to make it through. Now, within this framework, consider one of the better definitions of the relativity of time: "How long two minutes is, depends on which side of the bathroom door you're sitting on."

Now switch to the cosmic issue of God's public and last judgment on the quick and the dead. From the perspective of the oppressed, the Day of the Lord is a day of vindication, of pleasure. An early eleventh-century text from England notes: "Then will end the tyranny of kings and the injustice and rapine of reeves and their cunning and unjust judgments and wiles. Then shall those who rejoiced and were glad in this life groan and lament. Then shall their mead, wine, and beer be turned into thirst for them" (Byrhtferth 1972:242, lines 3–9). In a nineteenth-century account, a slave woman who had just seen her daughter unjustly lashed cried out: "There's a day a-coming! I hear the rumbling of the chariots, I see . . . white folks' blood a-running on the ground like a ribah and the dead heaped up that high! O Lord! Hasten the day when the blows and the bruises and the aches and the pains shall come to the white folks. . . . O Lord! Give me the pleasure of living to that day" (Scott 1992:5).

Within this framework, we should not be surprised that despite extensive strictures against it, Judaism, Christianity, and Islam all had their traditions around dating the advent of the end of the world, the apocalyptic scenario. Muslims had a term for those so engaged: "the exact persons" (*waqqatun*; Cook 2002:195nn16–17). No sooner did Maimonides denounce in the most urgent terms a false Messiah in Yemen than he revealed that his family had its own (owl's) tradition about dating the Messiah's advent (Maimonides 1952:xv). Chinese secret societies had their calendars and chronologies, and the many rebellions that punctuate Chinese history for the last two and a half millennia often arose when a countdown hit its term (Naquin 1976).

Now all this history may seem a bit distant from the nitty-gritty of time measurement, whether of the cyclical or the linear kind, but it actually permits us major insights into the Western passion for the subject. For indeed, Christians surpassed members of all the other apocalyptic traditions—monotheistic and, a fortiori, polytheistic—in their passion for dating the apocalypse. These apocalyptic dynamics enable us to understand why, of all religions and civilizations, Christianity alone adopted three major chronologies in succession and made these overhauls in the course of half a millennium.

And the key item in this fundamental cultural/intellectual difference between Christianity and all other traditions comes from the teaching known as the "sabbatical millennium" (Landes 1988). Exegetes (Barnabas, Irenaeus) combined Genesis and Psalms 89–90 to promise six thousand years of travail followed by a sabbatical millennium of peace and plenty. In 200 CE, Hippolytus, followed by Julius Africanus, dated the incarnation to 5500 (AM I). Without realizing it, these early chronographers set up a dance between roosters and owls that would play out over the next nine centuries and that included three different dating systems and three separate disappointments. Figure 2.2 illustrates the pattern graphically.

Generally, all apocalyptic discourse occurs along a continuum between roosters, who announce the imminent dawn of "The Day," and owls, who insist that the night is yet long. Owls have always been proven right, but roosters return periodically to dominate public discourse for brief if powerful moments, often in response to signs, wonders, prodigies, disasters, and eschatological calculations. But owls, by right of retrospective rewriting and archiving, dominate the written record. Owls are at the origins of the sabbatical millennium: they introduced and reinforced a chronology with several centuries still left before the target date, in effect saying, "Yes, a coming millennium, but not now" (i.e., nonapocalyptic chiliasm). Hippolytus established the first successful chronology around 200 CE (5700 AM I); Jerome/Augustine replaced it with the second around 400 CE (5600 AM II); and Bede replaced theirs around 700 CE (4652 AM III). Because it favors owls, the sabbatical millennium is the only form of millennial discourse with a relatively continuous written trace.

Any formal, date-based eschatological promises (chiliastic or not) are unusual: they are theologically illegitimate ("no one can know the time"), repeatedly denounced by serious theologians and preachers, and have always proven wrong. Moreover, given millennialism's volatile and profoundly political nature, a promise of "heaven on earth" at the end of the countdown rather than an eschatological scenario doubles the danger. Such "nonapocalyptic chiliasm" represents a kind of balloon mortgage in which clerics commit themselves in writing to repayment in full at a specific but distant time. One can imagine offering such calculations unofficially, orally, on an ad hoc basis, but to commit them explicitly to writing, and to have that written commitment taken up by a large number of clerics over many centuries, represents an extraordinary development unparalleled either in Judaism or in Islam.[3]

Roosters and their audiences do not favor such long-range calculations in times of waxing apocalyptic expectation, and their initial response to attempts to establish a chronology with a centuries-long waiting period was negative.[4] Roosters always favor something more immediate. With the (inevitable) disappointment, however, the chronological argument of the owls gained credence. As, over the generations and centuries, all other prophecies failed, the chronology became the single most convincing explanation for why Christ had yet to return. Thus, ironically, a chronology that may have had popularity only among an elite who preached it with limited success in its early centuries would, with the approach of its end date, exercise an ever-stronger attraction on roosters and their audiences. And just as roosters and their

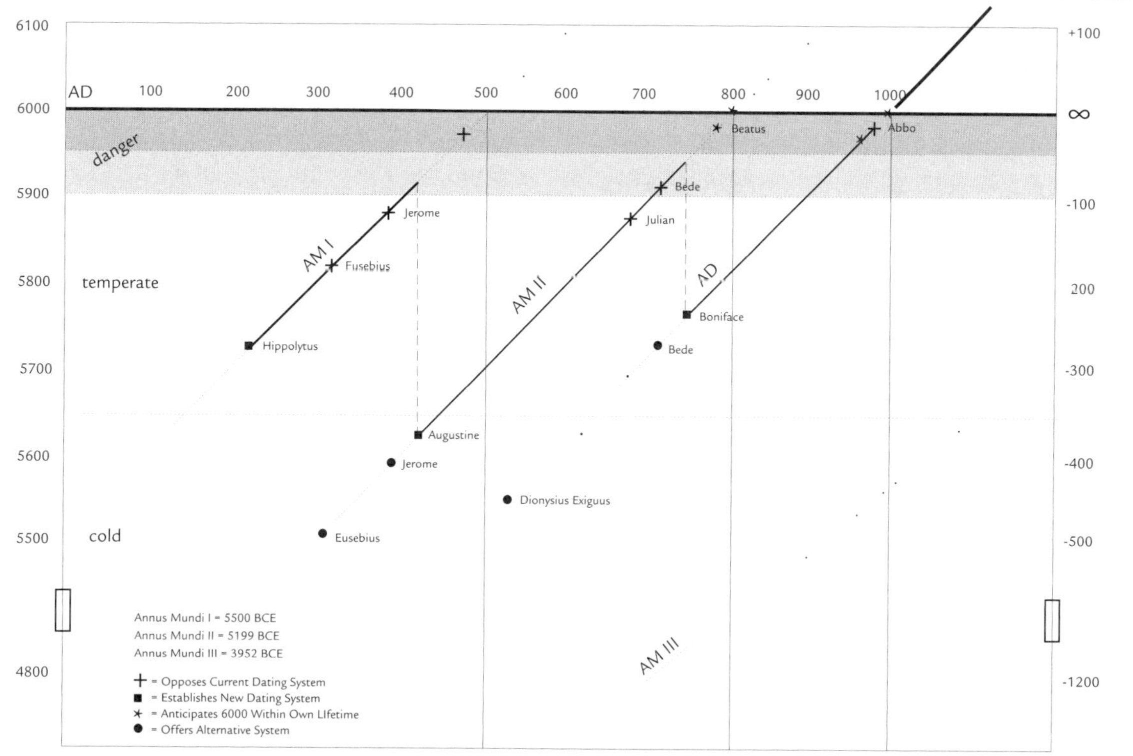

figure 2.2
Chronographical shifts and apocalyptic target dates. (Illustration by Richard Landes.)

audiences waxed enthusiastic about such chronologies as they approached term, owls would find the chronology increasingly problematic and try to disengage from it.

Each time the millennial date approached, then, the owls tried to replace the chronology in use with a new one. In the first two cases—AM I (500 CE = 6000) and AM II (801 CE = 6000)—they succeeded in substituting a new chronology that postponed the apocalyptic date by a few centuries, much as Hippolytus had with his introduction of AM I. This substitution is, by "normal" apocalyptic standards, unusual behavior. Most often redatings occur after the failure of a date and the ensuing "cognitive dissonance," as first analyzed by Leon Festinger (1956; Penton 1997). In this unusual case of nonapocalyptic chiliasm, however, it was not the roosters redating after disappointment, but owls redating before the advent of their balloon mortgage. We cannot know where and when the "new" chronology replaced the old or whether by the time the advent of 6000 occurred it had already been "dropped." We can only state with certainty that in the aftermath of both year 6000s (500 and 801 CE), the consensus moved rapidly toward the new chronology.

The result produced the homeostatic pattern evident in the graph: introduced in its temperate zone, any given era found favor in our sources (i.e., reasonably prominent clergy) as long as it remained in that zone. At the turn of the final century in the millennium, as the chronology "heated up," one finds the stirrings of a strong opposition within clerical circles that, growing over the course of the 5900s, replaced the old era with one that reset the clock. Bede's work provides the exception that proves the rule: his AM III dating was too far off (almost 1,350 years off—2048), but his Anno Domini (AD) date was just right (300 years off), and that was the dating system adopted by the Carolingians.

The year 1000 was the first eschatological target date that the Latin church could not/did not avoid. Unlike the previous two discarded chronologies, postmillennial Europe became still more committed to the Anno Domini dating system. No earlier culture has ever adopted and used a chronological system so pervasively.

For the purposes of this discussion, our focus lies not on the apocalyptic implications of this chronology for how we understand a generation's behavior at the moments of advent (500, 800, 1000), but rather on its unintended consequences for the history of timekeeping and its role in the tradition of "dating the end" that obviously still affects even modern, allegedly secular society. From this perspective, current speculation surrounding the Maya calendar's alleged "completion" in December of 2012 reflects a millennia-old pattern in Western timekeeping to invest big numbers with apocalyptic significance.[5] With each of the two great shifts under consideration here—either side of 500 and 800 and the sociocultural "mutation of 1000"—attention to and mastery of the chronological detail involved became ever more acute (Landes et al. 2003). This acuity developed not only at the time of the change from one system to the next (about a century before the date), but also at the approach of the era's implied apocalyptic term (countdowns) and at its passage (uniting around the new chronology).

Time of Change: The Early Sixtieth-Century AM (400–450, 700–750 CE)

It took a masterwork in chronology to replace the earlier *era mundi*: a critical part of the justification for such a striking move was the claim to greater accuracy. Thus, Eusebius's chronological work in the early fourth century (AM II) far outstrips in thoroughness and detail that of Hippolytus and Julius Africanus a century earlier (AM I). Similarly, Bede's *De temporum ratione*, with which he mainstreamed his alternative chronologies (AM III, AD), represents an extraordinary and extensive formulation of the concept of time measurement, from the smallest seconds to the interminable millennia, from dating Easter and the movements of the moon to not dating the apocalypse (Wallis 1999).

One might even argue that the Augustinian amillennial eschatology of the saeculum—we cannot know when it will end, but we *do* know that that end will not bring on the millennium—is taught and enforced by a regimen of knowledge about time (Markus 1989). And such regimens become especially important at the advent of the three millennial dates. It is no accident that both Charlemagne and Alcuin wanted a cadre of computists (Borst 1993:42–43). Hence, Bede's work represents a paradigm-setting scientific work in the Kuhnian sense and included the solution to the 532-year Easter cycle, that is, the period of time in which the moveable feast of Easter repeats its pattern of variation. Indeed, it is probably fair to say that time measurement is the only early medieval scientific field to actually advance over its antique predecessors (Wallis 1999). But at least at the time of its introduction, the separation had not yet taken place between the "pure" science of time measurement and the eschatological metaphysics of the fundamental Christian question: "When will Jesus return? When is the Parousia?"

Time of Approach: The Late Sixtieth-Century AM (Fifth, Eighth, Tenth Centuries CE)

Countdowns begin a century or more before the end date and initially reflect more of the owl's strategy: to point out the lengthy time between now and then. But with the passage of each generation, countdowns become increasingly problematic. Indeed, the approach of the date empowers the roosters. In the second countdown to 6000 (eighth century CE), this process culminated in Beatus of Liebana's spectacular illustrated commentary on the apocalypse in 5984 AM II. Buried in this otherwise largely derivative compilation of commentaries lies a strikingly original passage that gives an extensive treatment of chronology never before included in a commentary on Revelation, and it concludes with a countdown that establishes fourteen years remaining until the year 6000. Beatus concludes this exceptional passage with the following reflection, which begins with the cautious hoots of an owl and ends with a rooster's crow:

> The time remaining to the world is uncertain to human investigation. Our Lord Jesus Christ rejected every question on this matter when he said: "It does not belong to you to know the times or the moments which the father has put in his own power" (Acts 1:7); or again: "No one knows the day nor the hour—neither the angels of heaven, nor the Son, but only the Father" (Mark 13:32). *You should know the truth, the World will End in 6000 years, but who knows if they will not be shortened* (Landes 1988:193, emphasis added; see appendix 1, no. 17.)

Beatus, however, had few colleagues willing to take his direction.[6] On the contrary, the advent of the millennium in literary circles (our only source of evidence) met with the systematic resistance of trained timekeeping experts, whose knowledge of *computus* permitted them, on the one hand, to argue against the (AM II) roosters and, on the other, to use their authoritative knowledge to advance the single most important dimension of salvific timekeeping in this preapocalyptic saeculum, namely, liturgy.

Time of Advent: 6000 AM I (500, 801 CE)

How any given place or group experienced the passage of 6000—from those who did not know it was "happening" to those who thought it was the apocalyptic moment and the dynamics of interaction between everyone in between those two poles—remains invisible to the historian. As in all cases of the phenomenon, the apex of the curve of apocalyptic intensity also turns out to be the apex of cognitive dissonance, and the archival record, dominated by those who "knew" that 6000 did not bring either the end or the millennium, reflects the retrospective perfect of *ex post defectu*. I would like to suggest here, however, that the Carolingian imperial discourse that exercised its full hegemony on our AD documentation involves a channeling of apocalyptic hopes and fears—"l'espoir et l'effroi du moyen âge" as Michelet (1981 [1833]:231) called them—into organized, choreographed activities: from the liturgy at Saint-Riquier (Rabe 1995:111–137) to the architecture of Aachen (Heitz 1979:231–232) to the imperial coronation in Rome (Landes 1988:193–206; Van Meter 1996:131–170).

From Sabbatical Millennium to Millennium of the Incarnation and Passion

With each failure, there occurred a decoupling of the apocalyptic sense and the chronological mastery. In the sixth century, historians of the postimperial West showed an acute attention to chronology, universally using the new (nonapocalyptic) AM II (Landes 1988:186–191; appendix 1). In 809, Aachen hosted a major conference of chronographers, most probably brought there by a seriously disturbed emperor, who wanted his scholars to resolve a number of problems, including the increasingly "academic" question: what year was Annus Mundi? (Borst 2006:3:1034–1334; Palmer 2011). At the same time as, for Carolingian patriots, the millennial hopes of an imperial 6000 turned into the apocalyptic fears of imperial collapse, so did the focus on, indeed passion for, choreographing liturgy grow (Rabe 1995). Charlemagne's successor, Louis the Pious, oversaw a massive monastic reform, that of Benedict of Aniane, in which he could know at any time of prayer that monks everywhere throughout his empire were coordinated in their service of the Lord (Lackner 1972; Schmitz 1957). This was the early medieval version of Napoleon's desire to look at his watch and know what every French schoolchild was studying.

The year 1000 was the first eschatological target date that the Latin church could/did not avoid; with its advent, promises that clerics had made and repeated in public for over eight centuries reached their term.[7] The unique focus on 1000, its advent and passage in the sources, stands out, especially in the work of one of the key historians to live through it, Rodulfus Glaber.[8] And in the aftermath of the eschatological date embedded in its system, as we have noted, for the first time Latin Christendom did not discard the chronology, but rather became uniquely wedded to it.

As with other important eschatological target dates, the "failure" of 1000 to produce the millennium did not put an end to speculation. Quite the contrary, contemporaries (Glaber [*Quinque* 1.0.1, 1989:2] calls them "the wise and perceptive of the age") retargeted to the (still more Augustinian) date of the millennium of the Passion, 1033.[9] These two generations (preceding 1000 and 1033 CE) were steeped in an eschatological culture that historians have only begun to explore (Frassetto 2002; Landes et al. 2003). In particular, they mark an extraordinary intensification of previous practices (pilgrimage, relic cults, monastic reforms), as well as genuinely new phenomena: the (messianic) Peace of God movement, mass pilgrimage to Jerusalem, and demotic heresies (Landes 1995:285–332).

In one of the earliest and most remarkable expressions of approval for the "new" in traditionally conservative medieval ecclesiastical culture, Glaber (*Quinque* 1.1, 1989:3–4) speaks of the "new things [*nova*] that the Holy Spirit . . . brings about in this world . . . events that multiplied in unwonted fashion around the thousandth year of the humanity [Incarnation] of Christ, the Savior . . . and would continue until the last day."[10] Discovering how the members of that generation thought and spoke about these issues, how discourses between elites and commoners—roosters, owls, and the audiences they competed for—addressed the advent and passage of the first millennium since Christ is one of the major (and scarcely addressed) tasks of historians of the period around 1000.[11]

One aspect of this "millennial generation" concerns us here: the spread of Anno Domini dating, still somewhat limited in the tenth century, to virtually every aspect of the culture. Whereas previously, chancelleries had dated charters with AD only erratically (more often by the regnal year), the use of AD subsequently became so systematic that in French, AD dating is referred to as *le millésime*, from the Latin *millesimus* (thousandth).[12] From that time forward, no other culture ever adopted and used a chronological system with such fervor and unanimity and in so many areas of writing: chancelleries, charters, liturgy, architecture, historiography, epigraphy. In their near unanimity of usage and breadth of application, one might argue, Latin Christians went from counting down to a (failed) apocalyptic millennium—three times—to counting out indefinitely: an eschatological passion for time measurement had become a (proto)scientific one.[13]

Granted, the desire to count time precisely even after 1000 or 1033 had its apocalyptic dimension. From the turn of the millennium AD onward, most major apocalyptic predictions used the AD calendar as the foundation for calculations. After 1033, the next major date, which produced yet another massive pilgrimage to Jerusalem, was 1065, the first year of the new *annus magnus* of Easter dates according to Bede's Easter tables, the year that Good Friday and the Annunciation coincided.[14] The Joachite date of 1260, and the millennium-and-a-half marker of 1500/1533, produced intense focus on the date, a focus that, in the latter case, the advent of printing only intensified (Brendecke 1999:69–81; Reeves 1969:43–59; Schwartz 1995:61–68). Today, the AD system, often with the politically correct cognomen of Common Era, has become the dating system of global culture.[15]

All of which brings us back to David Landes's attempt to root the "objective" and precise timekeeping of the West in the monastic need for knowing the predawn hours for prayer. We have already seen the impact of the passage of 6000/800 on the work of Benedict of Aniane and his unhappy emperor Louis to reform and unify the liturgical activities of monasteries throughout the empire. This time the countdown to 1000 ignited the far more "bottom-up" monastic movement of Cluny, which approached the intensity of a 24-7 nonstop prayer cycle.[16] In the absence of a horizontal (i.e., historical) millennium that actually liberated the faithful, we find, among other social and religious phenomena, a monastic routine that aligned earthly society along a heavenly axis and reached out to the populace as a relic cult (Head 1992). This particular response to the messianic failure of collective salvation, where monasteries constituted the vertical alignment with God, placed particular pressure on liturgical deeds that, among their many functions, filled the void created between messianic hope and cold, hard reality, much as did the hard labor of Protestants in a later and still more dynamic era of individual watch-bearers who forged the modern world (Weber 2010).

Back in the more collective, but no less aroused, generation between the passage of the two Christian millennial dates from 1000 to 1033 (Fulton 2005:60–140; Landes 1995:285–332), Abbot Odilo of Cluny inaugurated the *omnium fidelium defunctorum* (All Souls' Day) somewhere between 1024 and 1033, under circumstances that suggest both an apocalyptic mindset and a concern for linking the liturgical calendar to the concerns of "all the faithful" (Paxton 2013). At virtually the same moment, at the approach of the year 1000 Anno Passionis, we also find the earliest evidence of the "Truce of God," which forbade warring during the weekly liturgical reenactment of the historical Resurrection (Easter Sunday, 1033). At Toulouges in 1027, the period lasted from Saturday morning to Monday morning (Head and Landes 1992:334–335). By 1041–1042, ambitions and scope had widened considerably: Glaber (*Quinque* 5.1.15, 1989:236–237) tells us that this *Treuga Dei* ("as the vulgus called it") spread from Aquitaine to all of Gaul and set aside the days from "sunset on Wednesday to sunrise on Monday" as free from vengeance on enemies.[17] This broad initiative of the bishops of Gaul to discipline war according to so extensively defined a liturgical calendar on a society-wide scale was unprecedented and owed its programmatic scope to none other than the most respected Christian in Europe at the time, Abbot Odilo of Cluny.[18]

That these efforts to pacify a belligerent aristocracy failed, which even contemporaries noted, should not blind us either to the collective energy that underlay such ambitious endeavors, nor to the role that liturgical spirituality played in communicating to, attracting the loyalty of, and disciplining a generation of commoners (Landes 2007). One finds similar dynamics at play in the emergence of free cities (communes) in Europe, perhaps the eleventh century's most "modern" product (Chédeville et al. 1998). And among those disciplines imposed by liturgy, one might place an awareness and a responsiveness to the measurement of time. Liturgical time, increasingly specific and omnipresent, yoked the apostolic Christian, wherever he might be a pilgrim in this terrestrial city, to the kingdom of heaven. In 1041, in the crisis of a failed millennial covenant, Odilo apparently believed that enough people were committed to this liturgical sacralization of time that they could successfully ban war and revenge killings for four days a week!

From this perspective, one might argue that the increasingly precise effort to count down to the end of the millennium and the beginning of the messianic era had an unanticipated consequence in the wake of its inevitable (but unthinkable) failure to arrive: the tradition of timekeeping developed in anticipation turned out to have other postapocalyptic uses, which, however unintended, nevertheless helped shape Western culture. Ironically, the failure of "the end of time" (or the millennium) in 1000 (and again in 1033) gave birth to millennium-long passion for precise time measurement and its use in organizing purposeful and (at least sometimes) peaceful and productive human activity.

NOTES

1 Note that when the Chinese discovered clocks, they eagerly adopted them to their cultural and ritual needs (Spence 1984). They did not, however, generate these precision micro timekeepers. On the uniqueness of the Western obsession with time in comparison with the Chinese, see Landes 1983:15–32.

2 This imagery, taken from a discussion in the Talmud (Sanhedrin, 98a), concerns roosters eager for the dawn of redemption and bats fleeing at the break of that day (Landes 2011:37–62).

3 The closest one comes in Islam is the *mujaddid* tradition of a "renewer" at the end of every century (Landau-Tasseron 1989). On 200 Anno Hegirae (815/816 CE), see Yücesoy 2009; on 1400 (1979/1980 CE), see Landes 2011:445–450. Perhaps the most striking case came in the year 1000 AH (1591/1592 CE) with its impact on Akhbar the Mogul emperor in India (Moin 2012). In Judaism, the year 6000 will coincide with 2240 CE.

4 The best documented case here is that of Bede in the early eighth century (i.e., the sixtieth century according to the chronology AM II, which he sought to replace with AM III; see Landes 1988:174–178).

5 Accordingly, the scenarios inspired by this apocalyptic peg in time run the gamut from (eschatological) cataclysmic destruction to a (millennial) transformative evolutionary leap.

6 See also the Kölner Notiz (798/5997), app. 1, no. 18.

7 In this sense, Abbo's experience "as a youth [*adolescentulus*]" with a preacher in Paris announcing the end of the world on 1000 underlines this dynamic (see app. 2, no. 2).

8 See appendix 2. For an extensive discussion of and precise references to these texts and their analysis, see Landes 2000b. No other year in the entire history of Christian historiography (until, perhaps, 1260) received this kind of attention at its advent.

9 Composing ex post defectu around 1040, Glaber, now an independent writer at Cluny, told the story of the advent of the second millennial date in 1033. For the text, see app. 2, no. 13; for analysis, see Landes 2000b:121–130.

10 The passage is longer and more complex than this citation and translation convey. For a discussion of the issue of novelty, see Landes 1995:265–268.

11 Despite a brief flurry of discussion around 2000, the majority of historians have chosen to reaffirm the older consensus that the year 1000 was not an apocalyptic date (e.g., Cushing 2005; Moore 2000:7–30; Riché 1999:11–26). See the nuanced discussion in Peters 2002. Fulton 2005:60–141 represents the most significant sustained exception to the general dismissal of a fundamental role for apocalyptic discourse at the turn of the millennium; see also the work of Holland (2008), a sophisticated nonspecialist.

12 Despite serious errors in Dionysius Exiguus's and Bede's calculations of the year of Jesus's birth (*Anno Incarnationis*), it was only at the advent of and passage of 1000 that historians showed any interest in correcting these matters. See Verbist 2010.

13 On the exceptional development of rational thought in this period, see Murray 1985. On the link between apocalyptic expectation and science, see Fried 2001.

14 On the calculations, see Annales altahenses maiores (1868:815). On this pilgrimage, see Joranson 1928 and, more recently, Jacoby 2005.

15 Even Hamas dated their founding and militantly Islamist charter to 1988 rather than 1408 Anno Hegirae. At least one Salafi pundit, noting the Christian origins of AD, considers this kind of "imitation" to be unworthy of Muslims (Ahmed 2012).

16 Lackner (1972:40–91) points out that one of the continuities between Cluny and Cîteaux (which is where others place the key dynamics leading to the mechanical clock) was the rigor with which both movements in their waxing early years pursued liturgical timing. By the mid-eleventh century, Peter Damian noted that at Cluny "the offices succeeded each other with such rapidity, that even in the long days of summer there remains only half an hour in which the brethren can talk in the cloister" (Lackner 1972:54). On Cluny and liturgy, see Iogna-Prat 1998:186–265, a provocative analysis of the extensive links between Cluniac spirituality and the world beyond the cloister. See also Carozzi 2002.

17 Glaber's account is corroborated by the text of a letter (1041–1042) from the bishops of Gaul to the bishops of Italy recommending the truce (Treuga Dei 1893). The letter explains the logic of the extra

days, that is, the key liturgical commemorations surrounding the Passion and Resurrection: the Ascension (Thursday, forty days after Easter), the Crucifixion (Friday), the entombment (Saturday), the Resurrection (Sunday). Note that this initiative is related less to warfare per se than the conduct surrounding feud and revenge killings (*homicidium*): "so that all people should be free of the fear of enemies" (Treuga Dei 1893:para. 2).

18 The letter from the bishops of Gaul is also signed by Abbot Odilo of Cluny, perhaps the most influential man of his day. On this Truce of God (Treuga Dei 1893), see Cowdrey 1970:52–53; Hoffmann 1964:81–88.

APPENDIX 1

Countdowns from the fifty-ninth and sixtieth centuries AM II (seventh and eighth centuries CE):

643: Et ab ascensione domini nostri Iesu Christi usque ad anno. XL. II. regnum gloriosissimi domni nostri Chlotari regis anni DC. I, qui faciunt in summa ab inicio mundi anni V milia DCCC. XX. C. IIII. Item et ab inicio mundi usque ad anno decimo Seggeberti regis anni V milia DCCCXX. C. IIII. *Et remanent dies sexto miliario, ut pro conpleatur anni CXXXX* [140 years]. Wurzburg, Mp. th. f. 28, 10–11th c.; *MGH SRM* VII, p. 493, no. 19:2; MS discussed p. 100.

673: A mundi principio usque ad diluvio sunt anni II. CC XLII. Item ab initio mundi usque ad passionem domni nostril Ihesu Christi sunt anni VCC XXVIII. Similiter ab initio mundi esque eo tempore, quand ciclo isto Victurius condedit sunt anni V DC XLVIII. In summa enim ab intio mundi usque in presente anno, id est sexto decomo anno regnante Clothario, filio Chlodoveo, sunt anni V DCCC. LXXIIII [5874]. Restant de sexto miliario anni CXXVI [126]; Krusch 1884:132.

673: Post hanc supputationem usque in annum praesentem, quo Chlotharius exercitum contra Wascones movit, id est quintodecimo regni ipsius, additi sunt anni LX. a tempore item Chlotharii abavi ipsius, in quo anno apostolicus vir domnus Chado episcopus et Barcelaicus abbas basilicae sancti Sulpicii a fundamento edificare ceperunt, circa illa priora ubi in corpore requiescunt. qui annus centesimus tercius decimus cum bissexto in ciclo Victorii reperitur et ab obitu sancti Sulpicii XX us VI us esse probatur, sunt ab exordio mundi usque ad predictum tempus anni VDCCCLXXIII [5873]. *de sexto vero miliario supersunt anni CXXVII* [127], quia idem Victorius a duobus Geminis et tempore passionis Christi cum consulibus in suo ciclo usque ad Constantinum et Rufum illius anni consules intimavit annos CCCCXXX et sine consulibus CII et ab initio mundi usque ad predictum annum, quo Constantinus et Rufus consules erant, titulavit in suo prologo annos esse VDCLVIIL peractis illis centum et duobus sine consulibus, quos ipse futuros scripsit, et CXIII de recapitulatione cicli invenies summam annorum superius intimatam, id est VDCCCLXXIII [5873]. BN Lat. 17544, 12th c.; *MGH AA* XI, p. 493, no. 7; MS 402n45.

674: Fiunt igitur ab exordio mundi usque in presentem eram, hoc est in annum quintum imperatoris Eraclii et quartum annum Siseboti principis gentis Gothorum anni VDCCCXII [5812 (*sic*)], post completionem huius temporis fluxerunt anni LX, scilicet a tempore Clodoarii usque ad XV tum annum regni itemque Clodoarii ac nepotis ipsius, quo ipse cum ingenti exercitu contra Danos abiit, estque a dormitione sancti Sulpicii XXVI tus in cyclo I vero Victorii CVIII cum bissexto, in aera ubi nomina consulum et Maximi habentur esse monstratur. hoc ideo dictum est, ut facilius deinceps reliqua ac futura summa annorum cognoscatur additis ergo his; LX annis fiunt a mundi initio anni VDCCCLXXIII [5873], *de sexto vero miliario fiunt residui anni CXXVII* [127]. Ambrosianus H 187 inf.; *MGH AA* XI, pp. 505–506, no. 6.

675: Ab inicio mundi usque quando passus est dominus Ihesus Christus anni VCCXXVIII. A passione domini nostri Ihesu Christi usque ad transitum Childeberti regis, in quo anno cyclus Victorii rurso ex passione dominica circulum annorum ad inicium rediit, sunt anni DXXXII. In summa ab inicio mundi usque in praedicto anno sunt anni VDCCLX. Ab eo anno usque primo anno regni Clotharii filii Chlodovei sunt anni LXXXVIIII. Ab inde usque transitum illius, quando Heldericus germanus suus tria hec regna Neustria Austria et Burgundia subiugavit, sunt anni quindecim et menses V. Hildericus regnavit in Neustria an. II et menses VI. Cui germanus suus Teodericus successit in regno ab eo anno, quando passus est dominus noster Ihesus Christus usque primo anno Teoderici regis anni sunt DCLXVIII [668]. Fiunt insimul ab inicio mundi usque in praedicto primo anno regni Teoderici incliti regis anni VDCCCLXX et VI [5876] *et restat de sexto miliario anni CXXIIII* [124]. *MGH AA* XI, p. 492.

699: In annum V. Childeberti regis Francorum, Pippino iubente, ab Adam sunt anni V milia DCCCC [5900]. Fuit pasca X. Kal. Aprilis, ascension Domini fuit Kal. Madias. Per cyclum annorum CXL. Repeticio a capite cychli. Krusch 1884:134.

715: + Invenit Lucerios presbeter monacos dom tum a . . . per ista croneca et per alia crone su. quod *septoagenta et quattuor ann[i ab ista compu]tavit quod sextus miliarios . . . [docit] esse explitos* conpotavit ipsos [anni LXXIIII et] an[o]s X in indiccione exsiente te[rtia decimo ann]o quarto Dagoberto rignante. BN lat. 10910, f. 184 (7–8th c.); *MGH SRM* II, p. 501; discussed in Wallace-Hadrill 1960:xlvii.

727: "Komputus im Berner Codex n. 611 aus dem Jahre 727 n. Chr." Edited by Krusch 1884:136; Borst 2006:373–374 (see following, no. 15).

736: [Continuator of Fredegar] Certe ab initio mundi usque ad passionem domini nostri Iesu Christi sunt anni 5228 et a passione Domini usque isto anno praesente, qui est in cyclo Victorii ann. 177. Kl. Ian. die dominica, ann. 735; *et ut istum miliarium impleatur, restant ann. LXIII* [63]. *MGH SRM* II, p. 176; Wallace-Hadrill 1960:92.

738: Sunt in summa ab exordio mundi usque ad praesente tempore anni quinque milia nongenti xxviii [5928, corr. 5929]. *Et remanent de sexto miliario anni LXXII* [72]. London, BM Cotton Nero A ii; Levison 1946:303–304. This manuscript updates this original calculation forty years later (see following at 778/5979).

742: usque DCLIIII vero eram, in qua beatus Isidorus cronicam suam condidit, id est quinto Eraclii imperatoris anno et quarto Sisebuti regis Gothorum, fiunt omnes ab initio anni VDCCCXV [5815]. deinde ergo usque his temporibus, in quo est era DCCLXXX [742 CE], creverunt anni CXXVI [126], qui additi ad superiorem summam faciunt omnes annos ab Adam usque in presentem eram, qui est DCCLXXX [780], VDCCCCXLI [5941]. posthac quippe supersunt usque ad finem sexte huius etatis vel *introitu septime etatis, in qua dominus in maiestate prestolatur adventus, anni* [59]. *MGH AA* XI, p. 169.

743: Invenimus collectum esse hanc coronicam sub era DCLXVI, sicut et in alia huius doctoris, quam prius edidit, repperimus cronica, per quam et hanc legimus eram. deinde a sequenti era DCLXVII usque in hanc presentem eram, que est DCCLXXI, creberunt anni CXVI, qui additi ad superiorem huius cronice summam faciunt simul omnes annos ab exordio mundi usque in hanc praefatam DCCLXXXI [*sic*] eram VDCCCCXLII [5942]. his itaque ita digestis fatemur denique difficile posse quemquam hominem annorum summam post etiam tempore incarnationis iam domini, pretermissa quoque era veritatis indice, per reges et principes conputare et in errore minime incurrere. proinde ergo necesse quippe est secundum morem prorsus eram illam partire, que suo reppererit tempore incurrere et quod partiendo invenerit, hoc quippe faciat ad ere maiore adicere, que ab Adam usque ad Christum noscitur percurrere; tunc sane probabitur quisque se scilicet ad veritatis potius tramite pervenisse. Vaticanus reginae n. 294; Venetus Marcianus Lat. II, 47; Vindobonensis 4831; *MGH AA* XI, p. 506.

745: a nativitate vero Iesu Christi domini nostri, quod fuit XL secundo anno imperii Octaviani Caesaris, usque in praesentem annum, id est tertio regni Ratchisi in Italia, sexto pontificatus Zachariae pape urbis Romae, indictio XV, sunt anni DCCXLV. *restant igitur de sexto miliario anni circiter LVII.* finitis vero sex aetatibus, id est sex miliaribus, seculi terminus est. si vero aliquid superstiterit (perstiterit), extra saeculum fit. Florence, Biblioteca Laurenziana, Plut. 20.54, 15v (Florence, s. xi), with the dating clause on fol. 15v, copied from a 10th c. exemplar, in Biblioteca Laurenziana, Conv. Soppr. 364; T. Mommsen, *MGH AA* IX, p. 158; Palmer 2011:1327n115.

748: a nativitate vero Iesu Christi domini nostri, quod fuit XL [suppl. in cod. 623] secundo anno imperii Octaviani Caesaris, usque in praesentem annum, id est tertio regni Ratchisi in Italia, sexto pontificatus Zachariae pape urbis Romae, indictio XV, sunt anni DCCXLV. *restant igitur de sexto miliario* *anni circiter LVII* [57]. *finitis vero sex aetatibus, id est sex miliaribus, seculi terminus est.* si vero aliquid superstiterit, extra seculum fit. *MGH AA* IX, p. 158, no. 5; cod. 623 f. 116 = cod. 54 f. 15.

767/778: Sunt in summa ab exordio mundi usque ad praesente tempore anni quinque miliano genti viginti octo [5928] *et remanent de sexto miliario anni lxxii, subtractus xl remanent xxxii* [32]. Qui vult contendere, requirat Iulio Africano et Eusebio Cesare episcopo, Hyreonimo presbito erudissimo viro, Agustino Ypporegense [regiae] episcopo, Orosio presbito, Victorio qui cy[. . .] Hispalense episcopo et Gregorio catholico et erudito viro Tononense episcopo velalius quam pluris quod longum est enarr[a] tore per singulae perplexo stilo has sentencias helicuit, qui uult contendere apud ipsus conferat quia superna illa civitas ex angelis et hominibus constat. London, British Library, Cotton Nero A ii, fols. 34v–36 (France or Lombardy, 767), at fol. 35v; Palmer 2011:1327n117. Palmer connects this calculation to the *Computus of 727* (Borst 2006:1:374), which also gives thirty-two years to complete the sixth millennium. According to Levison (1946:304), the mistakes (5928/727 rather than 5939/738) obscure the actual date, which he considers to be 778.

775: Ab exordio mundi usque ad nativitatem Christi numerantur anni VCXCVI. A nativitate Christi usque ad passionem anni XXXII et menses tres. A passione domini usque in anno presente, hoc est in anno VII Caroli regis inditione XIII, in quo anno celebratur pascha VII k. Apr., luna XVIIII, implentur anni DCCXLVIII [748], ita ut in hoc pascha incipiat annus DCCXLV [745], fiuntque insimul ab exordio ann. VDCCCCLXXI [5971]. Greci computant ab initio usque in praesente supra scripto anno VICCLXXX [6280]. distat vero in hac subputatione inter Grecos Latinosque an. CCCVIII [308]. secundum argumenta Egiptiorum computant Greci ab exordio mundi an. VICCLXXXIII [6283], Latini dicunt VDCCCCLXVIII [5968]: distant inter se ann. CCCVIII [308]. *MGH AA* XI, p. 492.

784: Beatus Commentary, second recension: Colligitur omne tempus ab Adam usque ad Christum anni V milia CCXXVII; et ab adventu Domini nostri Iesu Christi usque in praesentem eram, DCCCXXIV, sunt anni DCCLXXXVI. Computa ergo a primo homine Adam usque in praesentem eram DCCCXXIV, et invenies annos sub uno V milia DCCCCLXXXVI [5986]. *Supersunt ergo anni de sexto miliario XIIII* [14]. Finiebit quoque sexta aetas in era DCCCXXXVIII. Residuum saeculi tempus humanae investigationis incertum est. omnem enim de hac re quaestionem Dominus noster Iesus Christus abstulit dicens: non est vestrum scire tempora vel momenta, quae pater posuit in sua potestate. et alibi: de die autem illa et hora nemo scit, neque angeli caelorum, neque filius, nisi pater solus. Et quia diem dixit et horam, aliquando pro tempora dicuntur, aliquando vero simpliciter intelleguntur. Ut sciatis, *in veritate sexto millesimo anno finiendus erit mundus,* utrum impleantur, an minuentur, soli Deo cognitum est. Ceterum ut suspicemus, de septimo non possumus, quia nec scriptum invenimus. Sex diebus operatus est Deus, quos invenimus a mane usque ad vesperam esse finitos; de septimo non legimus, nisi requievisse; et ut per sex dies sex milium annorum figuram ostenderet, quibus saeculi istius aetas deducitur, indicaret in septimo resurrectionem omnium sanctorum. Beatus 1930:IV:5:367–368.

798: Kölner Notiz (Additamenta): Ab Adam usque ad passionem Christi anni [5228]. Sunt anni ab initio mundi secundum veritatem Hebreorum, ut transtulit Hieronimus, usque ad istum annum XXXI. regni Karoli regis—ipse est annus, quando hospites accepit de Saxonia tertiam partem populi et quando missi venerunt de Grecia ut traderent ei imperium—anni VDCCCCLXXXXVIII [5998], secundum vero LXX anni VICCLXVIII [6268], ab incarnatione Domini [798]. Cui vero sic non placet, sudet et legat et melius numeret. *Additamenta Coloniensia,* I.2, *MGH SS;* 1864:730; Borst 2006:781–782.

800: A quinto vero anno Eraclii imperatoris usque ad vigisimo octavo Liuprandi regis Langobardorum et mea infelicis Ioh. miseroli nativitate subputantur ann. CXXIIII. dehinc usque ad XXVII cristianissimi et piissimi dom. Carolo [sic] magni regis Francorum et Langobardorum in Italia inveniuntur anni sexaginta duo. colleguntur ergo *omnes anni ab Adam usque ad presentem* octava indictione, quo pascha Hebreorum idus Aprilis, pascha vero nostrum tertio decimo kl. Martias, principium mundi quarto id Aprilis, luna condita id. Aprilis, homo ad imaginem dei factus septimo decimo kl. Martias, *anni sex milia.* Codex Mutinensis Estensis ord. I, no. 11; *MGH AA* XI, p. 400, no. 32; citation from p. 491.

Specific mentions of the year 1000 Anno Domini or Anno Passionis in the contemporary documentation:

1. **968:** mille anni a nativitate Christi [one thousand years from the birth of Christ; thirty-two-year mistake based on thinking that the Easter Tables recorded Anno Passionis rather than Anno Domini]. *Annales sancti Floriacensis, MGH SS* 2, p. 255.
2. **994–996:** Abbo's letter to Hugh and Robert: De fine quoque mundi coram populo sermonem in Ecclesia Parisiorum adolesceotulus audivi, quod statim finito mille annorum numero Antichristus adveniret, et non longo post tempore universale judicium succedere . . . [I heard a sermon in the church in Paris that as soon as the number of a thousand years should come, the Antichrist would come, and not long thereafter, the Last Judgment would follow]. Abbo of Fleury, *Apologeticus ad hugonem et rodbertum reges francorum*, London, British Library, MS 10972, fol. 22v, PL 139:471–472; dated before 996 by Vidier 1965:105–107; to 994–995 by Mostert 1987:48–51.
3. **999:** Iste fuit Girbertus, tempore cuius inpletus est annus millesimus ab incarnatione Domini [This was Girbertus (Pope Sylvester), during whose time the thousandth year since the incarnation of the Lord was completed]. *Annales pragenses, MGH SS*, 3, p. 120.
4. **1000:** Anno dominicae incarnationis 1000, ind. 13. epacta 12. concurrente 1. termino pascali 9. Kal. Apr. 4 Kal Apr. feria 6, terraemotus . . . generali et vasto tremore totius orbis magnitudo passim contremuit, ut cunctis fieret manifestum, quod ore veritatis fuerat ante promissum. His namque et aliis signis quae praenuntiata fuerunt opere completis, hinc iam fit nostra spes certior omni visu, de his quae restant ordine complendis [In the year of the incarnation of the Lord, 1000, indiction 13, epacts 12, concurrents 1, paschal term 9, on Good Friday March 29 (4 of the calends of April), there was a great earthquake. . . . It was not as one often occurs . . . but rather the whole magnitude of the earth shook everywhere with a general and vast tremor, so that that which had been promised before by the mouth of truth might be made manifest to everyone. With these and other signs that had been predicted having been completed by divine operation, hence our hope is made more certain in the sight of all regarding those that remain to be completed in due order]. *Annales elnonenses maiores, MGH SS*, 5, p. 12.
5. **1000:** Tertio Ottone imperante. Millesimus annus supercrescens statute computationis numerum, secundum illud quod legitur scriptum: Millesimus exsuperat et transcendit omnia annus [With Otto III ruling. The thousandth year surpassed the number of established reckoning, according to what we read written, "The thousandth year surpasses and transcend all years"]. *Annales hildesheimenses* III, Preface, *MGH SS*, 3, pp. 91–92.
6. **1000:** Interea millesimus ab incarnatione Domini annus feliciter impletus est et hic est annus archiepiscopi 12 [Meanwhile the thousandth year from the incarnation of the Lord was happily completed and this is the twelfth year of the archbishop]. Adam of Bremen, *History of the Archbishops of Hamburg-Bremen*, xlii (40), *MGH SS rer. Ger.*, 2, p. 101; 1917; 2002.
7. **1001:** Data mense augusto, regnante Rotberto rege, anni ab Incarnatione Domni nostri Jesu Christi usque in presentem diem mille et I [Dated the month of August, with Robert reigning as king, in the year from the Incarnation of our Lord Jesus Christ until the present day, one thousand and one]. Saint Hilaire charter Poitiers, AD Vienne, G 591 C 3 no. 52, CPR no. 30; http://www.cn-telma.fr/originaux/charte1184/.
8. **ca. 1015:** Post salutiferum intemerate virginis partum millenarii numeri linea consummata et in quinto cardinalis ordinis loco et in eiusdem quarte ebdomade inicio clarum mane illuxit seculo [After the sequence of the one thousand count since the salvific birth through the immaculate virgin was completed, and in the fourth place of the sequence of cardinal numbers and in the beginning of the fifth week, a brilliant dawn illuminated the world]. Thietmar of Meerseburg, *Chronicon*, 6.1.1; *MGH SS rer. Germ. N.S.*, 9, p. 275; 1935.
9. **ca. 1025:** Non secius ergo que dicuntur quin solito multiplicius circa millesimum humanati Christi Salvatoris contigerunt annum [This was certainly true of the events that multiplied in unwonted fashion around the thousandth year of the humanity (Incarnation) of Christ, the Savior]. Glaber, *Quinque* 1.0.1; 1989:2.

10. **ca. 1025:** Quod presagium Iohannis prophetie congruit, quia dixit Sathanan solvendum: et expletis mille annis [All this accords with the prophecy of John who said that Satan would be freed after a thousand years]. Glaber, *Quinque* 2.12.23; 1989:92; see Landes 2000a:26–43.

11. **ca. 1025:** Nunc igitur, quoniam de priorum gestis aliqua retulimus, ab illo et infra, ut spopondimus, anno, videlicet millesimo nati cuncta vivificantis verbi, tercii sumamus incitamen huius operis libelli [Now, having reported things preceding, we begin the third book of this work where we will tell, as promised, the events that marked the thousandth year of the Verb]. Glaber, *Quinque* 3.0.1; 1989:94.

12. **ca. 1031:** Ipsius [William of Volpiano's] namque imperio, maxima iam ex parte eventuum ac prodigiorum, quae circa et infra Incarnati Salvatoris annum contigere millesimum, descripseram [For at his command I had already described most of the events and prodigies that occurred around and during the thousandth year of the Incarnation of our Savior]. Glaber, *Vita sci. Guillelmi*, 13; 1989:294.

13. **ca. 1040:** Post multiplicia prodigiorum signa quae tam ante quam post circa tamen annum Christi Domini millesimum in orbe terrarum contigere, plures fuisse constat sagaci mente viros industrios, qui non his minor propinquante euisdem domincae passionis anno millesimo fore predixere quod utique evidentissime contigit [After the many prodigies which had broken upon the world before, after, and around the millennium of the Lord Christ, there were plenty of able men of penetrating intellect who foretold others, just as great, at the approach of the millennium of the Lord's Passion, and such wonders were soon manifest]. Glaber, *Quinque*, 4.1; 1989:170.

14. **Later eleventh century:** Habemus hujus pietatis exemplum nostris pene temporibus declaratum Roberto apud Merovingiam, quæ alio nomine dicitur Francia, tenente jus regium, post mille a passione Domini volumina annorum. Ipso milenarii impleto anno, cum peracta quadragesimali observatione, sanctæ Parasceves dies advenisset, visæ sunt multis per loca multa in aëre igneæ acies, prodigioso visu corda se intuentium perterrentes. Extemplo fama (malum, quo non aliud velocius ullum mobilitate viget) multorum perculit aures. Cui quot sunt plumæ, dicuntur & infore linguæ. [After the passage of a thousand years since the Passion of the Lord, in that very year of the millenary fulfillment, when the observation of Lent was done and the day of holy preparation—that is, the day of the Crucifixion of Christ—had arrived, long columns of fire were seen high in the air in many places and by many people. This prodigious vision terrified the hearts of those who contemplated it. Immediately this rumor—and disaster thrives with a rapidity that nothing else exceeds—reached the ears of the multitudes, of whom it is said there are as many tongues in them as their feathers.] *Miracula s. Agili abbatis*, 1, 3; *AA SS Augustii*, vol. 6, p. 588.

REFERENCES CITED

Primary Sources

Adam von Bremen (later eleventh century)

1917 *Gesta Hammaburgensis ecclesiae pontificum.* Monumenta Germaniae Historica, SS rer. Germ. 2.2. Impensis Bibliopolii Hahniani, Hannover.

2002 *History of the Archbishops of Hamburg-Bremen.* Translated by Francis Joseph Tschan. Columbia University Press, New York.

Additamenta Coloniensia (Kölner Notiz) (798)

1864 *Annales Sancti Petri Coloniensis.* Monumenta Germaniae Historica, scriptores 16. Impensis Bibliopolii Hahniani, Hannover.

Annales altahenses maiores (eleventh century)

1868 Monumenta Germaniae Historica, scriptores 20. Edited by George Pertz, pp. 772–824. Impensis Bibliopolii Hahniani, Hannover.

Beatus of Liebana (late eighth century)

1930 *Commentarius in Apocalypsin.* Edited with commentary by Henry A. Sanders. Papers and Monographs of the American Academy in Rome 7. American Academy in Rome, Rome.

Byrhtferth (early eleventh century)

1972 *Byrhtferth's Manual.* Edited by S. J. Crawford. Early English Text Society, Original Series 177. Humphrey Milford, London.

Glaber, Rodulfus (1025–1044)

1989 *Rodulfi Glabri historiarum libri quinque* [The Five Books of the Histories]. Edited and translated by John France. Oxford University Press, Oxford.

Hamas Founding Charter

1988 Translated by Avalon Project. Yale Law School, New Haven, Conn.

Maimonides (1172)

1952 *Epistle to Yemen: The Arabic Original and Three Hebrew Versions.* Edited by Abraham Halkin. Translated by Boaz Cohen. American Academy for Jewish Research, New York.

Miracula s. Agili abbatis (later eleventh century)

1743 *Acta Sanctorum, Augustii,* vol. 6. Société des Bollandistes, Antwerp.

Thietmar of Meerseburg (early eleventh century)

1935 *Chronicon.* Edited by Robert Holtzmann. Monumenta Germaniae Historica, SS rer. Germ. n.d., 9. Impensis Bibliopolii Hahniani, Hannover.

Treuga Dei, Archidioecesis Arelatensis (ca. 1037–1041)

1893 Monumenta Germaniae Historica, leges, constitutiones et acta publica imperatorum et regum 1, *911–1197.* Edited by Ludwig Weiland, pp. 596–597. Impensis Bibliopolii Hahniani, Hannover.

Secondary Sources

Ahmed, Iftakhar

2012 Muslims Adopting the Anno Domini (A.D.) Calendar. Tawheed Movement. Electronic document, http://tawheedmovement.com/2012/08/05/muslims-adopting-the-anno-domini-a-d-calendar/, accessed September 29, 2013.

Aveni, Anthony F.

1980 *Skywatchers of Ancient México.* University of Texas Press, Austin.

Borst, Arno

1993 *The Ordering of Time: From the Ancient Computus to the Modern Computer.* University of Chicago Press, Chicago.

2006 *Schriften zur Komputistik im Frankenreich von 721 bis 818.* Hahnsche Buchhandlung, Hannover.

Brendecke, Arndt

1999 *Die Jahrhundertwenden: Eine Geschichte ihrer Wahrnehmung und Wirkung.* Campus Verlag, Frankfurt.

Carozzi, Claude

2002 Apocalypse et calendrier au Moyen Age. In *L'attente des temps nouveaux: Eschatologie, millénarisme, et visions du futur du Moyen Âge au XXe siècle,* edited by André Vauchez, pp. 23–28. Brepols, Tournhout.

Chédeville, André, Jacques Le Goff, and Jacques Rossiaud (editors)

1998 *La ville en France au Moyen Âge.* Point, Paris.

Cook, David

2002 *Studies in Muslim Apocalyptic.* Darwin Press, Princeton, N.J.

Cowdrey, H. E. J.

1970 The Peace and the Truce of God in the Eleventh Century. *Past and Present* 46:42–67.

Cushing, Kathleen

2005 *Reform and the Papacy in the Eleventh Century: Spirituality and Social Change*. Manchester University Press, Manchester.

Dohrn-van Rossum, Gerhard

1992 *History of the Hour: Clocks and Modern Temporal Orders*. Translated by Thomas Dunlap. University of Chicago Press, Chicago.

Festinger, Leon

1956 *When Prophecy Fails: A Social and Psychological Study of a Modern Group that Predicted the Destruction of the World*. Harper, New York.

Frassetto, Michael (editor)

2002 *The Year 1000: Religious and Social Response to the Turning of the First Millennium*. Macmillan, New York.

Fried, Johannes

2001 *Aufstieg aus dem Untergang: Apokalyptisches Denken und die Entstehung der modernen Naturwissenschaft im Mittelalter*. C. H. Beck, Munich.

Fulton, Rachel

2005 *From Judgment to Passion: Devotion to Christ and the Virgin Mary, 800–1200*. Columbia University Press, New York.

Guenée, Bernard

1980 *Histoire et culture historique dans l'occident medieval*. Aubier Montaigne, Paris.

Head, Thomas

1992 The Judgment of God: Andrew of Fleury's Account of the Peace League of Bourges. In *The Peace of God*, edited by Thomas Head and Richard Landes, pp. 219–238. Cornell University Press, Ithaca, N.Y.

Head, Thomas, and Richard Landes (editors)

1992 *The Peace of God: Social Violence and Religious Response around the Year 1000*. Cornell University Press, Ithaca, N.Y.

Heitz, Carol

1979 Retentissement de l'Apocalypse dans l'art de l'époque carolingienne. In *L'apocalypse de Jean: Traditions exégétiques et iconographiques, IIIe au XIIIe siècles*, edited by Renzo Petraglio, pp. 217–243. Librairie Droz, Geneva.

Hoffmann, Hartmut

1964 *Gottesfriede und Treuga Dei*. Schriften der MGH 20. A. Hiersemann, Stuttgart.

Holland, Tom

2008 *Millennium: The End of the World and the Forging of Christendom*. Little, Brown, London.

Iogna-Prat, Dominique

1998 *Ordonner et exclure: Cluny et la société chrétienne face a l'hérésie, au Judaïsme, et à l'Islam, 1000–1150*. Aubier, Paris.

Jacoby, David

2005 Bishop Gunther of Bamberg: Byzantium and Christian Pilgrimage to the Holy Land in the Eleventh Century. In *Zwischen Polis, Provinz und Peripherie: Beiträge zur byzantinischen Geschichte und Kultur*, edited by L. Hoffmann, pp. 267–285. Mainzer Veröffentlichungen zur Byzantinistik 7. Harrassowitz Verlag, Wiesbaden. Electronic document, http://www.mgh-bibliothek.de/dokumente/a/a139920.pdf, accessed September 29, 2013.

Joranson, Einar

1928 The Great German Pilgrimage of 1064–1065. In *Crusades and Other Historical Essays Presented to Dana Munro*, edited by L. J. Paetow, pp. 3–43. F. S. Crofts, New York.

Krusch, Bruno

1884 Zur Einführung des griechischen Paschalritus im Abendland. *Neues Archiv* 9:99–169.

Lackner, Bede K.

1972 Benedict of Aniane and Post-Carolingian Monachism. In *The Eleventh-Century Background of Citeaux*, pp. 1–39. Consortium Press, Washington, D.C.

Landau-Tasseron, Ella

1989 The "Cyclical Reform": A Study of the Mujaddid Tradition. *Studia islamica* 70:79–118.

Landes, David

1983 *Revolution in Time: Clocks and the Making of the Modern World.* Bellknap Press, Cambridge.

Landes, Richard

1988 Lest the Millennium Be Fulfilled: Apocalyptic Expectations and the Pattern of Western Chronography, 100–800 CE. In *The Use and Abuse of Eschatology in the Middle Ages*, edited by W. Verbeke, D. Verhelst, and A. Welkenhuysen, pp. 137–211. Leuven University Press, Leuven.

1995 *Relics, Apocalypse, and the Deceits of History: Ademar of Chabannes, 989–1034.* Harvard University Press, Cambridge, Mass.

2000a The Birth of Popular Heresy: A Millennial Phenomenon. *Journal of Religious History* 24(1):26–43.

2000b The Fear of an Apocalyptic Year 1000: Augustinian Historiography, Medieval and Modern. *Speculum* 75:97–145.

2007 Economic Development and Demotic Religiosity: Reflections on the Eleventh-Century Takeoff. In *History in the Comic Mode: The New Medieval Cultural History*, edited by Rachel Fulton and Bruce Holsinger, pp. 101–116. Columbia University Press, New York.

2011 *Heaven on Earth: The Varieties of the Millennial Experience.* Oxford University Press, New York.

Landes, Richard, Andrew Gow, and David Van Meter (editors)

2003 *The Apocalyptic Year 1000: Religious Expectation and Social Change, 950–1050.* Oxford University Press, New York.

Levison, Wilhelm

1946 *England and the Continent in the Eighth Century.* Oxford University Press, Oxford.

Markus, Robert

1989 *Saeculum: History and Society in the Theology of St. Augustine.* Cambridge University Press, Cambridge.

Michelet, Jules

1981 [1833] *Le Moyen Âge.* Edited by Claude Mettra. Laffont, Paris.

Moin, A. Azfar

2012 *The Millennial Sovereign: Sacred Kingship and Sainthood in Islam.* Columbia University Press, New York.

Moore, R. I.

2000 *The First European Revolution, c. 970–1215.* Blackwell, Oxford.

Mostert, Marco

1987 *The Political Thought of Abbo of Fleury.* Verloren, Hilversum.

Murray, Alexander

1985 *Reason and Society in the Middle Ages.* Clarendon, Oxford.

Naquin, Sue

1976 *Millenarian Rebellion in China: Eight Trigrams Uprising of 1813.* Yale University Press, New Haven, Conn.

Palmer, James

2011 Calculating Time and the End of Time in the Carolingian World, c. 740–82. *English Historical Review* 126:1311–1320.

Paxton, Frederick

2013 *The Death Ritual at Cluny in the Central Middle Ages.* Studia Monastica, Series Fontes. Brepols, Turnout.

Penton, M. James

1997 *Apocalypse Delayed: The Story of Jehovah's Witnesses.* 2nd ed. University of Toronto Press, Toronto.

Peters, Edward

2002 Mutations, Adjustments, Terrors, Historians, and the Year One Thousand. In *The Year 1000: Religious and Social Response to the Turning of the First Millennium*, edited by Michael Frassetto, pp. 9–28. Palgrave/Macmillan, New York.

Rabe, Susan A.

1995 *Faith, Art, and Politics at Saint Riquier: The Symbolic Vision of Angilbert.* University of Pennsylvania Press, Philadelphia.

Reeves, Marjorie

1969 *The Influence of Prophecy in the Later Middle Ages: A Study in Joachimism.* Oxford University Press, Oxford.

Riché, Pierre

1999 *Les grandeurs de l'an mille.* Hachette, Paris.

Schmitz, P., OSB

1957 L'influence de Saint Benoît. In *Il monachesimo nell'alto medioevo e la formazione della civiltà occidentale*, pp. 401–415. Settimane di Studio del Centro Italiano di Studi sull'Alto Medioevo 4. Presso la Sede del Centro, Spoleto.

Schwartz, Hillel

1995 *Century's End: An Orientation Manual towards the Year 2000.* Doubleday, New York.

Scott, James

1992 *Domination and the Arts of Resistance: Hidden Transcripts.* Yale University Press, New Haven, Conn.

Spence, Jonathan D.

1984 *The Memory Palace of Matteo Ricci.* Penguin, New York.

Van Meter, David

1996 The Empire of the Year 6000: Eschatology and the Sanctification of Carolingian Politics. PhD dissertation, Department of History, Boston University, Boston.

Verbist, Peter

2010 *Duelling with the Past: Medieval Authors and the Problem of the Christian Era (c. 990–1135).* Brepols, Turnhout.

Vidier, A.

1965 *L'historiographie à Saint-Benoît-sur Loire.* Paris.

Wallace-Hadrill, J. M.

1960 *The Fourth Book of the Chronicle of Fredegar.* Thomas Nelson and Sons, London.

Wallis, Faith

1999 *Bede, The Reckoning of Time.* Liverpool University Press, Liverpool.

Weber, Max

2010 *The Protestant Ethic and the Spirit of Capitalism.* Edited and translated by Stephen Kalberg. Oxford University Press, Oxford.

Yücesoy, Hayrettin

2009 *Messianic Beliefs and Imperial Politics in Medieval Islam: The 'Abbāsid Caliphate in the Early Ninth Century.* University of South Carolina Press, Columbia.

3

Ecumene Time, Anecumene Time

Proposal of a Paradigm

ALFREDO LÓPEZ AUSTIN

Worldview as System

It has been said that the concept of linear time is one of the cornerstones of Judeo-Christian thought. Its importance lies beyond its continuity: the concept of linear time generates relationships that penetrate and mold all areas of thought and action. In short, linear time in the Judeo-Christian tradition is not only an enduring concept, but also one that shapes and structures. In the theoretical framework of Fernand Braudel (1958:727, 1974:64, 1980:27), this concept corresponds to the rhythm of the longest time span of the *longue durée*, a rhythm that seems to be static. In what follows, I propose a paradigm for the Mesoamerican concept of time, which, like Judeo-Christian linear time, also shapes and structures. It forms part of the enduring nucleus of the Mesoamerican tradition (López Austin 2001) and is equivalent to what J. Eric S. Thompson (1954:137, 1992:196), though referring only to the Maya, called "the philosophy of time."

It is possible to formulate such a paradigm because time pertains to worldview, and worldview can be characterized as a collective, historical, rational, and systematic creation. Its collective character implies that it is a product of the aggregate, day-to-day activity of all members of a human group, who, communicating among themselves, gradually elevated their thought to various levels of abstraction. It is historical because it is subject to social transformations, thus it is in perpetual creation, always mutable, never definitive. It is rational, though not necessarily the result of deliberate reflection, as it derives from the thought and action of human beings, which are communicated by way of expression rooted in logic and subject to continuous processes of abstraction. And finally, it is systematic because increasing levels of abstraction necessitate consistency, although admittedly this consistency is relative and is affected by all sorts of social contradictions.

Based on these suppositions, a paradigm can be constructed (López Austin and López Luján

2009:33–36), in this case, for a concept of time in a tradition that goes back to the early days of agriculture (2500 BCE) in territory currently part of Mexico and that endures to this day within indigenous communities in Mexico and Central America. This paradigm is based on abundant documentary and ethnographic information; therefore, the citations offered here are merely examples for consideration. This study seeks to synthesize the central concepts of the enduring nucleus and to link them consistently for heuristic purposes, in the hope that this paradigm will serve as a source of hypothetical orientation for researching particular issues and will be continuously perfected in the face of specific data.

Ecumene and Anecumene

One of the main concepts in the enduring nucleus of Mesoamerican tradition is the primary division between two kinds of space-time, which I am calling *ecumene* (of this world) and *anecumene* (not of this world). The immediate space-time of believers—that is, the constructor-users of this worldview—is the ecumene, which is the space-time of creatures (human beings, animals, plants, minerals, elements, astral bodies, and so on), as opposed to the anecumene, which is the space-time of supernatural beings (deities and forces) and is forbidden to creatures. The ecumene has the following characteristics: (1) it houses these creatures, along with a multitude of supernatural beings that inhabit it either temporarily or permanently (Figure 3.1); (2) it is bounded by the world's creation (the Maya 13.0.0.0.0 4 Ahau 8 Cumku) and destruction (the Maya *butic*; Casas 1967:2:507), that is, its beginning and end are established by the gods; and (3) it is a dependent product of the anecumene, from which it receives elements that dynamize it, for everything that exists in the space-time of creatures was produced in the world of the gods, thus the ecumene's movement and sustenance come from the space-time of supernatural beings. The anecumene, on the other hand, has the following characteristics: (1) it does not house or admit creatures, except for certain parts of them consisting only of subtle substance; (2) it created the ecumene; and (3) it has no marked beginning or end. Although the ecumene

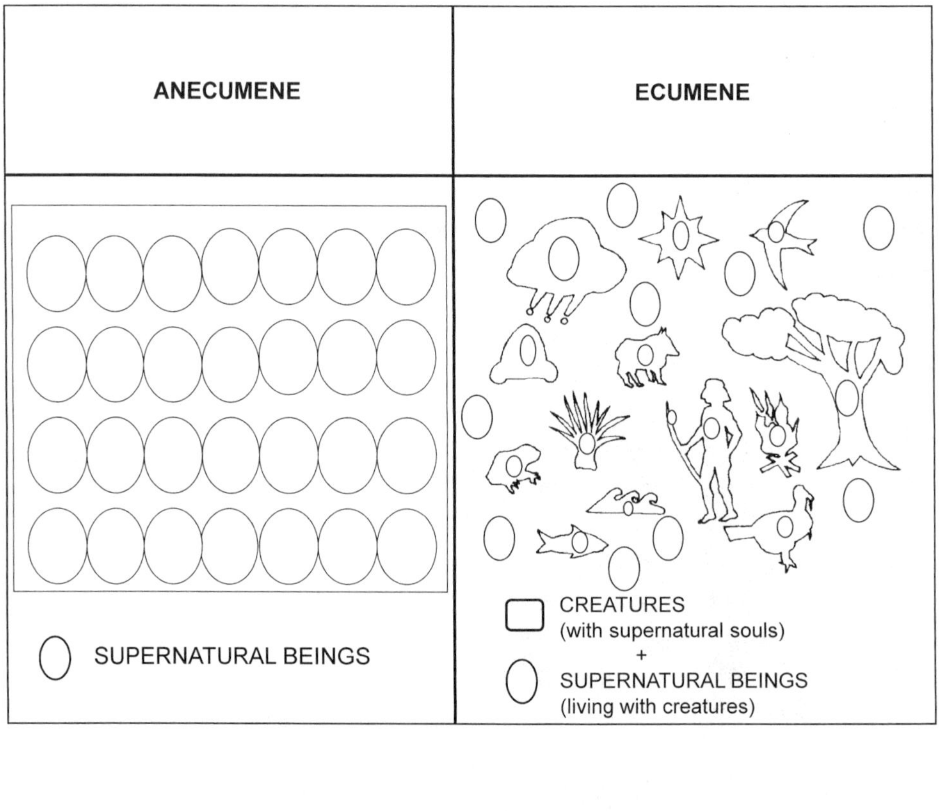

figure 3.1
The domains of anecumene and ecumene. (Illustration by Alfredo López Austin.)

and the anecumene are two very different dimensions, the circulation of deities, forces, and subtle substances between the two space-times is allowed through certain thresholds and portals. Transit through these thresholds and portals, however, is only permitted at precise times and under particular circumstances.

As the Greek philosopher Xenophanes of Colophon well noted, believers construct the supernatural from their intimate experience, projecting their reality when imagining the other world. For example, the Ixil of El Quiché, Guatemala, say the gods invented incense and candle offerings, and the people learned these rituals in order to communicate with them (Colby and Colby 1981:40–42, 1986:53–55). This projection, however, also implies contrasts that differentiate believers from the supernatural to explain their own condition. The gods, whether they are conceived in human form or not, are just like human beings; they are cognitive beings with volition, capable of acting on their thoughts. Moreover, the gods have social values: feelings of love, hate, or reciprocity and an inclination toward justice or vengeance. Even Xenophanes's god, without human form, was cognitive, morally perfect, and able to listen to mortals. Using Mesoamerican projections, I intend to contrast the concept of ecumene time with what believers attribute, with differentiating characteristics, to anecumene time. Ecumene temporal reality, therefore, is both a model and antimodel of anecumene temporal reality.

Mythic Processes of Creation

The anecumene space-time dimension is perceived to be divided further into two dimensions: one of divine leisure and one of myth. The mythic dimension has a preparatory processual zone and a liminal zone, as the latter is in contact with the ecumene space-time dimension (López Austin 1975:289, 1993:41–43, 1996:63–65; Figure 3.2). Nothing further can be said about the divine leisure space-time, for a state that is void of consequences is inconceivable. It is simply a point of reference in human reflection. The mythic dimension, however, corresponds to transformations, in which deities acquire the

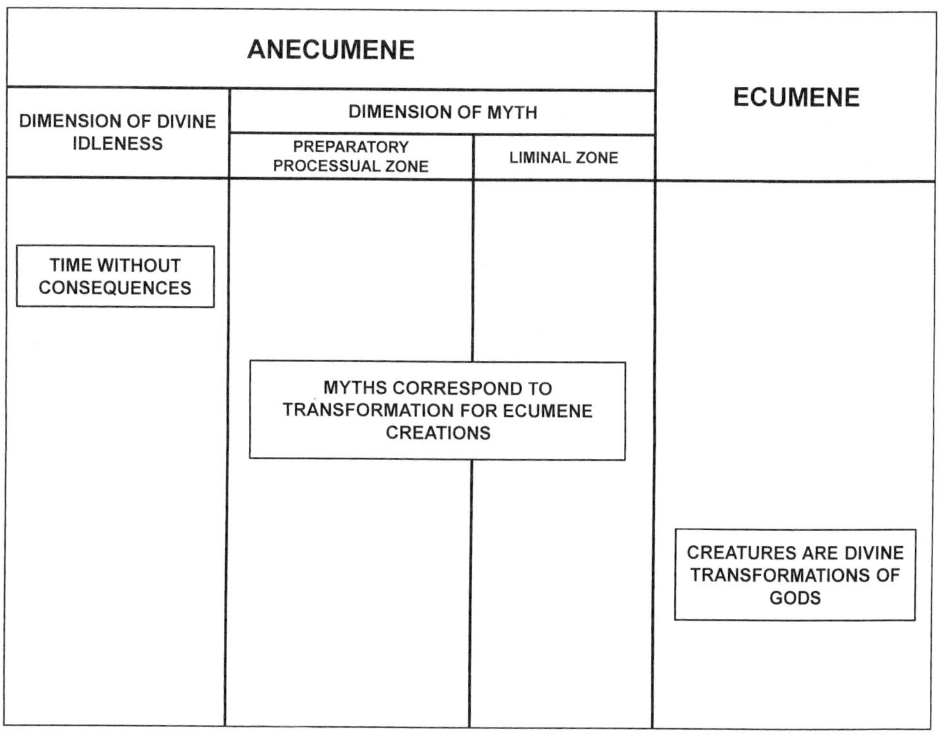

figure 3.2 Division of anecumene and ecumene. Subdivision of the domain of anecumene. (Illustration by Alfredo López Austin.)

necessary traits for the moment of the ecumene's creation and the final installation of creatures in primogenial time. Accounts of these divine processes, which combine many diverse elements, are expressed in mythic narratives such as the adventures of gods. Their characters take on human, animal, mineral, and other forms and are heterogeneous beings with humanlike powers, able to communicate fully with each other and to change form up until the culminating moment when they obtain their definitive characteristics.

This dimension makes it necessary to distinguish between cosmic processes (nodal aspects of myth) and their narrative expression as the adventures of gods (heroic aspects of myth). It should be noted that each myth is independent of other myths, although all of them belong to a common myth complex. For example, two different myths about the Sun and Moon may coexist if one refers to their different temporal cycles and the other to their different degrees of brightness. Because of this mythic autonomy, it is not rare to find the same character playing different roles in different narratives: one being may have, and be represented by, two or more mythic personalities; kin relationships between gods may vary in different myths; and so on. A linear interrelation, much less a historical succession, should not be sought in the corpus of mythic narratives (López Austin 1993:43, 1996:66). The mythic context is like the bar of a comb that holds many parallel teeth, each independent of the others (Figure 3.3)

Earlier I said that the Mesoamerican mythic dimension has a preparatory processual zone and a liminal zone. The former ends with the culmination of myths, which are followed, in the latter, by the first sunrise and the initiation of the Sun's course. From this moment on, the earth's muddy surface solidifies, the gods stop transforming, the mountains no longer move from place to place, the new types of beings lose their ability to communicate with each other, and so on.

The well-known ancient Nahuatl myth of the birth of the Sun at Teotihuacan from the immolation of Nanahuatzin (Sahagún 1950–1982:7:3–8, 2000:2:694–697) clearly describes the liminal zone. Nanahuatzin perishes in the fire and descends into the underworld. After his journey through the

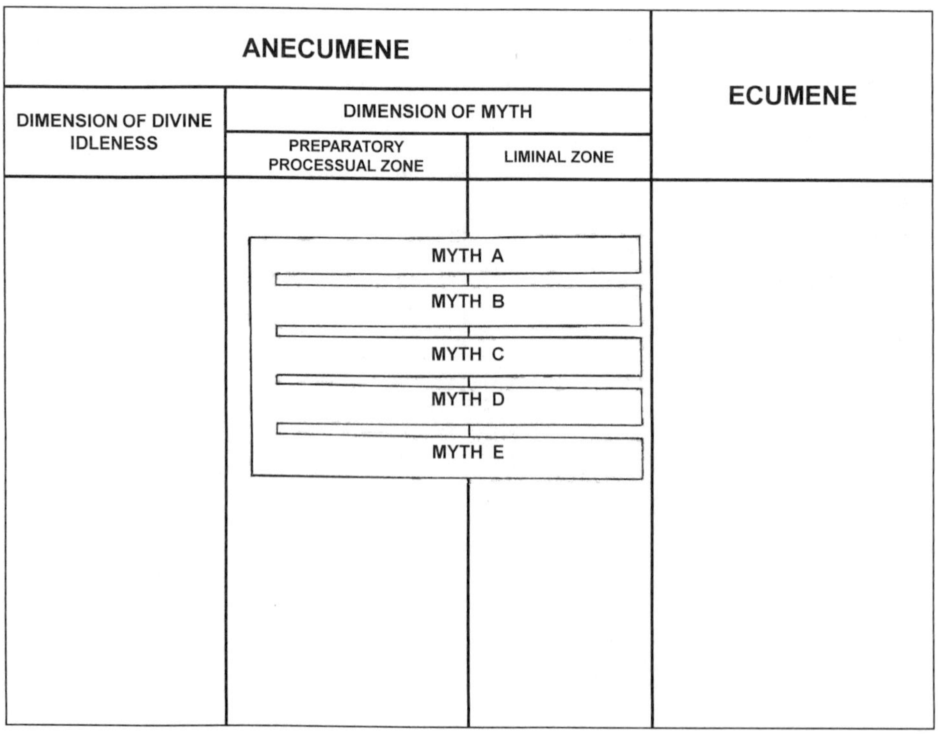

figure 3.3
Each myth is independent of other myths. (Illustration by Alfredo López Austin.)

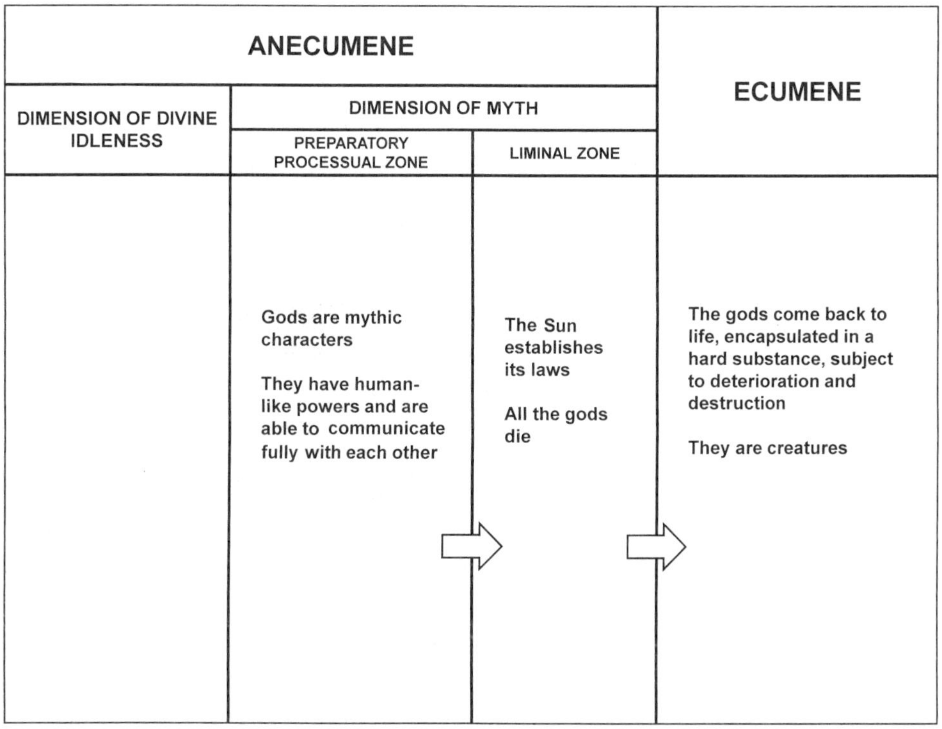

figure 3.4 The transformation of the gods. (Illustration by Alfredo López Austin.)

bowels of the earth, he emerges in the east and orders all the other gods to immolate themselves as he did. The last one to do so is Xolotl, whose concealment in the water after his sacrifice turns him into the animal called axolotl. In other words, the death of the gods in the liminal zone sends them to the underworld, where they acquire their worldly nature consisting of a hard and heavy substance that envelops the subtle substance of the deceased deities and converts them into creatures of this world. The god Yappan, for example, is transformed into a scorpion when he is decapitated by the god Yaotl, who becomes the species of locust called "head carrier" (Ruiz de Alarcón 1953:176–177, 1984:205–206). This process is the origin of ecumene beings, those who had a different nature in the time that Hernando Ruiz de Alarcón calls "aquel siglo" ("that century"; 1953:55–58), the anecumene space-time. Jacinto de la Serna (1953:231) says it was believed that trees had rational souls because of their "human" condition in the other century. In sum, the gods of myth are transformed by death in the liminal zone into creature-creators. The creatures are gods encapsulated in a heavy, hard, and perceptible corporeal substance, subject to deterioration and destruction in a life-death cycle of existence (Figure 3.4) They are like the Sun, whose heat and fire constitute a cover that dies each day in the west, passes through the subterranean world of death, and comes back to life each morning. This transformation explains the belief in Mesoamerican tradition, still present today, that all creatures of this world possess a rational soul within them.

Laws of the Sun

In the preparatory processual zone, laws are not lacking. As an Ixil diviner asked, "How could the gods be without law?" (Colby and Colby 1981:140, 1986:155). In an ancient Nahuatl myth, child gods were expelled from the Tamoanchan anecumene for violating the orders of the Father and the Mother. The sentence itself was a mandate that established a tacit destiny: the punished children would go and populate the surface of the earth and the underworld (Codex Telleriano-Remensis 1995:13r; Codex Vaticanus A 1996:27v). The order implied the beginning of the ecumene with the inclusion of the expelled deities in the life-death cycle. In Maya accounts, the delegation of creative action is more explicit. An ancient Guatemalan myth clearly states that the children received the mission of creating the things of this world (Casas 1967:2:505–506). Although laws were not lacking in the process of creation, violence escalated to the point that the situation was unbearable on the eve of the first sunrise. In that muddy and shadowy realm, the gods residing on earth—the creatures' antecessors—practiced cannibalism and devoured their own children and siblings; they ate raw flesh and despicable things; they paired incestuously; they raped and killed each other; and so on. Today these protocreatures are known as "gentiles," demons, monkeys, jaguars, or Jews (Garza 2002:64): "Les decían *tsitsimimej*; eran los anteriores [They called them *tsitsimimej*; they were the anterior ones]" (Taller de tradición oral 2009:40). The rule of the Sun, light, and the establishment of suitable laws for the ecumene space-time that was born became indispensable.

When the gods' characteristics were set with the Sun's first rays, the cosmic apparatus that would serve to receive the orderly flow of cycles was complete. A green ceiba tree rose in the center as a symbol and memorial of the annihilation, a column that provided support between the earth and the sky (the plate and the cup), and the site where the *k'atuns* would govern (Books of Chilam Balam 1948:155). Four columns also rose, and these would sequence the periods of time to successively pour onto the earth (Books of Chilam Balam 1948:114), ruled by a calendar "composed of myriad overlays of cycles of differing lengths and portents" (Tedlock 1982:1). Many accounts refer to the establishment of cycles that would rule in the ecumene, among them the day-night mytheme of the "Musicians of the Sun" (Histoyre du Mechique 1905:32–33, 1965:111–112; Mendieta 1945:1:86), the myth of the "Birth of Huitzilopochtli" (Sahagún 1950–1982:3:1–5, 2000:1:300–302), and the figures found in the central lower portion of page 52 of the Codex Vindobonensis (1992).

The Sun established its laws. In order to differentiate between the previous time and the one now legislated, a Nahuatl myth refers to the latter with the term *tlaneltoca* (Dakin 1977:63), which can be translated as "[when] what was established [was] followed." With the first dawn, the mortal destiny of the reborn gods was determined, for their hard shells deteriorated and were destroyed. One of the gods said: "Todos los que íbamos a amanecer teníamos que morir después [All of us who went out at dawn later had to die]" (Taller de tradición oral 2009:45). The ability of creatures to intercommunicate as the gods had in the preparatory processual zone, where all beings spoke, even cooking implements and reed mats, was suppressed (Fagetti 2003:71). Each creature exercised its function on earth, but their work wore them down and they had to be reinvigorated with food. Where the animals lived and which ones could be eaten was established (Popol Vuh 1950:85–86, 1964:26–27). Human beings merited special legislation: they were charged with worshiping and feeding the gods (Popol Vuh 1950:165, 1964:103). For human sustenance, a contract was struck with the Earth in which she provided food in return for human cadavers (Shaw 1972:58–59). Cannibalism was forbidden to humans, and prohibitions against incest were established. In sum, the Sun—illuminator, creator, and guardian (Colby and Colby 1981:38, 1986:52)—became "the initial and primary symbol of ethical, spatial, and temporal order" (Gossen 1972:136). For "cuando apareció el día, todo estuvo preparado para resolver, todo se preparó antes de que hubiera vida como ahora [when the day appeared, everything was prepared to settle, everything was prepared before there was life as today]" (Boege 1988:90). In this way, the concepts of creation and all processes in the liminal zone served as the archetypes that explained the daily transits of the gods in the cosmos and the prayers and offerings that humans delivered to them.

Cosmic Flows

The ecumene exists thanks to flows that emanate from the anecumene. It was said that many gods were transfigured during creation to give rise to creatures. Some, perhaps by hiding in rivers or mountains when the general immolation took place

(Colby and Colby 1981:167, 1986:183), remained as guardians. Others received the mission of traveling between the anecumene and the ecumene, causing transformations—some cyclical—and hierophanies. Upon crossing over to the ecumene, deities could enter a creature and occupy it until it died. In this sense, the heavy shell that covered each creature could accommodate multiple contents. Human beings, the prototypical creatures, contain in their bodies a multitude of heterogeneous animate entities (deities that give them life, new characteristics, thought, special abilities, and so on) that are subject to various dangers and, on top of that, do not always coexist harmoniously (López Austin and López Luján 2009:105, cuadro 2). One of these entities, which M. Esther Hermitte (1964:122, 1970:84) has described as a co-essence, is the *nahual* that a human shares with another being (see *nawal* in Brown, this volume). Another entity, which confers the ability to travel through portals, has received the name *way* among the Maya (Houston and Stuart 1989); Barbara Tedlock (1982:53, 110) has found a similar entity called *coyopa* among contemporary K'iche'. Ancient kings who were deified when they ascended to power could gain animate entities, as their souls acquired additional portions of the substance of one or more gods. Subsequently, when the king died, these divine parts returned to their sources of origin. Therefore, rulers do not become gods when they die, only while they are alive.

It should be noted that in both the liminal zone and the ecumene the laws of the Sun dictate that activity causes deterioration that must be counteracted with food. This law also affects the deities that travel from one realm to the other; therefore, humans must feed them with prayers, rites, offerings, and even their own bodies. In Guatemala, the Achi' say that Jesus Christ himself complains when they do not feed him: "Ya no me recuerdan con candelas y por eso estoy sufriendo mucha hambre (siendo que el humo de la candela es mi comida); ya no me dan de beber; ya no me dan de comer [They no longer remember me with candles and therefore I am suffering much hunger (the smoke from the candle being my food); they no longer give me anything to drink; they no longer give me anything to eat]" (Shaw 1972:49).

Of course, the transit in either direction is not entirely free for gods, forces, the dead, creatures, offerings, or prayers. Portals are located at precise points, and passage occurs at precise times in the calendar (Albores 1997:429; Durán 1971:159–160, 1984:1:85; Neff Nuixa 2001:356), for the cycles are in force in both the ecumene and the liminal zone, where they were first created and installed. Therefore, some portals only open once a year. A large opening, such as at a large festival, can be dangerous because beings from the underworld could attempt an invasion (Knab 1991:54; Lupo 1995:162). The cyclicity of the opening also requires that offerings be delivered at precise dates and times (Villa Rojas 1990:751). They must be made on the day that the deity "comes back to life" (*resucita*; Reyes García 2001:167; Schultze Jena 1954:49, 75)—that is, the date when it crosses the barrier of the ecumene or when it reaches the liminal zone. If it is in the preparatory processual zone, it may be summoned to enter the liminal zone, as the K'iche' do with *kumpisión*, a call urging the sacred person to come the following day to receive the offering (Bunzel 1952:301–302). In order to confirm its efficacy, the K'iche' believer waits for the reply of divine acceptance (*rétal*) to know whether the place, time, quality, and quantity of the offering were adequate (Schultze Jena 1954:93, 95). Some transits are instantaneous; others are slow, like the crossing of the dead to the anecumene; and some are divided into segments, like the seven gates of the Sierra Norte Nahuas of Puebla (Lupo 1995:70).

Conformation of Temporal Units

Sylvanus G. Morley (1946:223–224, 1972:219, 221) proposes two fundamental characteristics of Maya divinities: the possibility of two or more fusing together to form one unit and the ability of one god to serve as patron for two or more temporal units. We are indebted to J. Eric S. Thompson (1954:138–139, 1970:199, 1975:248, 1978:96, 1992:197) for studying the divinization of time among the ancient Maya. The British researcher identified

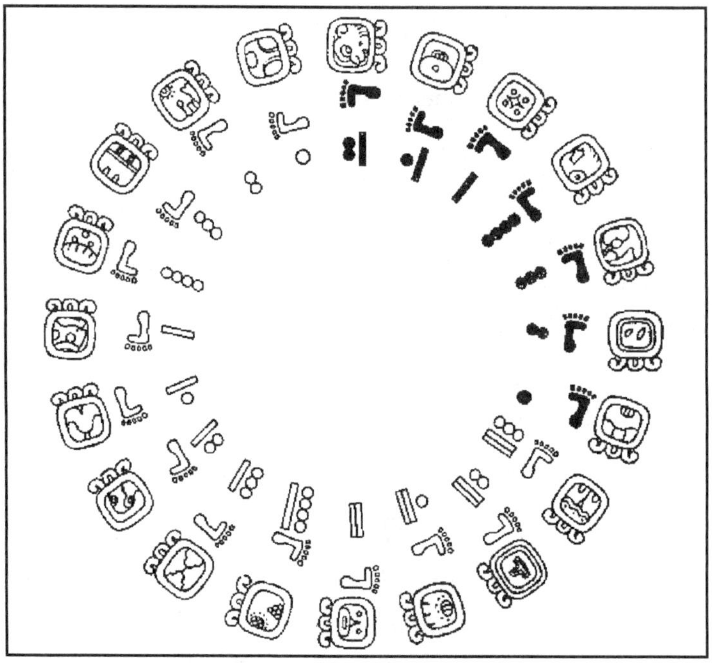

figure 3.5
The birth of the month. (Illustration by Alfredo López Austin.)

the personality of the day-gods, the composition of the temporal unit of the day by two different deities, the divine character of other units, as well as how they were worshiped in antiquity. Moreover, based on works of Bishop Núñez de la Vega, Oliver La Farge, Douglas Byers, and J. S. Lincoln, he insisted on the historical continuity of these concepts in current beliefs and rituals dedicated to time in some highland Maya communities.

Although Thompson's proposals were sound, they did not all achieve general acceptance. A subsequent study has closely reviewed beliefs concerning the divine character of the various temporal units. Its author, María Eugenia Gutiérrez González (2008), approached the problem from iconographic, linguistic, historical, ethnographic, and epigraphic perspectives and concluded that Maya Initial Series temporal cycles were personified and deified in Classic-period religious thought. She acknowledged, however, that data for the *bak'tun* and longer periods were insufficient (Gutiérrez González 2008:105). It is also important to add that ethnographic information considerably enriches the panorama. Authors like Oliver La Farge and Douglas Byers (1931:172–173), Ruth Bunzel (1952:267), Leonhard Schultze Jena (1954:49), Mary Shaw (1972:104–105), Robert Carmack (1979:381–382), Barbara Tedlock (1982:107), and Benjamin N. Colby and Lore M. Colby (1981:62–63, 65–66, 67, 289–290, 1986:76–77, 80, 81, 294–295) have provided material of great value.

Other researchers have continued to study the composition of the gods, considering them divine powers with abilities including fusion, fission, dividing their substance, and, thus, potentially occupying two or more places simultaneously, including the interior of their images (López Austin 1983, 1993:135–138, 1996:176–179). These divine faculties are also attributed to temporal units (León-Portilla 1968:88–90, 1973:82–83). We think, for example, that the numerical element in the composition of the days has a value as important as its figurative counterpart. This interpretation is clearly reflected in the Maya myth of the birth of the month, in which a divine personage steps in the footprints left by another unnamed personage (Figure 3.5). The union of each foot in each footprint forms a combination of thirteen numerals and twenty figurative signs. In this manner the twenty units are completed (Book of Chilam Balam of Chumayel 1967:116–119, 1973:97–101). On the other hand, book 4 of the Florentine Codex (Sahagún 1950–1982, 2000:1:343–432) reiterates that the numeral 1 confers so much force to its figurative counterpart that it becomes the head of the *trecena* and as such alters the action of the twelve days it commands.

When breaching the ecumene, the temporal units occupy and filter into all that exists as a great transforming force. This event has posed a dilemma among scholars as to whether this force exerts absolute fatality or simple influence. Some authors have leaned toward the former, saying, for example, that "people had no control over their fate" (Bricker 1981:7, 1993:27). Others, in contrast, have emphasized the power of humans to confront the determination of time (Garza 1975:105–106; León-Portilla 1963:119–121, 1974:198–199).

It should be noted here that effects of destiny and the deterministic force of temporal units may vary considerably. For example, the fatal character of k'atun 8 Ahau only held for the Itzá, and the years 1 and 7 Rabbit (*tochtli*) only held for the Toltec and the Mexica. Moreover, the force of the temporal units may differ: the fate of the day 1 Death (*miquiztli*) was almost ineludible (Sahagún 1950–1982:4:33–36, 2000:1:367–368), although destiny generally is avoidable and not a fatal force. As Bunzel (1952:275) has stated, destiny is "a vast concatenation of forces." The array of units simultaneously arriving to the ecumene struggle with each other and confront other divine forces already there. Faced with this onslaught of forces, human beings defend themselves with knowledge, foresight, proper conduct, discretion, penance, prayers, and offerings, for the devout act also persuades invading powers (Sahagún 1950–1982:4:33, 2000:1:367; Schultze Jena 1954:65). Destiny, as one reads in the Books of Chilam Balam (1948:180), "como puede no suceder, puede suceder [just as well may not happen as happen]."

Glimpses of the Anecumene

Except for mythic references to the shadowy, humid, and soft ambience of the moment before the pristine sunrise, few ancient or contemporary texts provide detailed descriptions of the anecumene. Some accounts, however, are of great value. In the first place, and this is obvious, the basis for imagining the anecumene is experience in this world (López Castro and Ruiz Medrano 2010:241), although it has been said that the former is an inverted duplicate of the latter (Lok 1987:219). This duplicate, however, is not exact in all respects, for the anecumene is imagined to differ considerably from the perceptible world. Human beings use this contrast to explain their worldly reality.

This difference between the anecumene and the ecumene is often measured by the effects of transit between them. Various animate entities in a human being travel to the anecumene, leaving the body's heavy matter behind in this world. The most common means of travel is in dreams, through which gods and the dead may be visited. Those who ingest psychotropics, as well as those who possess abilities to manage the occult, also travel. Sometimes the threshold is crossed by order of the gods or merely by accident. Prayers and the aromas of offerings also travel, but the passage distorts the nature of things (López Austin 1993:70–71, 1996:97–98): smoke from this world turns into tortillas for thunderbolts (*rayos*) in the anecumene (Carrasco 1960:107), food from the anecumene becomes oak leaves or rotten sticks when it crosses into the ecumene (Davis 1963:202–203), and gold extracted in the anecumene turns into excrement here.

Interestingly, the most important transformations occurring during the passage involve space-time and what are clearly different dimensions. For example, Tim Knab (1991:35, 37, 40, 47), referring to Nahuas in the sierra of Puebla, describes an underworld space anchored by a center with four fixed boundary points, where differentiated portals may be used by the specialized traveler to precisely locate a desired site. Thus, traveling at will, in the shadows, from one place to another, does not require one to follow a fixed route.

In terms of time, the traveler who spends an extended period in the anecumene may find upon returning that only an instant has passed in the ecumene; or, conversely, after spending a very short time in the anecumene, one may return to find many years have transpired in the ecumene (Figuerola Pujol 2010:301; Schultze Jena 1977:30–31; Técnicos Bilingües 1985:33). A good example of this phenomenon is the story of two traders who stop at a store to buy cigarettes. One of them goes in, but he and the store disappear before the eyes of his astonished companion. A year later, the companion returns to the same place and sees the store again. He goes in, finds his lost friend, and tells him to leave immediately. The friend, unaware of the passage of time outside, protests because he has not gotten his cigarettes yet (López Palacios 1998:21–24). Similarly, divine beings from the anecumene seem to have an erroneous perception of the passage of time in the ecumene (Durán 1984:2:220–221, 1994:218–219). The explanation is clear: ecumene and anecumene times and spaces have different qualities. For example, in the region of death a deceased Mixtec tells his widow: "Sé que me estás viendo como si fuera yo misma, como si estuviera viva, pero te pido que no me abraces, porque ya no soy parte de tu vida. Estoy en otra parte . . . es otra forma del tiempo [I know you are looking at me as if I were the same, as if I were alive, but I ask you not to embrace me, because I am no longer part of your life. I am in another land . . . ; it is another form of time]" (López Castro and Ruiz Medrano 2010:241). Generally, in the anecumene the traveler communes with the occult, with time past, and with time yet to come, which are likewise perceived by those who consume psychotropics. It was said that *ololiuhqui* revealed where to find a runaway wife (Serna 1953:236). Francisco Hernández (1959:1:105) recorded that Nahuas who used *poyomatli* could see the future, just as Mazatecs claim that those who consume mushrooms can, but they add that those transported to the remote past encounter Adam and Eve (Boege 1988:92). In this total presence, the traveler may observe inverse causal processes, like the regression of years in the mythical Aztlan recorded by Fray Diego Durán (1984:2:221–222, 1994:220).

An important characteristic of the other time can be found in therapeutic practice. According to Hernando Ruiz de Alarcón (1953:163, 1984:190), when a Nahua medic treated a fractured bone, he confronted the infirmity in both the somatic and supernatural realms. In addition to applying a splint to the patient, he would address the fracture in an incantation as "la codorniz señorial originaria del lugar del alboroto que daña el hueso del Mictlan [the lordly quail from the place of the disturbance that damages the bone from Mictlan]" (Ruiz de Alarcón 1953:163, 1984:190). The context is obvious: the medic is situated in the myth where Quetzalcoatl descends to the region of the dead, asks the god Mictlantecuhtli to give him the bones necessary for creating human beings, and when returning with them is disturbed by quail. The startled Quetzalcoatl falls with the bones, and they break into pieces. The god gets back up and gathers the fragments, which he will use to form humans. The medic, in his therapeutic action, travels to the anecumene to access the myth located in the preparatory processual zone, but he reduces the context merely to the episode of the bone fracture. His precise action will have consequences for his patient (López Austin 1975:292–295, 1993:53, 1996:76). His presence occurs at the exact site in the anecumene of a particular mythic process. Likewise, an Ixil transported to the anecumene is unable to gather firewood because he arrives at a mythic moment when trees have not yet spread over the world (Colby and Colby 1981:184, 1986:200).

Sometimes the sources merely provide isolated phrases, brief flashes, that can be quite revealing. For example, a discourse for greeting a king speaks of "ca omito in topan, in mictlan [what was said in the sky, in the region of the dead]" as "ca omocuap[p]ano in tlatolli [the word (that) already crossed the bridge]," and Josefina García Quintana (1980:72–73, 87) interprets this text as the determination of the destiny of humans in the sky and in the underworld, which reaches the ecumene across the liminal space. This expression is in accordance with the Nahuatl text of Cristóbal del Castillo (1991:190–191), who wrote that the figurative sign of the day 1 Water (*atl*) is the word of the great Tlaloc ("in iquechol atl oncan tlatoa in huey Tlaloc").

Time in the Preparatory Processual Zone

Time in the mythic dimension is usually called "the time of darkness," "when there still was no light," or "before the dawn." Based on the aforementioned information, however, two types of

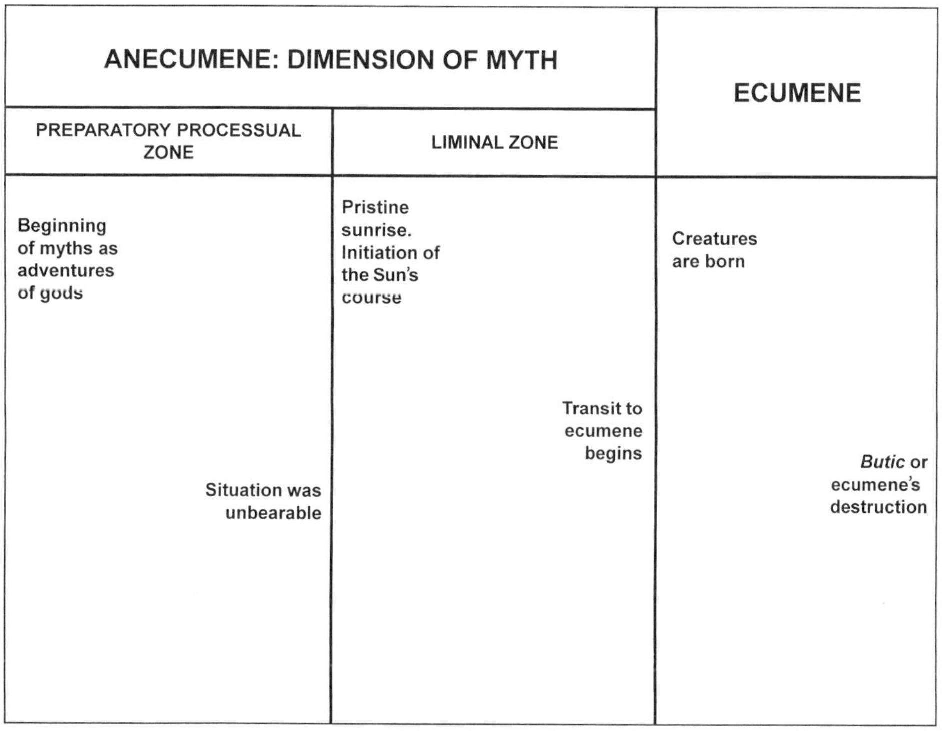

figure 3.6
The limits between both mythic zones and between the liminal zone and ecumene. (Illustration by Alfredo López Austin.)

time may be distinguished in this dimension: one in the preparatory processual zone and the other in the liminal zone (Figure 3.6). The Chol, for example, say about the former: "Hubo un tiempo antiguo, 'Tiempo sin tiempo' . . . cuando el mundo fue formado en esta oscuridad líquida por el padre-madre, antes aún de que el Sol fuera a su vez concebido. En este tiempo 'en el que las palabras carecían de significado y las cosas aún no tenían nombre' [There was an ancient time, a 'time without time' . . . when the world was formed in this liquid darkness by the mother-father, even before the Sun eventually was conceived. In this time 'words lacked meaning and things still had no name']" (García de León 1988:35). Researchers have referred to this time in the preparatory processual zone in different ways—"tiempo estático primordial [primordial static time]" (Garza 2002:53), "the totality of time" (López Austin 1993:53, 1996:76)—or have compared it to eternity (López Austin 1993:61, 1996:87; Thompson 1954:138, 1992:199). Let us analyze these qualifiers.

Mythic time begins, according to Nahuas, with the sin that occasions the expulsion of the gods from Tamoanchan to the surface of the earth and the region of the dead and, in Maya mythology, when the creator forms the k'atuns and gives them the task of creating the things of the world (Montoliu Villar 1989:29). Both myths specifically refer to the processes of creature formation. The characters of the myth (the creator deities) are the elements that will be transformed to produce the creatures of the ecumene. The product may be an individual being (e.g., a star), a kind of being (e.g., maize), part of the cosmic apparatus (e.g., the four trees), a worldly process, a temporal unit, a cycle, a rite, an institution, and so on.

The preparatory processual zone's time is dynamic, for in it the deities acquire the forms and characteristics that the different creatures will possess. The elements in play (the characters in their adventures) fuse and divide. Each mythic process is autonomous (like the teeth of a comb), as the myths have no singular context of succession linking them together (Figure 3.7). Each myth is formed in stages that have a consecutive logical order; however, in this temporal dimension, all of the stages exist simultaneously: they are perennial, since they

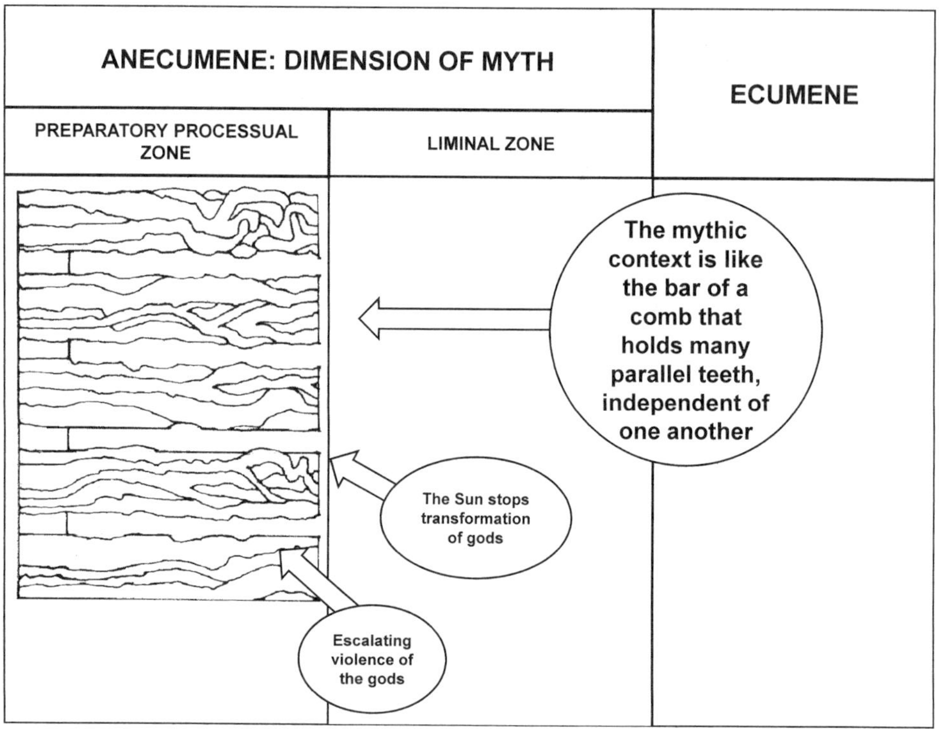

figure 3.7
Five autonomous myths up to the pristine sunrise. (Illustration by Alfredo López Austin.)

are free of the deterioration, death, and existential impermanence that the Sun prescribes as law in its dominions (López Austin and López Luján 2009:235–236). Myth complexes have an absolute presence; they include all possible processes. The preparatory processual zone includes all that was, is, and will be in the ecumene, but also processes never realized but possible. Thus, here, in an eternal present that may be surmised in an immense imaginable past or future, is found everything that transcends the limits of ecumene reality existing between the world's creation and butic.

If cycles are created in these mythic processes in the preparatory processual zone, how can these same processes be measured calendrically? As we saw in the Maya myth of the birth of the month, creation is a strange game: the days are being created with footsteps upon footprints, but at the same time, things in the ecumene are being created in a calendrical cycle that is barely being born. "On 1 Chuen he raised himself to his divinity, after he had made heaven and earth. On 2 Eb he made the first stairway. It descended from the midst of the heavens, in the midst of the water, where there were neither the earth, rocks nor trees" (Book of Chilam Balam of Chumayel 1967:117, 1973:98). The creation occurred in a calendrical sequence that was itself being created (López Austin 1993:45–48, 1996:67–70). For this reason, creatures may have secret names of the day of their creation: trees and objects made of wood are called 1 Water (atl), deer are called 7 Flower (*xochitl*), and so on. And the gods have as their names the dates of myths in which they act—one or more names, according to the number of their participations. So why are some myths measured in years? How can the "Historia de los mexicanos por sus pinturas" (1965:25) say that six hundred years had passed before the ordering of the world? I believe this statement is merely an extrapolation rendering the immeasurable measurable. A few lines later, this source contradicts itself when declaring that "todo lo susodicho fue fecho y criado sin que en ello pongan cuenta de año, sino que fue junto y sin diferencia de tiempo [all of the above was made and created without them giving the year count there, but rather together without any difference in time]" (Historia de los mexicanos por sus pinturas 1965:27).

Other issues arise. Guilhem Olivier (2004) has wondered what to make of the gods appearing in myths and rites with different ages, as if they were subject to the effects of time, and why these ages vary. I think the answer is in the autonomy of processes in the anecumene: a god may play the role of a child in one mythic adventure and the role of a mature or older adult in others. Moreover, some myths, as they are processes, include the age progression of some deities, which can become reality in ecumene time. For example, Mazatecs believe that Chumaje, the mother who feeds the maize field with her milk, is a girl in the morning and becomes older as the day progresses (Boege 1988:151). Federico Navarrete Linares (2004:42) raises a second problem when stating that the gods of the ancient Nahuas are subject to the cycles. His opinion is sound but must be limited to the dominions of the Sun and its laws. In the preparatory processual zone—specifically in myth—the process of closing a temporal unit can occur, but the actual cyclical recurrence that traps the gods is established with the first appearance of the Sun in the sky.

Time in the Liminal Zone and Its Transit into the World

Although the preparatory processual zone and the liminal zone belong to the mythical dimension of the cosmos, and although in some myths it is difficult to perceive the boundaries between the two, their respective qualities of time greatly differ. In order to function, the world needs a cosmic apparatus with passageways and portals that permit the flow of forces and gods in both directions between the anecumene and ecumene space-time dimensions. The principal components of this apparatus are the *axis mundi* and the four cosmic trees. From bottom to top, the axis mundi is basically composed of the underworld or region of death, the sacred mountain, and the great central tree. This trilogy is projected toward the horizontal extremes of the cosmos, reproduced in the four directions (Figure 3.8). The most important transits from anecumene to ecumene and vice versa take place via this triple axis mundi and its four projections and are ruled by the calendrical sequence of the Sun's world.

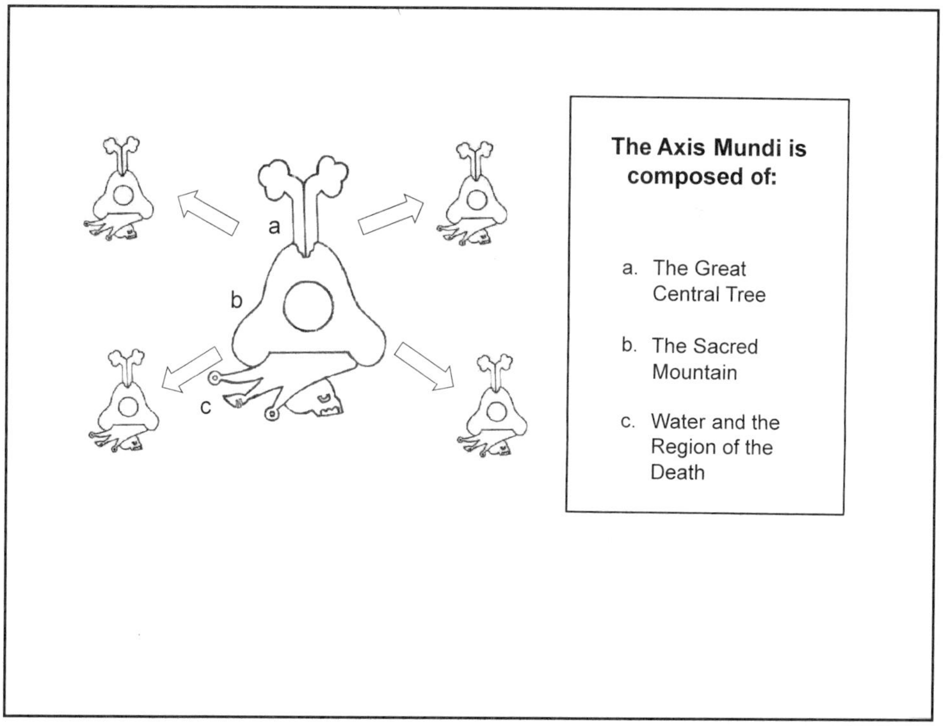

figure 3.8
The axis mundi projected toward the four corners of the world. (Illustration by Alfredo López Austin.)

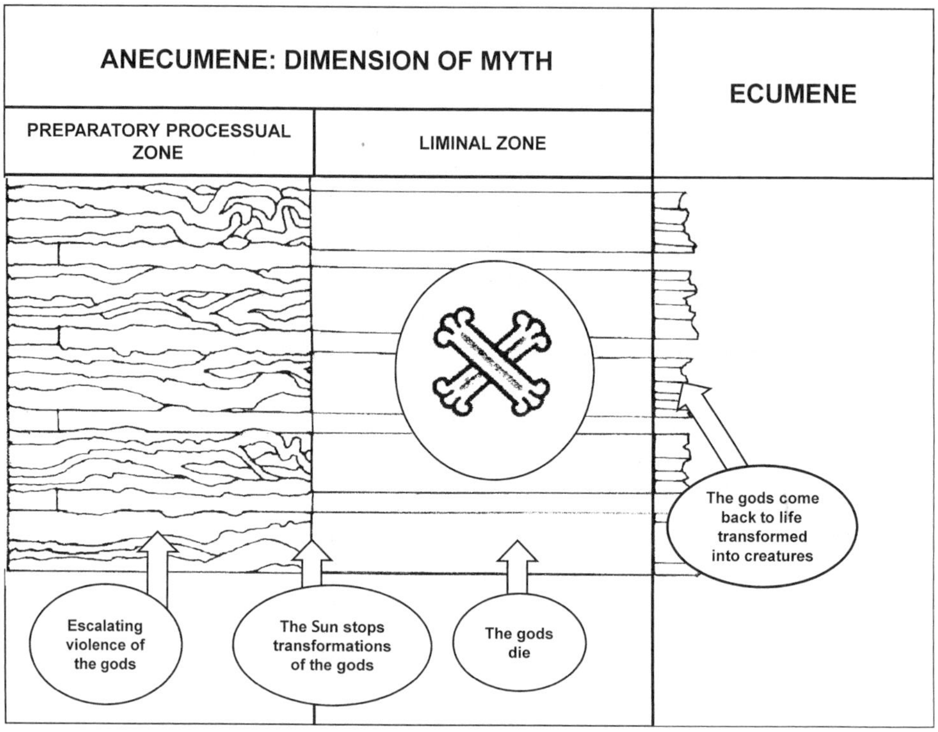

figure 3.9
Five autonomous myths up to the birth of creatures. (Illustration by Alfredo López Austin.)

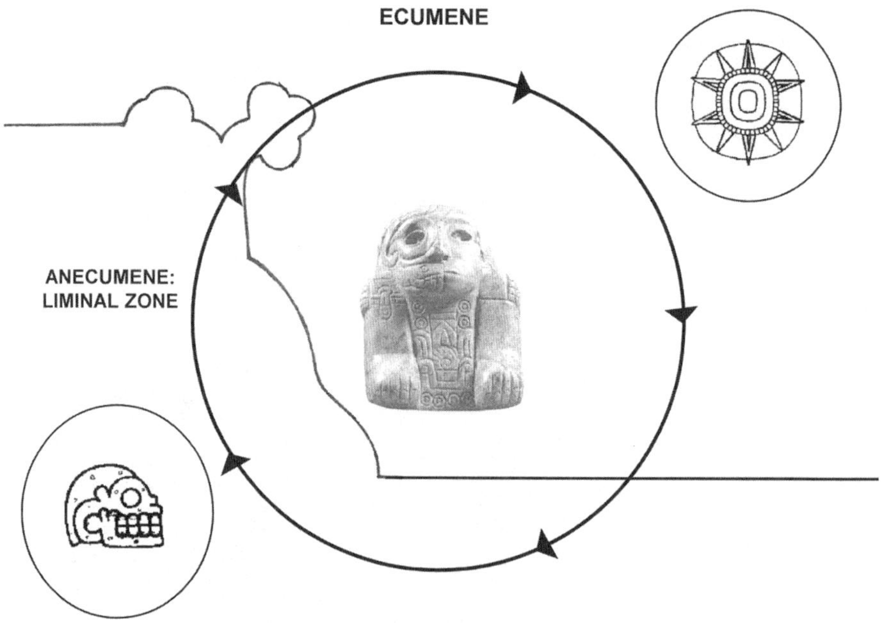

figure 3.10
The life-death cycle of existence. (Illustration by Alfredo López Austin.)

The liminal space-time begins with the pristine appearance of the Sun on the horizon and includes the end of the protean state of the gods: the death of the gods sent into the world, their resurrection, and their preparation for crossing into the ecumene, now trapped in a cycle. It concludes precisely with the transit between the two space-time dimensions (Figure 3.9). We must also consider that not only are all of the cycles created and propelled in the liminal zone, but that half of each cycle occurs there, including the half of the life-death cycle that houses the dead during the time that transpires between what was a worldly life and what will be their return (Figure 3.10).

The processes of creation, as I have said, became the archetype that explains the daily transits in the liminal zone. Under this logic, forces and deities arrive at the liminal zone where they come together and acquire (1) their location in the great cosmic apparatus where the laws of the Sun rule, (2) their sequence, (3) their name, (4) their hierarchy, including (5) their cyclicity, and (6) a covering for crossing into the ecumene. This covering is indispensable for them to work in the world. It may be nearly imperceptible or quite noticeable.

Let us look at an example of the location of time gods in a cycle from the aforementioned Maya myth of the birth of the month, "when the world was not yet created":

> Then occurred the invention of the word of our Lord God, when there was no word in heaven, when there were neither rocks nor trees. Then they went to consider [what they were], and [the voice] spoke as follows: "Thirteen entities, seven entities, one." So it spoke when the words came forth, at the time when there was no word. Then the reason was sought by the first ruling day (the first day Ahau) why the meaning of the word to them was not revealed so that they could declare themselves. Then they went to the center of heaven and joined hands. Then the following names were set up in the middle of the land: the Burners, four of them: 4 Chicchan, the Burner; 4 Oc, the Burner; 4 Men, the Burner; 4 Ahau, the Burner. (Book of Chilam Balam of Chumayel 1967:118, 1973:100)

The specificity ("thirteen entities, seven entities, one") signifies that the month finished integrating its twenty days: the first thirteen were formed with the first series of thirteen numerical signs (the feet); to which were added the seven figurative signs that were missing (the footprints), with a second round of numbers, until arriving at seven. Before the first dawn, the days had not manifested their voice ("when there was no word"). Then they formed a ring in the center of the sky and joined hands to order their succession one by one. For their successive appearance four time gods were installed on the earth as burners, that is, passageways for the twenty days to come out one by one in levorotatory sequence. Their respective names were the tree of the east, the tree of the north, the tree of the west, and the tree of the south and they had

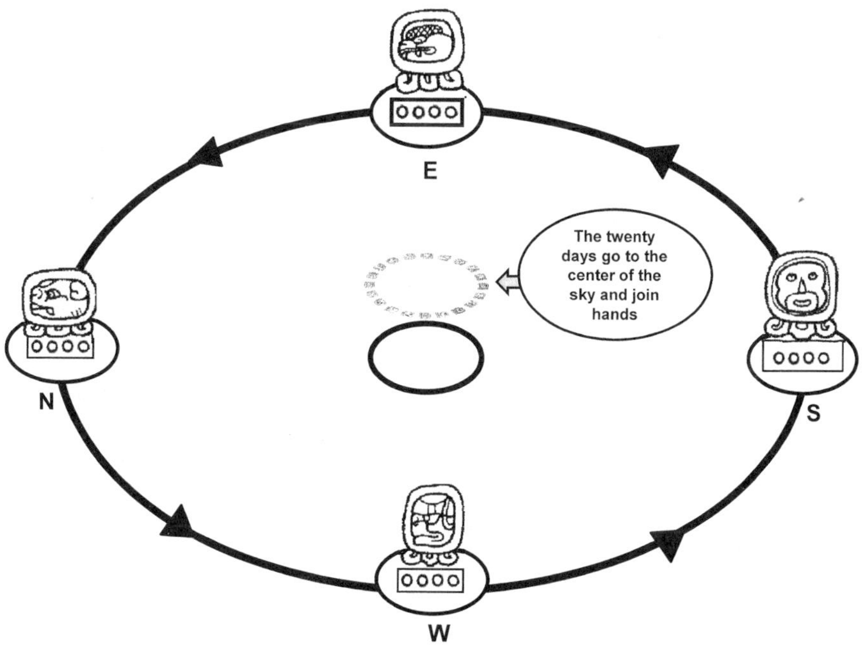

figure 3.11
Calendrical names of the four cosmic trees. (Illustration by Alfredo López Austin.)

the same figurative signs as the four Nahua trees in the "Historia de los mexicanos por sus pinturas" (1965:32; Graulich 1983:579; López Austin 1993:256–257, 1996:328–329): Serpent, Dog, Eagle, and Flower or Lord (Figure 3.11).

Another text refers to the place where the k'atuns were situated to operate in the world: "Se alzará también Yaax Imixché, Ceiba-verde, en el centro de la provincia como señal y memoria del aniquilamiento. Ella es la que sostiene el plato y el vaso; la estera y el trono de los katunes por ella viven [Yaax Imixché, Green Ceiba, will also rise in the center of the province as a symbol and memorial of the annihilation. It supports the plate and the cup; the mat and the throne of the katuns reside there]" (Books of Chilam Balam 1948:155). The text speaks of the most important tree, the central one, the green one, which is one of the elements of the axis mundi. The plate-cup *difrasismo*, or metaphoric couplet, refers to the underworld and the sky. The mat-throne difrasismo signifies "government," equated with the period of influence of each k'atun on the ecumene.

In summary, the time deities are organized sequentially on the world axis to act upon the creatures. The formation of hierarchies includes the temporal units under the rule of a larger unit, which conditions the action of its subordinates or servants. They will be its burden, its retinue (Books of Chilam Balam 1948:4, 226). In the same manner, other larger units are constituted, such as the cycle called *uudz katunob*, or doubling-back k'atuns (Books of Chilam Balam 1948:91). In the succession of governments, while one k'atun is on the throne, the next one resides inside the ruling house as a guest, lavished with attention by the reigning k'atun (Book of Chilam Balam of Chumayel 1967:89, 1973:39; Books of Chilam Balam 1948:204). In this way cyclical orders of temporal action upon the ecumene are established—irruptions that, when not the product of a single time god, will not occasion identical repetitions, but rather a cyclicity of influences over a linear course. This is where a type of movement that some authors have compared to a spiral similar to thread around a spindle (Garza 2002:62; Read 1998:94–96, 106) is produced.

Situated on its throne, each temporal unit is named; it shows its face, "the face that dominates the sky," and emits the "voice of dominion" it was given at the moment of its composition and placement in the cosmic apparatus. With its face

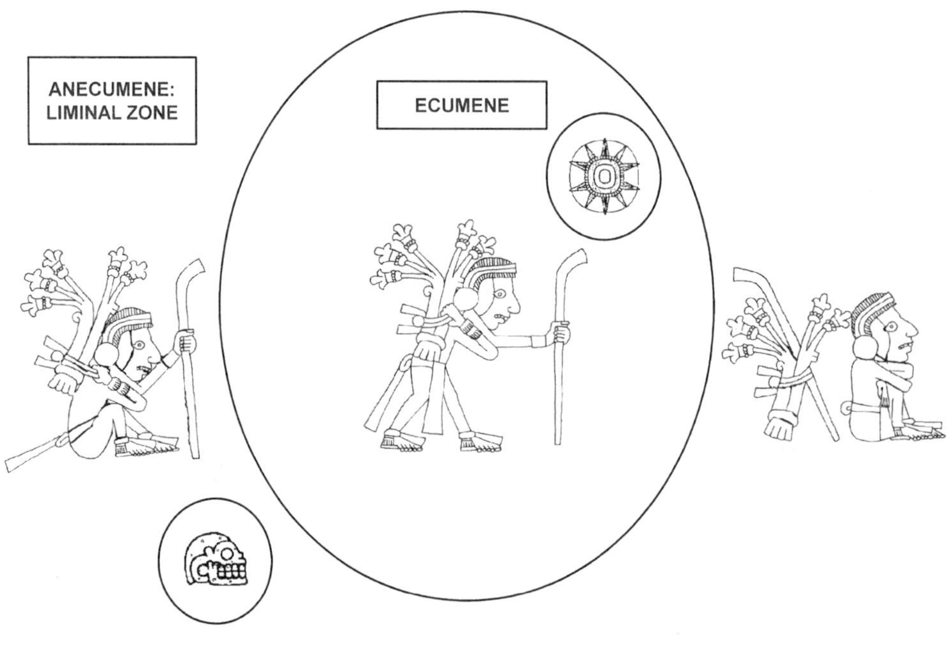

figure 3.12
The burden's transit. (Illustration by Alfredo López Austin.)

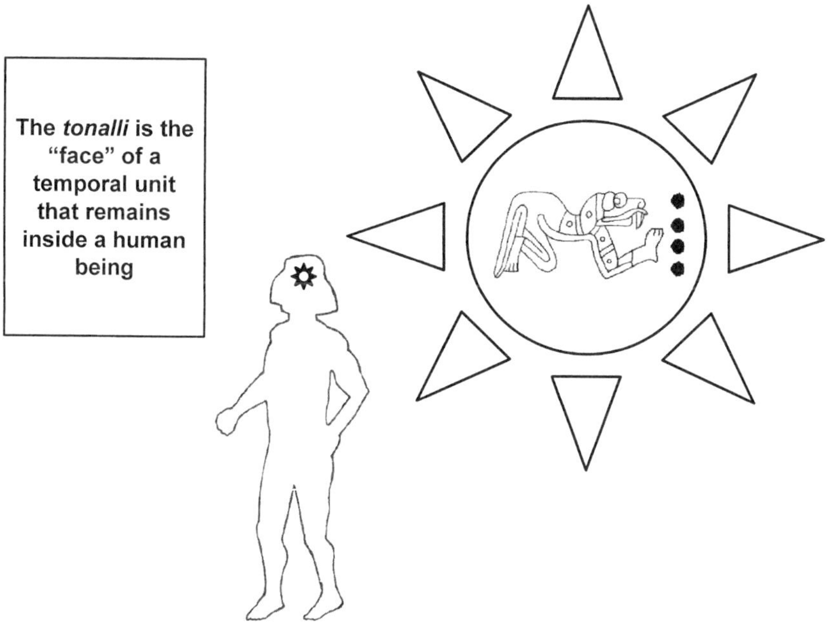

figure 3.13
The *tonalli* as an animate entity. (Illustration by Alfredo López Austin.)

"it becomes present"; this is the nature of its influence. Its voice "reveals the content of its burden" (Books of Chilam Balam 1948:69, 95–96). This voice mentioned in the Maya texts is the one in Nahua texts that was "said in the sky and in the region of death" so that "the word crosses the bridge." Such expressions are repeated today in sacred texts. Bunzel (1952:287) has recorded them in a prayer by a diviner from Chichicastenango who ritually confronts the time gods: "Let me invoke their name and faces, lords." Tedlock (1982:110) says that among the K'iche', the "face of his day" (*uwäch uk'ij*) is compared to a nahual (nawal) that penetrates the child born on the date of the time god's rule and thus produces what should be understood as a co-essence.

In order to cross from anecumene to ecumene a ritual is necessary, as one observes in a study Bruce Love (1994:18–25) conducted of the k'atuns in the Codex Paris. Love was concerned with the nature of the process. Is it a bloody ritual? Is it necessary for the gods, and specifically the time gods, to die so that they may govern in the world? The mythic archetype of the transformation of deities destined for the ecumene says that they previously died. Must the gods be sacrificed in order to travel to the ecumene? The answer is still pending.

The crossing initiates the work of the gods in the ecumene. Ancient Nahua texts say that time rises (*moquetza*; Sahagún 1950–1982:bk. 4; 2000:1:342–432). Both in antiquity (Thompson 1954:138, 1992:197) and in our times (Bricker 1980), crossing into the ecumene is compared to the journey a carrier or bearer makes with a load on his back (Figure 3.12). The load or burden is destiny. Both bearer and burden fuse together to form a unit of time while performing their function. They are then one god, as is characteristic among Mesoamerican deities. The work is tiring, and humankind must contribute to divine recuperation with prayers, rites, and food. The worship must be opportune, whether in the time of arrival or the time of ruling in the ecumene (Books of Chilam Balam 1948:227). These are the resources that humankind possesses to influence this will of the gods and mitigate the force of their "words." In the anecumene, the gods lay down their burden, recover their lost energy with food and rest, and are able to return to their dimension of origin.

The previously mentioned myth of the expulsion of the gods from Tamoanchan says: "And because of this some came to earth and others to hell, and these are the ones who terrify them" (Codex Telleriano-Remensis 1995:13r, 266). This myth may

figure 3.14
The three different types of time. (Illustration by Alfredo López Austin.)

ANECUMENE: DIMENSION OF MYTH		ECUMENE
PREPARATORY PROCESSUAL ZONE	LIMINAL ZONE	
TIME AS ABSOLUTE PRESENCE	TIME AS PROCESSES OF TRANSIT	TIME AS TRANSPIRING PRESENT

be interpreted as the capture of the gods within cycles, rotating between the ecumene and the liminal zone of the anecumene. Some of them, however, remain on earth and function as guardians and meteorological phenomena (Knab 1991:52). In the liminal zone—whether in the region of the dead, inside the sacred mountain, or in the passageways for leaving the world again via the cosmic tree—the creator deities are in their cyclical death phase of existence or are guardians and meteorological phenomena in repose. Accounts of journeys to the interior of the sacred mountain or to the region of death abound in which the aforementioned time distortion experienced by travelers is often found. The inversion of space that Rossana Lok (1987:219) mentions has its correspondence in time, for in the underworld day/night and wet/dry season relationships are the inverse of those that occur on the surface of the earth (Grigsby n.d.; Knab 1991:52). Spatial inversion is produced by the orbital course of the dead Sun, whereas temporal inversion is a product of an earlier cyclical stage in the anecumene and the predecessor of the stage that follows. As for the inhabitants of the liminal zone, why were they "the ones who terrify them"? An idea existed in antiquity and persists in our day that the beings from the time before creation are savages, primitives, cannibals—that is, anterior to the laws of the Sun. They are the dangerous *tzitzimime* (Sahagún 1950–1982:6:37, 2000:2:507). They resent that they were displaced by the definitive creatures and await the right moment to bring an end to them and occupy the surface of the earth. They have been at the point of destroying human beings with the help of the Virgin of Mercy and Jesus Christ, but fortunately the good Saint Thomas intervened and impeded their attack (Villa Rojas 1990:748–749).

If the destiny of the deities expelled from Tamoanchan is to remain prisoners of the cycle, what happens to the temporal units? Recall the image of the bearers. When concluding their task in the ecumene, they are so fatigued that they must replenish their forces. Thompson (1954:138–139, 1992:197–198) observes how time deities, tired at the end of their journey, laid down their burden. Among the Ixil, the Sun concludes its work sweaty and tired and is received at the end of the day by people on their knees (Colby and Colby 1981:38, 1986:51). This information suggests that leaving, as well as entering, the ecumene requires a ritual in the liminal zone.

But time does not entirely leave the ecumene. Some pieces of time remain in the world, accumulating inside humans as the years pass, warming the hearts of the elderly. Other pieces remain, also within people, as co-essences or animate entities, such as *tonalli* (Figure 3.13). Time imparts its nahual (nawal) to them, as Tedlock (1982:110) records. Even after a person dies, a piece of time remains in the form of a name, for we see that a vestige of its ancient identification survives in the K'iche' word *k'ix*, which means "sunlight," "day," and also "name" (Bunzel 1952:285–286). This name does not go away. Among ancient Nahuas, the deceased had an "earthly name" (*tlalticpactoca*) that could "be borne" (*quitocamama*) by a grandson to "enhance the tonalli" (*quitonaleoa*) of the grandfather (Sahagún 1950–1982:6:203–204, 1979:2:173r, 2000:2:646).

Conclusion

When we are comparing Mesoamerican concepts concerning time in the anecumene and in the ecumene, three different types can be distinguished. The preparatory processual zone of the anecumene has processes that, although divided logically into successive stages, do not expire in the sense of one giving way to the next. They are all perennial. It is a dynamic in which stages remain, unaffected by the laws of the Sun, to mark the transition from existence. Anyone who willfully visits this zone is situated in the proper moment and stage of the desired myth. This type of time may be described as *absolute presence*.

In the liminal zone of the anecumene, the laws of the Sun establish the cycles and the order of the sequences. Transits to the ecumene are prepared, and gods concluding their stay in the ecumene are received and thus continue in the anecumene part of the cycle. The gods are perennial—they are not destroyed—but they are subject to existential changes in life and in death. This time may be described as *processes of transit*.

Finally, in the ecumene, time exists only sequentially. Its appearance is strictly determined by the laws of the Sun and lasts only for the interval established for its passage. Neither the past nor the future exists here. This type of time may be described as the *transpiring present* (Figure 13.14).

Acknowledgments

I would like to thank those who helped me in the preparation of this work, especially Leonardo López Luján, for his detailed, critical reading of the text, and Scott Sessions, translator of the original Spanish into English.

REFERENCES CITED

Albores, Beatriz
1997 Los quicazcles y el árbol cósmico del Olotepec, estado de México. In *Graniceros: Cosmovisión y meteorología indígenas de Mesoamérica*, edited by Beatriz Albores and Johanna Broda, pp. 379–446. El Colegio de Mexiquense, Zinacantepec, and Instituto de Investigaciones, Universidad Nacional Autónoma de México, Mexico City.

Boege, Eckart
1988 *Los mazatecos ante la nación: Contradicciones de la identidad étnica en el México actual.* Siglo Veintiuno, Mexico City.

Book of Chilam Balam of Chumayel
1967 *The Book of Chilam Balam of Chumayel.* Translated and edited by Ralph L. Roys. New ed. University of Oklahoma Press, Norman.

1973 *Libro de Chilam Balam de Chumayel.* 3rd ed. Translated and edited by Antonio Mediz Bolio. Universidad Nacional Autónoma de México, Mexico City.

Books of Chilam Balam

1948 *El libro de los libros de Chilam Balam.* Translated and edited by Alfredo Barrera Vásquez and Silvia Rendón. Fondo de Cultura Económica, Mexico City.

Braudel, Fernand

1958 Histoire et sciences sociales: La longue durée. *Annales: Économies, sociétés, civilisations* 13(4):725–753.

1974 La larga duración. In *La historia y las ciencias sociales*, translated by Josefina Gómez Mendoza, pp. 60–106. 3rd ed. Alianza, Madrid.

1980 History and the Social Sciences: The *Longue Durée.* In *On History*, translated by Sarah Matthews, pp. 25–54. University of Chicago Press, Chicago.

Bricker, Victoria Reifler

1980 El hombre, la carga y el camino: Antiguos conceptos mayas sobre tiempo y espacio, y el sistema zinacanteco de cargos. In *Los zinacantecos: Un pueblo tzotzil de los altos de Chiapas*, edited by Evon Z. Vogt, pp. 355–370. Reprint of the 1st ed. Instituto Nacional Indigenista, Mexico City.

1981 *The Indian Christ, the Indian King: The Historical Substrate of Maya Myth and Ritual.* University of Texas Press, Austin.

1993 *El Cristo indígena, el rey nativo: El sustrato histórico de la mitología del ritual de los mayas.* Translated by Cecilia Paschero. Fondo de Cultura Económica, Mexico City.

Bunzel, Ruth Leah

1952 *Chichicastenango: A Guatemalan Village.* Publications of the American Ethnological Society 22. University of Washington Press, Seattle.

Carmack, Robert M.

1979 *Historia social de los quichés.* Ministerio de Educación, Guatemala City.

Carrasco, Pedro

1960 Pagan Rituals and Beliefs among the Chontal Indians of Oaxaca, Mexico. *Anthropological Records* 20(3):i–v, 87–115.

Casas, Bartolomé de las

1967 *Apologética historia sumaria.* Edited by Edmundo O'Gorman. 2 vols. Instituto de Investigaciones Históricas, Universidad Nacional Autónoma de México, Mexico City.

Castillo, Cristóbal del

1991 *Historia de la venida de los mexicanos y otros pueblos, e Historia de la conquista.* Translated and edited by Federico Navarrete Linares. Instituto Nacional de Antropología e Historia, Proyecto Templo Mayor, GV, and Asociación de Amigos del Templo Mayor, Mexico City.

Codex Telleriano-Remensis

1995 *Codex Telleriano-Remensis: Ritual, Divination, and History in a Pictorial Aztec Manuscript.* Facsimile ed. and study by Eloise Quiñones Keber. University of Texas Press, Austin.

Codex Vaticanus A

1996 *Códice Vaticano A 3738.* Facsimile ed. by Ferdinand Anders, Maarten Jansen, and Luis Reyes García. Akademische Druck- und Verlagsanstalt, Graz, and Fondo de Cultura Económica, Mexico City.

Codex Vindobonensis

1992 *Códice Vindobonensis Mexicanus I.* Facsimile ed. by Ferdinand Anders, Luis Reyes García, Maarten Jansen, and Gabina Aurora Pérez Jiménez. Sociedad Estatal Quinto Centenario, Madrid; Akademische Druck- und Verlagsanstalt, Graz; and Fondo de Cultura Económica, Mexico City.

Colby, Benjamin N., and Lore M. Colby

1981 *The Daykeeper: The Life and Discourse of an Ixil Diviner.* Harvard University Press, Cambridge, Mass.

1986 *El contador de los días: Vida y discurso de un adivino ixil.* Translated by Juan José Utrilla. Fondo de Cultura Económica, Mexico City.

Dakin, Karen

1977 Pedro Cuaresma and Other Nahuatl Stories. *Tlalocan* 7:47–66.

Davis, Marjorie

1963 Cuicatec Tales about Witchcraft. *Tlalocan* 4(3):197–203.

Durán, Diego

1971 *Book of the Gods and Rites and the Ancient Calendar.* Translated and edited by Fernando Horcasitas and Doris Heyden. University of Oklahoma Press, Norman.

1984 *Historia de las indias de Nueva España e islas de tierra firme.* 2 vols. 2nd ed. Porrúa, Mexico City.

1994 *The History of the Indies in New Spain.* Translated and edited by Doris Heyden. University of Oklahoma Press, Norman.

Fagetti, Antonella

2003 *Los que saben: Testimonios de vida de médicos tradicionales de la región de Tehuacán.* Instituto de Ciencias Sociales y Humanidades, Benemérita Universidad Autónoma de Puebla, and Comisión Nacional para el Desarrollo de los Pueblos Indígenas, Puebla.

Figuerola Pujol, Helios

2010 *Los dioses, los hombres y las palabras en la comunidad de San Juan Evangelista Cancuc en Chiapas.* Universidad Nacional Autónoma de México, Mérida.

García de León, Antonio

1988 *Resistencia y utopía: Memorial de agravios y crónica de revueltas y profecías acaecidas en la provincia de Chiapas durante los últimos quinientos años de su historia.* 2nd ed. Era, Mexico City.

García Quintana, Josefina

1980 Salutación y súplica que hacía un principal al *tlatoani* recién electo. *Estudios de cultura náhuatl* 14:65–94.

Garza, Mercedes de la

1975 *La conciencia histórica de los antiguos mayas.* Centro de Estudios Mayas, Coordinación de Humanidades, Universidad Nacional Autónoma de México, Mexico City.

2002 Origen, estructura y temporalidad del cosmos. In *Religión maya*, edited by Mercedes de la Garza Camino and Martha Ilia Nájera Coronado, pp. 53–81. Trotta, Madrid.

Gossen, Gary H.

1972 Temporal and Spatial Equivalents in Chamula Ritual Symbolism. In *Reader in Comparative Religion: An Anthropological Approach*, edited by William A. Lessa and Evon Z. Vogt, pp. 135–149. Harper and Row, New York.

Graulich, Michel

1983 Myths of Paradise Lost in Pre-Hispanic Central Mexico. *Current Anthropology* 24(5):578–588.

Grigsby, Thomas L.

n.d. The Tepemaxtla as the Nagual of the Transfigured Christ. In press.

Gutiérrez González, María Eugenia

2008 El paso del katún: La personificación del tiempo entre los mayas del Clásico. MA thesis, Facultad de Filosofía y Letras, Universidad Nacional Autónoma de México, Mexico City.

Hermitte, M. Esther

1964 Supernatural Power and Social Control in a Modern Mayan Village. PhD dissertation, Department of Anthropology, University of Chicago, Chicago.

1970 *Poder sobrenatural y control social en un pueblo maya contemporáneo.* Translated by Carmen Viqueira. Departamento de Antropología, Instituto Indigenista Interamericano, Mexico City.

Hernández, Francisco

1959 *Historia natural de Nueva España.* 2 vols. Obras completas 2–3. Universidad Nacional Autónoma de México, Mexico City.

Historia de los mexicanos por sus pinturas

1965 In *Teogonía e historia de los mexicanos: Tres opúsculos del siglo XVI*, edited by Ángel M. Garibay K., pp. 21–90. Porrúa, Mexico City.

Histoyre du Mechique

1905 Histoyre du Mechique, manuscrit française inédit du XVIe siècle. Edited by Édouard de Jonghe. *Journal de la Société des Américanistes* 2:1–41.

1965 Historia de México. In *Teogonía e historia de los mexicanos: Tres opúsculos del siglo XVI*, edited by Ángel M. Garibay K., pp. 91–120. Porrúa, Mexico City.

Houston, Stephen D., and David Stuart

1989 *The Way Glyph: Evidence for "Co-essences" among the Classic Maya.* Research Reports on Ancient Maya Writing 30. Center for Maya Research, Washington, D.C.

Knab, Tim J.

1991 Geografía del Inframundo. *Estudios de cultura náhuatl* 21:31–57.

La Farge, Oliver, and Douglas S. Byers

1931 *The Year Bearer's People.* Department of Middle American Research, Tulane University, New Orleans, La.

León-Portilla, Miguel

1963 *Aztec Thought and Culture: A Study of the Ancient Nahuatl Mind.* Translated by Jack Emory Davis. University of Oklahoma Press, Norman.

1968 *Tiempo y realidad en el pensamiento maya: Ensayo de acercamiento.* Instituto de Investigaciones Históricas, Universidad Nacional Autónoma de México, Mexico City.

1973 *Time and Reality in the Thought of the Maya.* Translated by Charles L. Boilès and Fernando Horcasitas. Beacon Press, Boston.

1974 *La filosofía náhuatl estudiada en sus fuentes.* 4th ed. Instituto de Investigaciones Históricas, Universidad Nacional Autónoma de México, Mexico City.

Lok, Rossana

1987 The House as a Microcosm. In *The Leiden Tradition in Structural Anthropology: Essays in Honor of P. E. de Josselin de Jong,* edited by R. de Ridder and Jan A. J. Karremans, pp. 211–223. E. J. Brill, Leiden.

López Austin, Alfredo

1975 Algunas ideas acerca del tiempo mítico entre los antiguos nahuas. In *Historia, religión, escuelas: XIII Mesa Redonda,* edited by Jaime Litvak King, pp. 289–298. Sociedad Mexicana de Antropología, Mexico City.

1983 Nota sobre la fusión y la fisión de los dioses en el panteón mexica. *Anales de antropología* 20(2):75–87.

1993 *The Myths of the Opossum: Pathways of Mesoamerican Mythology.* Translated by Bernard Ortiz de Montellano and Thelma Ortiz de Montellano. University of New Mexico Press, Albuquerque.

1996 *Los mitos del tlacuache.* 3rd ed. Universidad Nacional Autónoma de México, Mexico City.

2001 El núcleo duro, la cosmovisión y la tradición mesoamericana. In *Cosmovisión, ritual e identidad de los pueblos indígenas de México,* edited by Johanna Broda and Félix Báez-Jorge, pp. 47–65. Consejo Nacional para la Cultura y las Artes and Fondo de Cultura Económica, Mexico City.

López Austin, Alfredo, and Leonardo López Luján

2009 *Monte sagrado–Templo Mayor: El cerro y la pirámide en la tradición religiosa mesoamericana.* Universidad Nacional Autónoma de México and Instituto Nacional de Antropología e Historia, Mexico City.

López Castro, Hermenegildo F., and Ethelia Ruiz Medrano

2010 *Tutu Ñuu Oko: Libro del Pueblo Veinte; Relatos de la tradición oral mixteca de Pinotepa Nacional, Oaxaca.* Centro de Investigaciones y Estudios Superiores en Antropología Social and Instituto Nacional de Lenguas Indígenas, Mexico City.

López Palacios, Juan

1998 *Relatos de Xoxocotla.* Instituto de Investigaciones Antropológicas, Universidad Nacional Autónoma de México, Mexico City.

Love, Bruce

1994 *The Paris Codex: Handbook for a Maya Priest.* University of Texas Press, Austin.

Lupo, Alessandro

1995 *La tierra nos escucha: La cosmología de los nahuas a través de las súplicas rituales.* Consejo Nacional para la Cultura y las Artes and Instituto Nacional Indigenista, Mexico City.

Mendieta, Gerónimo de

1945 *Historia eclesiástica indiana.* 4 vols. Salvador Chávez Hayhoe, Mexico City.

Montoliu Villar, María

1989 *Cuando los dioses despertaron: Conceptos cosmológicos de los antiguos mayas de Yucatán estudiados en el Chilam Balam de Chumayel.* Instituto de Investigaciones Antropológicas, Universidad Nacional Autónoma de México, Mexico City.

Morley, Sylvanus G.

1946 *The Ancient Maya.* Stanford University Press, Stanford, Calif.

1972 *La civilización maya.* Translated by Adrián Recinos. 2nd ed. Fondo de Cultura Económica, Mexico City.

Navarrete Linares, Federico

2004 ¿Dónde queda el pasado?: Reflexiones sobre los cronotopos históricos. In *El historiador frente a la historia: El tiempo en Mesoamérica*, edited by Virginia Guedea, pp. 29–52. Universidad Nacional Autónoma de México, Mexico City.

Neff Nuixa, Françoise

2001 La Lucerna y el Volcán Negro. In *La montaña en el paisaje ritual*, edited by Johanna Broda, Stanislaw Iwaniszewski, and Arturo Montero, pp. 353–373. Instituto de Ciencias Sociales y Humanidades, Universidad Nacional Autónoma de México, Consejo Nacional para la Cultura y las Artes, and Instituto Nacional de Antropología e Historia, Mexico City.

Olivier, Guilhem

2004 También pasan los años por los dioses: Niñez, juventud y vejez en la cosmovisión mesoamericana. In *El historiador frente a la historia: El tiempo en Mesoamérica*, edited by Virginia Guedea, pp. 149–180. Universidad Nacional Autónoma de México, Mexico City.

Popol Vuh

1950 *Popol Vuh: The Sacred Book of the Ancient Quiché Maya.* Translated and edited by Delia Goetz, Sylvanus G. Morley, and Adrián Recinos. University of Oklahoma Press, Norman.

1964 *Popol Vuh: La antiguas historias del quiché.* Translated and edited by Adrián Recinos. 7th ed. Fondo de Cultura Económica, Mexico City.

Read, Kay Almere

1998 *Time and Sacrifice in the Aztec Cosmos.* Indiana University Press, Bloomington.

Reyes García, Luis

2001 *¿Cómo te confundes? ¿Acaso no somos conquistados?: Anales de Juan Bautista.* Centro de Investigaciones y Estudios Superiores en Antropología Social, Biblioteca Lorenzo Boturini, Insigne, y Nacional Basílica de Guadalupe, Mexico City.

Ruiz de Alarcón, Hernando

1953 Manual de ministros de indios. In *Tratado de las idolatrías, supersticiones, dioses, ritos, hechicerías y otras costumbres gentílicas de las razas aborígenes de México*, by Jacinto de la Serna et al., edited by Francisco del Paso y Troncoso, vol. 2, pp. 17–130. Fuente Cultural, Mexico City.

1984 *Treatise on the Heathen Indian Superstitions That Today Live among the Indians Native to This New Spain, 1629.* Translated and edited by J. Richard Andrews and Ross Hassig. University of Oklahoma Press, Norman.

Sahagún, Bernardino de

1950–1982 *General History of the Things of New Spain: Florentine Codex.* Translated and edited by Arthur J. O. Anderson and Charles E. Dibble. 13 vols. School of American Research and University of Utah, Santa Fe, N.Mex.

1979 *Códice Florentino: Manuscrito 218–20 de la Colección Palatina de la Biblioteca Medicea Laurenziana.* 3 vols. Archivo General de la Nación, Secretaría de Gobernación, Mexico City.

2000 *Historia general de las cosas de Nueva España.* Edited by Alfredo López Austin and Josefina García Quintana. 3 vols. Consejo Nacional para la Cultura y las Artes, Mexico City.

Schultze Jena, Leonhard

1954 *La vida y las creencias de los indígenas quichés de Guatemala.* Translated by Antonio Goubaud Carrera and Herbert D. Sapper. 2nd ed. Ministerio de Educación Pública, Guatemala City.

1977 *Mitos y leyendas de los pipiles de Izalco.* Translated by Gloria Menjivar Rieken and Armida Parada Fortin. Cuscatlán, San Salvador.

Serna, Jacinto de la

1953 Tratado de las supersticiones y costumbres gentílicas que oy viuen entre los indios naturales de esta Nueva España. In *Tratado de las idolatrías, supersticiones, dioses, ritos, hechicerías y otras costumbres gentílicas de las razas aborígenes de México,* edited by Francisco del Paso y Troncoso, vol. 1, pp. 47–368. Fuente Cultural, Mexico City.

Shaw, Mary

1972 *Según nuestros antepasados . . . : Textos folklóricos de Guatemala y Honduras.* Instituto Lingüístico de Verano, Guatemala City.

Taller de Tradición Oral

2009 *Tejuan tikintenkakiliaj in toueyitatajuan: Les oíamos contar a nuestros abuelos.* Taller de Tradición Oral de la Sociedad Agropecuaria del Centro de Estudios y Promoción Educativa para el Campo, Puebla, and Instituto Nacional de Antropología e Historia, Mexico City.

Técnicos Bilingües

1985 *Agua, mundo, montaña: Narrativa nahua, mixe y popoluca del Sur de Veracruz.* Técnicos Bilingües de la Unidad Regional de Acayucan, Premiá, and Dirección General de Culturas Populares, Tlahuapan, Puebla.

Tedlock, Barbara

1982 *Time and the Highland Maya.* University of New Mexico Press, Albuquerque.

Thompson, J. Eric S.

1954 *The Rise and Fall of Maya Civilization.* University of Oklahoma Press, Norman.

1970 *Maya History and Religion.* University of Oklahoma Press, Norman.

1975 *Historia y religión de los mayas.* Translated by Félix Blanco and Arturo Gómez. Siglo Veintiuno, Mexico City.

1978 *Maya Hieroglyphic Writing: An Introduction.* 3rd ed. University of Oklahoma Press, Norman.

1992 *Grandeza y decadencia de los mayas.* Translated by Lauro José Zavala. 3rd ed. Fondo de Cultura Económica, Mexico City.

Villa Rojas, Alfonso

1990 *Etnografía tzeltal de Chiapas: Modalidades de una cosmovisión prehispánica.* Consejo Estatal para el Fomento a la Investigación y Difusión de la Cultura, Gobierno del Estado de Chiapas, Tuxtla Gutiérrez.

4

When Pre-Sunrise Beings Inhabit a Post-Sunrise World

Time, Animate Objects, and Contemporary Tz'utujil Maya Ritual Practitioners

LINDA A. BROWN

> The Past, the Present and the Future present a unified whole as healthy as it is possible to present under the circumstances in which Atitecos[1] have lived for the past four centuries. It is, I think, possible to show that such a binding of all times into one Time represents a strong and permanent current of feeling and thinking in Atitlán.
>
> —E. Michael Mendelson, "Religion and World View in Santiago Atitlán"

For contemporary Maya ritual practitioners, concepts of time and the past can take overtly material and innately permeable forms. In the highlands of Guatemala, Tz'utujil ritual practitioners routinely collect small objects—including antiquities—for use in ceremonies, a practice long noted among previous Mesoamerican peoples as well (e.g., Coe 1966; Coe and Diehl 1980; Hamann 2002; Miller 1973; Miller 1986; Smith 2003; Umberger 1987). Carefully stored on altars and in personal divination bundles, these items hold a unique ontological status as they are considered animate manifestations of the past in the present (Figure 4.1).

In this essay, I discuss insights from research I conducted with contemporary ritual practitioners who routinely engage with animate objects in the course of daily life. The observation that many of these items were Pre-Columbian artifacts especially intrigued me, so I developed a project to explore what these items meant to their indigenous curators and how they used them. When I began this endeavor, I approached sacred assemblages as collections of things. But over time, I realized that to understand what these materials meant to their caretakers, I needed to shift my perspective from viewing them as collections of things to viewing them as assemblies of beings. They are gatherings of potent actors—pre-sunrise beings—who come from a previous era prior to the creation of human beings and our present sun.

As an inseparable conjunction of ancestral being, ancestral power, and ancestral thing, animate sacred materials dissolve a number of

figure 4.1
Contents of an open bundle containing archaeological and natural sacred objects; the bundle was recently fed tobacco as indicated by the presence of three cigarette butts and ash. (Photograph by Linda A. Brown.)

modernist assumptions about the world. The two I approach in this essay are (1) the framing of time as a linear, one-way progression of quantifiable, equal units (in this scenario, the past is over and the future is not yet here as "time marches on," forming a backdrop against which other activities take place); and (2) a rational dualist perspective that separates living beings from inert things. Much more will be said about both of these themes during the course of this essay, but it is evident that the "binding of all times into one Time," noted by the ethnographer Michael Mendelson (1957:497) in his research in Santiago Atitlán during the 1950s, is still prevalent among Tz'utujil ritual practitioners today.

A central concern for me throughout this research has been my ability as a Western scholar thoroughly entrenched in my own cultural frameworks to understand and adequately communicate what ritual practitioners were telling me: that the past emerges in the present through engagement with potent animate objects. In this process, I would be remiss if I did not give credit to my friend and Tz'utujil research collaborator Dolores, with whom I have worked for many years. Dolores is from a family that includes some powerful ritual practitioners. Through our mutual friendship, she opened the pathway for the interviews I discuss here. She showed great patience when she encountered my own questioning rationalism and thoughtfully explained to me just how things really work. Moreover, she served a key role in translating Tz'utujil during interviews.

A significant aspect of the experience of time emerges in the practice of engaging with "found" animate materials. Although I use the word *found*, it does not capture an individual's experience of this practice, as there is nothing random in these encounters. Instead, ritual practitioners understand that sacred objects are placed or, more appropriately, place themselves in their pathways. Typically, individuals begin to encounter sacred materials when they are undergoing divine election. In fact, the appearance of such an object plays a key role in validating one's destiny to walk a sacred path and in confirming that one has a divine mandate for such a calling.

In the following pages, I open with a discussion of the interviews and the methods I employed. I then present a framework for understanding animate objects by discussing how Tz'utujil ritual practitioners conceptualize ancestral beings and personhood. From there, I turn my attention to the central importance of dreams in communications with powerful pre-sunrise beings. In the final section of the essay, I present a more detailed discussion of two specific types of animate objects and examine the pre-sunrise beings associated with each, discussing how individuals incorporate them into their ritual practices. Finally, using the ethnographically rich perspectives of ritual practitioners, I turn to the material record to suggest new insights into the ancient Maya past.

Interviews with Tz'utujil Ritual Practitioners

Tz'utujil ritual practitioners collect a wide variety of objects as sacred materials. Home altars and personal bundles house assortments of stones, seeds, and unique cultural items that hold special resonance for their curators. I (Brown 2000) initially became interested in such "ritual collecting" when I noticed that archaeological objects were frequent components of sacred assemblages, an observation that prompted many questions. What did these artifacts mean to the individuals who routinely

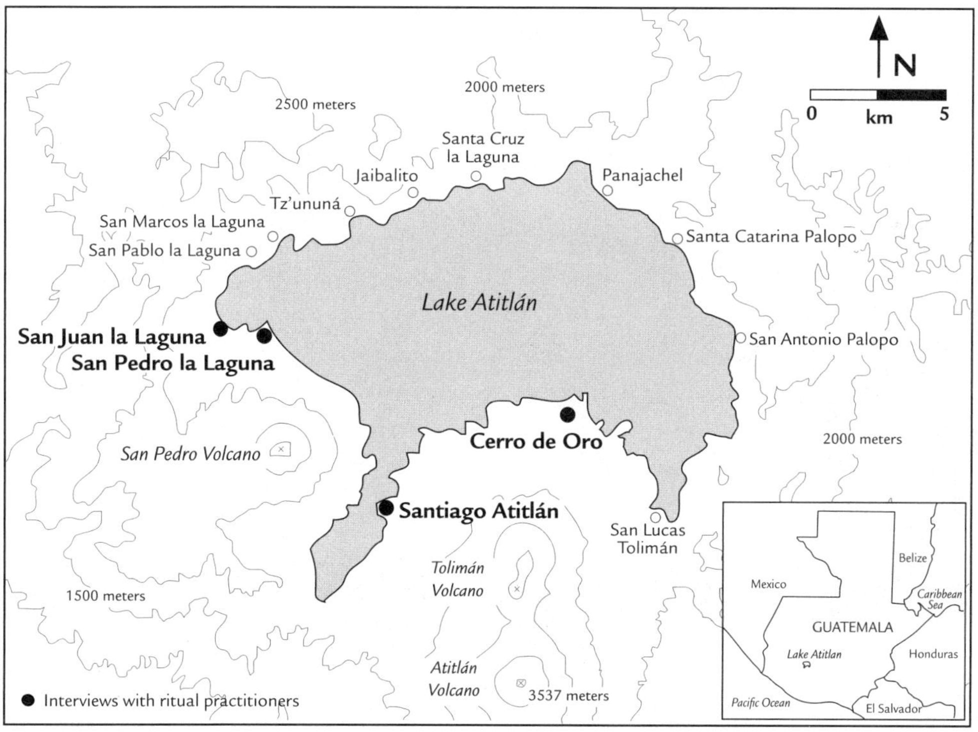

figure 4.2 Map of Lake Atitlán, showing Tz'utujil communities where we conducted interviews. (Map by Linda A. Brown.)

included them in their ceremonial practices? Which objects were collected, and in what manner were they used? How might contemporary meanings enrich and inform our ideas about archaeological materials as well as "indigenous archaeologies" and collecting practices in the ancient past?

With these questions in mind, I developed an archaeologically focused short-term ethnographic project in the Tz'utujil-speaking communities along the southwestern side of Lake Atitlán. I primarily spoke with people from the town of Santiago Atitlán, but I also conducted interviews with several individuals from the nearby communities of Cerro de Oro, San Pedro la Laguna, and San Juan la Laguna (Figure 4.2). Today, most residents of these communities identify as orthodox Catholics or evangelical Protestants, although a few individuals still retain knowledge of the traditional ritual practices known collectively as *costumbres*,[2] and these were the individuals I interviewed. Project participants held various ritual specializations including *aj'kun* (hunter),[3] *ajq'ij* (day-keeper), *iyom* (midwife), *rukoy b'aaq* (bonesetter), *yo'ol chay* (curative bloodletter), *ruki kik 'om* (spider-bite curer), *telinel* (caretaker of Rilaj Mam),[4] and *nabeysil* (overseer of communal sacred bundles). Often, a single person held more than one role.

Over the course of several field seasons, I interviewed twenty-four different individuals. My interview methods involved the seemingly unlikely combination of cognitive anthropology and archaeologically focused lines of inquiry. As I was concerned about being intrusive with questions that might be too direct, I began the interviews with a very general, open-ended question: Could you tell me the kinds of things one might find and use as

figure 4.3
Ritual toolkit created for pile sort questions. (Photograph by Linda A. Brown.)

figure 4.4
A project participant conducting a pile sort. (Photograph courtesy of Linda A. Brown.)

sacred materials? Much to my surprise, most people took me to their altars or opened their divination bundles and said, "These kinds of materials." Thus, at the end of intitial interviews I had a fairly good idea about the varieties of objects collected.

Based upon the initial responses, I decided to follow up with a second interview using a technique known as the pile sort. Pile sort questions seek to delineate a cognitive "map" by articulating how individuals classify items belonging to a specific cultural domain, in this case, sacred materials. To employ this technique, I created a model ritual toolkit containing some of the types of objects I observed in personal bundles or altar assemblages. The toolkit included natural and cultural items as well as modern objects and archaeological artifacts borrowed from a museum display case at a local hotel (Figure 4.3). During the pile sort interview, I handed the toolkit to a participant and asked them to group similar items together in piles (Figure 4.4). Once all the objects were sorted, I asked individuals to provide a word or phrase for each of their groupings.

After the fieldwork, I subjected the interview data to various analytical techniques using the software ANTHROPAC (Borgatti 1996). But prior to discussing specific results, I would like to turn my attention to Tz'utujil understandings of ancestral beings and personhood, as these contextualize what ritual practitioners had to say about sacred materials and the experience of time.

Ancestral Beings and Personhood

The interviews generated a wealth of information concerning the meanings and uses of sacred objects. Pertinent to the topic of this volume, ritual practitioners told me that the efficacy of sacred objects was directly linked to their origins—that is, they are animated by beings from the deep past. Following Margaret Bruchac (2005:56), I use the phrase "deep-time narratives" to describe oral traditions that convey how things were in a far distant time and place. One Mesoamerican deep-time narrative, known from at least the colonial period forward, concerns the existence of multiple worlds or eras prior to the one we live in today (e.g., Christenson 2007; López Austin, this volume; Monaghan 1995:32; Redfield and Villa Rojas 1934:330–331; Tedlock 1985; Tozzer 1907:153). These earlier worlds are often envisioned as existing in liminal darkness and situated in an anecumenical cosmic time and space, as Alfredo López Austin notes in this volume, during the period before the creation of our present "sun." In our interviews, several Tz'utujil ritual practitioners referred to this earlier existence and explicitly linked their sacred animate materials to this deep-time past. Here is what Pablo[5] said about this relationship:

> The earth has changed many times in the past. It is always changing and creating. These sacred materials are part of the earth; they come from there. They are the first beings to be born on the earth. They came before us, before the creation of this sun. They come from the darkness. These stones have a lot of talent, knowledge, and medicine. Even if it's a very small piece—it is powerful!

As Pablo makes clear, the sacred materials curated by ritual practitioners are not understood to be lifeless inert things but, instead, are powerful ancestral beings who were born in a previous time and place. In Santiago Atitlán, these Early Power People, as Nathaniel Tarn (1997) called them, are known collectively as *nawales*.[6] Nawales manifest themselves in a variety of forms, but a common attribute is their connection to a magical ancestral past that continues to empower the present (e.g., Carlsen 2009:312; Mendelson 1957:484–485; O'Brien 1975:43; Stanzione 2003:21). The ancient nawales had amazing abilities and knowledge that allowed them to divine the future, write the sacred songs and dances, cure illness, combat witchcraft, create idols that could speak and move, and control the forces of earth and sky.

During a previous era, the nawales walked on the surface of the earth like we do today. At the end of their time, they departed to live in the mountains surrounding Santiago Atitlán, where they still exist as AchiJab (Rain Men), the lightning-throwing rain warriors (Carlsen 2009:312; Stanzione 2003:58). As they retreated, they left behind physical remnants in the form of certain stones and relics that embody their powerful essence; these are the objects collected and valued by ritual practitioners today. Atitecos describe these objects as being *puro-nawal* (pure nawal), meaning they are infused with nawal essence (Carlsen 2009:312). Not many objects hold such potency, but those that do physically embody deep-time beings.

For ritual practitioners, objects possessing nawal essence are active participants in how the past is known in the present. Central to this process is an understanding that these sacred materials possess significant aspects of personhood. Historically for Maya peoples, the category of personhood has not been synonymous with human beings but instead transcends us (e.g., Astor-Aguilera 2010; Gossen 1994, 1996; Hendon 2012; Houston and Stuart 1998; Looper 2003; McAnany 1998; Meskell and Joyce 2003; Monaghan 1998). Personhood can be extended to diverse elements, and membership does not require the exterior appearance of our species. Trees, rocks, wild animals, houses, segments of time, corn, saints, lightning, stelae, crosses, and the dead, just to name a few, can display attributes of personhood (Astor-Aguilera 2010; Brown and Emery 2008; Carlsen 1997; Houston and Stuart 1998; Looper 2003; McAnany 1998; Meskell and Joyce 2003; Monaghan 1998; O'Neil 2012; Stanzione 2003; Wilson 1990). Similarly, some beings who do look like humans are not persons at all but antisocial creatures, such as transforming witches (e.g., Carlsen and Prechtel 1994; Monaghan 1998). Thus, personhood is not a fixed a priori attribute possessed by human beings, or any other types of beings for that matter. Instead, personhood is relational; it is a status that is constituted in daily social interactions among various types of persons: human and other-than-human

(Monaghan 1998:140). All persons acquire significance and power through interactions in which they learn how to "act as a person" in mutually constitutive social relationships (Hallowell 1976 [1960]:363).

We might ask, then, what is it to "act as a person"? To act as a person, one is expected to be sociable and engage in positive reciprocal relationships. These engagements may include the sharing of food and drink; entertainment with music and dance; personal caretaking such as dressing and bathing; attendance at important community events; communication with others; sharing of knowledge, insight, and advice; and being all-around responsible community members. In other words, personhood derives from engaging in personlike interactions (e.g., Hallowell 1976 [1960]; Harvey 2006; Monaghan 1998).

Puro-nawal objects act as persons when they participate in daily interactions with their human curators. Reciprocal exchange is an important aspect of these interactions as nawal and ritual practitioner provide vital necessities to one another. From the perspective of a ritual practitioner, nawales act as persons when they present humans with the life-giving rains, agricultural growth, human fertility, and the wild plants and animals of the forest, among other things. In addition, the nawales bestow individual ritual practitioners with esoteric gifts in the form of cures, divinatory insights, and ongoing mentoring in dreams, an important point to which I shall return shortly. In exchange, ritual practitioners act as persons when they remember and praise the ancestral nawales like all good humans should. In prayers, they plead with the nawales to come to their altars and partake of their offerings of fire and heat, as there are none in their worlds. During these rites, the nawales are provided with burning candles, smoldering copal smoke, pungent tobacco, and firewater liquors to help keep them warm (O'Brien 1975:44). In the midst of such interactions, two types of persons emerge—nawal and ritual practitioner—as the requirements for mutual existence are forged in exchanges between human and other-than-human entities (e.g., Hallowell 1976 [1960]).

As the sacred objects cared for by ritual practitioners possess qualities of personhood, they are understood to experience personlike emotions such as fatigue or jealousy. During interviews, Manuel gave voice to a common understanding when he said: "They [puro-nawal] get tired when they do a ceremony. When they get tired I give them some rum to drink and I talk to them. I tell them that it's time to sleep and then I cover them so they won't feel cold." As Manuel makes quite apparent, puro-nawal objects are thoroughly socialized household members. Like other household members, they are provided sustenance, spoken to kindly, encouraged to rest when weary, and sheltered against the cold. In fact, during interviews ritual practitioners often referred to them as spouses, a perspective noted by Martín Prechtel (Tarn 1997:120), who stated that an "*aj'kun's* art is thought of as being his wife and his wife is then something like his mistress."[7] These perspectives underscore the personal and intimate nature of the relationship between human caretakers and *puro-nawales*.

Communication forms a key component in the relationship and reciprocal exchanges among ritual practitioners and nawal-permeated sacred objects. As humans call upon and remember the nawales in rituals and prayers, the nawales communicate with ritual practitioners in dreams, as Pablo describes:

> In the old days, the sky was on the ground, and those guys [puro-nawales] they lived in the dark. The black dancers, they represent the night, they started to dance. They have a kind of a dance that they do, and they lifted up the sky and then the sun came up, the light came out. Nowadays we have the light of the sun, and it's OK, but mostly in the old days they were living in the dark. Before the sun rose, all of these guys [puro-nawales] were there. They are from the beginning. All those rocks [puro-nawales] were there, and they were talking to each other. Then, when the Maya people came, they talked to them in dreams. They talked to each other first, but after the day dawned and the sun first rose they started to talk to the Maya people in dreams.

As Pablo makes clear, the sacred materials curated by Tz'utujil ritual practitioners are communicating objects. In fact, this communication is how an individual knows whether any particular object is puro-nawal. If it is, then it communicates in dreams. Collections of puro-nawal objects function as dream bundles. They play an active role in nightly knowledge exchange as they direct important information to their human caretakers. In dreams, the nawales visit to provide answers to problematic divinations, share special medicines and prayers, or help diagnose sick clients. Moreover, they actively mentor ritual practitioners in dreams; in fact, many people told me that such dream revelation was the way they learned their craft. In dream consultations, nawales share wisdom from the ancestral deep-time past, which is subsequently incorporated into ceremonies and cures performed in the present during waking hours. In traditional Atiteco epistemology, these dream consultations are one of the most powerful means through which the past is known and the wealth of ancestral knowledge revealed. Given the prominence of dreams in local epistemology about the past, I turn briefly to Tz'utujil concepts of dreaming in the following section.

Dreaming

From the perspective of Atitecos, dreams are associated with a very specific part of the human soul. Each individual contains various souls, one of which is called the *q'aaq'aal*, or the fire-spirit. The q'aaq'aal can temporarily leave the body without harm,[8] and it does so at night when we dream. During interviews several people told me that if one sees an emberlike spark shooting up from a house roofline at night, it is the q'aaq'aal of one of the inhabitants leaving on such a nocturnal journey.

Once the q'aaq'aal leaves to wander outside the corporeal body, it takes on a form called the *ajelbal*, literally, the "transformation" (see Mendelson 1957:398–403). Usually, the ajelbal is an animal, but it can also be a celestial body or meteorological phenomenon. As the ajelbal roams the darkened world it meets and interacts with various other forces and beings. Dreams are taken to be the actual experiences of our wandering ajelbal, and the dream characters with whom we interact during sleep are real beings that our ajelbal actually meets during these nightly sojourns. Thus, for the Tz'utujil, as also noted among traditional Plains peoples, "there is no separation between the world-as-dreamed and the world-as-lived" (Irwin 1994:236).

It is the ajelbal who serves as the intermediary in nightly dream communications between ritual practitioner and nawal. Just as the human fire-spirit takes the form of an ajelbal when wandering at night, a nawal will take the form of a very particular dream character—its ajelbal, so to speak—to convey important messages. Generally, the nawal takes the form of a human in dreams, although it can also appear as an animal, plant, unusual object, or celestial phenomenon. Regardless of the form, the retention and interpretation of nightly communications from nawales is a central skill in the repertoire of a ritual practitioner. In fact, dream communications are so important that Robert Carlsen and Martín Prechtel (1994:97) noted that the dream world "is in many ways the *aj'kun's* workshop."

The importance of dreams in accessing ancestral ritual knowledge is not limited to the Tz'utujil Maya. In the 1950s, a Tzotzil elder in highland Chiapas told Calixta Guiteras-Holmes (1961:149), "I know everything by my dreams." In another Tzotzil community, William Holland noted the following about a ritual practitioner:

> Instead of a medical education, he learns the secrets of the curing art in revelations in his dreams. If the ancestor gods have selected a particular person to be a curer, he will dream that his spirit has gone to the sacred mountain of his companion animal. There he may see an Indian boy, a Ladino, or practically any human form, which he interprets as having been sent to instruct him how to cure. He is shown all the details of how to build an altar, the proper candles, the herbs, the prayers, and related rituals of the curing ceremonies he is to practice. (Holland 1962:233)

Among the Tzeltal, Esther Hermitte (1964:108) reported that dreams tell individuals which particular medicines can be used for curing and whether a client will get better or die. Meanwhile, Q'eqchi' elders in Alta Verapaz noted that the mountain spirits talk to people in dreams, and "dreams are the principal fount of knowledge itself" (Wilson 1990:74).

Similar to Tz'utujil puro-nawal objects, the well-known communicating crosses in Quintana Roo, Yucatan, are understood to possess qualities of personhood and communicate with their human caretakers (Astor-Aguilera 2010). And here again is a connection with dreams. Miguel Astor-Aguilera (2010), in a seminal study focusing on communicating crosses, interviewed a recent widow whose husband had been the caretaker of such a cross. She told him that "the cross-objects never physically spoke to the husband but that their communications were transmitted within his dreams. After the death of the husband, the cross-objects began to communicate with the widow and her sons" (Astor-Aguilera 2010:100).

Traditionally, dreaming has been given strong ontological priority and regarded as a primary source of knowledge and power among various indigenous peoples (e.g., Carlsen and Prechtel 1994; Flannery and Chambers 1985; Furst 2006; Hallowell 1976 [1960]; Irwin 1994; Tedlock 1981, 1992). Yet, as Lee Irwin (1994:233) notes, "the analytic tendency even in contemporary research has been to devalue dreaming and dream contents because they seem to lack the socially conditioned logic of the everyday."

Interestingly, puro-nawal objects and science come together in recent dream research conducted by Dr. Deirdre Barrett at the Harvard Medical School. Barrett (2001, 2007, 2011; Barrett and McNamara 2007) understands dreams as thinking in a different biochemical state of consciousness. Thus, the same issues and problems that concern us in a waking state will inform and preoccupy our dreaming minds as well. Due to the loosening of rational and culturally normative thought patterns during rapid eye movement (REM) sleep, however, dreams can be particularly helpful for creative problem solving and gaining new insights. Barrett encourages us to guide dreams toward such ends by employing the technique of dream incubation, which involves concentrating on a problem before sleep to help induce a dream about it. Notably, she suggests as part of this process that we place material objects associated with the problem by the bedside as these help to direct dreams toward solutions (Barrett 2011:32).

In her investigations, Barrett (2011:33) found that when research participants directed dreams in this manner, solutions or creative insights were presented to them by a dream character. Advances in brain-imaging technology demonstrate why. Positron-emission tomography (PET) scans show that the area of the cortex associated with self-volition is far less active during a REM dream state than it is when we are awake. Thus, Barrett (2011:33) reports that "one computer programmer received repeated nocturnal lessons from Albert Einstein—and when people had answers to problems presented to them via a dream character they had trouble taking full credit for what their dreaming mind had done. This tendency fits brain findings for REM sleep in which the dorsolateral prefrontal cortex, associated with perception of volition, is less active."

Similarly, the primary agent involved in any Tz'utujil divination, curing, or dream revelation is a nawal and not the ritual practitioner. The ritual practitioner is a conduit and partner in these encounters and thus would not take credit for the insights imparted by the nawales during nightly communications. Whether such dream mentoring comes in the form of an ancestral nawal or Albert Einstein is only a matter of one's cultural filter.

Sacred Objects as Nawales

Who are these pre-sunrise nawales who continue to empower the present with ancestral knowledge from the past, and what is their connection to the types of materials collected and curated today? Although a complete discussion of all the nawales whom interviewees identified with specific materials is beyond the scope of this essay, in the following pages I present two—obsidian and the female figurine—as examples. First, I discuss how ritual practitioners describe and use each material as well as their associated nawales and attributes. Then, using interview data as a springboard, I examine a related material aspect of ancient Maya culture to explore how we might use contemporary understandings to arrive at new interpretations of the archaeological past.

Obsidian as Lightning Rain Warriors

Virtually all interviewees linked the obsidian artifacts with lightning. Lightning is an aspect of the AchiJab, "Rain Men," the ancient nawales who reside in the mountaintops surrounding Santiago Atitlán. To illustrate the association between lightning and obsidian, I begin with a story shared by Manuel, who told me about his divine call to become an aj'kun:

> When I was twelve years old, I had bad pains in my head, a pain that felt like lightning. You know when you see lightning when it rains? They look like snakes, a very bright snake up in the sky. That's how I felt in my head. One night I went to sleep, and I dreamed. A small man [a nawal] came to me and he said, "This is the kind of tool you are going to find, and you are going to help people. You are not going to die, but this is your job." That's what the small man told me in a dream.
>
> During this time, I was a Catholic [*acción católica*], but then I changed my religion. When I was twelve, thirteen, fourteen, fifteen years old, I was studying the Bible and I was preaching, but after that I changed my religion. I had to go back to do this work [aj'kun] because I am not a preacher. This is the work I have to do! Then, after the dream, I found the medicine tool [puro-nawal] in the ground. It was one meter deep in the ground when I found it. I was digging to plant coffee. The stone is what the little man showed me in the dream.

The medicine tool revealed to Manuel in his dream was intimately tied to his suffering from

***figure** 4.5*
Altar with a large obsidian polyhedral core puro-nawal in the center foreground. (Photograph by Linda A. Brown.)

sharp headaches. The tool Manuel encountered when planting coffee that day was a very large obsidian polyhedral core (Figure 4.5). For Manuel, being presented with such a potent puro-nawal object during the time that he was undergoing an intense degree of inner conflict about which religious path to follow was a clear sign that he was destined to be an aj'kun. Once he accepted his role, the lightning-snake headaches ceased.

Lightning is a powerful phenomenon that possesses clear agency as it illuminates the darkened night, ignites fires, slices though trees, and instantly kills. Atitecos understand obsidian to be created from a lightning strike[9] and, as such, to be a form of materialized lightning, a concept shared by other Maya and Mesoamerican peoples as well (e.g., Alcorn 1984:141; Aulie and Aulie 1978:61; Foster 1945:184–185; Miller 1956:246; Monaghan 1995:109; Schele and Miller 1986:73).

Interestingly, Tz'utujil ritual practitioners link obsidian to a very specific role—lightning as protection against malevolent beings who can attack and harm humans. Interviewees told me that the AchiJab monitor the affairs of Atitecos from their mountaintop perches around Santiago Atitlán. If they see that a dangerous underworld being has infiltrated the community, they will throw a lightning bolt, killing the creature on the spot. This protective strike creates physical remnants in the form of obsidian, especially the polyhedral cores and polished stone axes[10] encountered in the fields. For this reason, ritual practitioners often describe obsidian cores as "bullets"; they are lightning materialized in the form of a formidable weapon.[11] The association of obsidian with defensive actions taken by the nawales is further reinforced by its dark color, which links it to dreams, the night, and death. Given the role of lightning as a potent weapon used against denizens of the underworld, objects created by lightning are particularly powerful medicine against curses and spiritual attacks.

During interviews, individuals described two general forms of malicious attacks. Curses can be thrown as a type of entanglement that entraps the vital power of a person, or they can be sent via a process that some *aj'kuna* referred to as "blowgun," the insertion of a foreign substance into the body of a victim. In either situation, ritual practitioners harness the power of obsidian as a lightning weapon to cut free any knots ensnaring victims or spiritually extract intrusive objects that may be sickening their clients. Additionally, the glistening, smooth surface of lightning-powered obsidian can protect individuals against such attacks by deflecting curses and refracting them back upon the sender. Interestingly, the understanding of lightning as an active agent in protecting humans from various types of malevolent activities is not unique to the Tz'utujil as it is known among other Maya groups including the Tzotzil (Guiteras-Holmes 1961:291), Tzeltal (Hermitte 1964:37, 1970:91; Nash 1970:16, 141, 1973:207), Chol (Aulie and Aulie 1978:46), Jacaltek (Montejo 1984), Ch'orti' (Fought 1972:413–414, 436; Wisdom 1940:396), Yucatek (Tozzer 1907:156), Lacandon (Bruce 1975:334), Q'eqchi' (Thompson 1970:274), and Huastec (Alcorn 1984:140–141) as well as more broadly in Mesoamerica (e.g., Monaghan 1995:109–110; Munch 1983:370, 373; Spero 1987). A Ch'orti' Maya rainmaker made this association overt in the following prayer: "If there is somewhere an evil spirit at work, today, conjure your sword; conjure your thunderbolt. Do not let them rule" (Fought 1972:413–414).

In addition to its role in protection, ritual practitioners also link obsidian to divination. Individuals call upon obsidian lightning to help diagnose gravely ill patients and to communicate the causes of such illnesses during dreams. Some people also physically utilize obsidian artifacts for scrying divinations, in which they gaze upon an obsidian artifact with a lit candle nearby and interpret the images that emerge on its dark, reflective surface. As obsidian contains the power to heal and protect individuals from spiritual attacks, it is especially effective for divining whether a client is sick from a curse and, if so, whether they will recover or die from this dark assault.

Obsidian is also associated with another type of lightning and divination. Storms occurring in Pacific lowlands can produce displays of sheet, or heat, lightning across the dark Atitlán nightscape. Sheet lightning is related to the divination

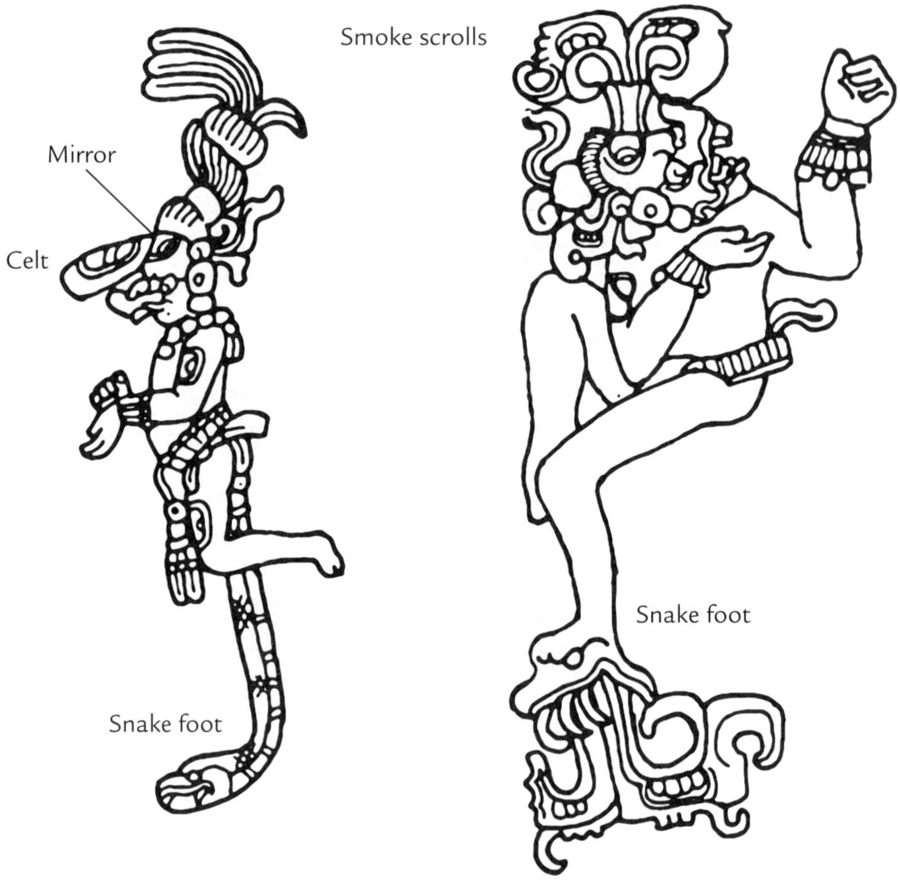

figure 4.6
Two illustrations of the god K'awiil as a scepter. (Drawing by Linda Schele, © David Schele, courtesy of the Foundation for the Advancement of Mesoamerica Studies, Inc. [Schele 2000; Schele and Miller 1986:fig. 33].)

technique known as "lightning in the blood" (Tedlock 1982:53). With lightning in the blood, communications from nawales physically surge through the body and are experienced as an involuntary flicking, jumping, or snakelike movement. Persons with this gift are attentive to these embodied messages and interpret them based on the area of the body that experienced the spontaneous lightninglike movement (see Tedlock 1982:53).

The widespread association of lightning and its material manifestations—obsidian, chert, and axes—with protection from malicious beings and malevolent acts suggests an interesting avenue for thinking about the past. For the ancient Maya, lightning was strongly associated with the enigmatic god K'awiil. During the Classic period, K'awiil was personified as the flaming lightning axe carried by the rain god Chahk (Coggins 1988). Having a repertoire of lighting-related symbols, K'awiil was often depicted with a smoking celt, burning torch or cigar, a single serpent leg,[12] and an obsidian forehead mirror (Taube 1992:79; Figure 4.6).

For the ancient Maya, K'awiil had multiple associations and was linked with fertility, abundance, and the rebirth of the corn god (see Taube 1992:60–79; Thompson 1970:289). Iconography and hieroglyphic texts also connect K'awiil with ancestral authority, royal ancestors, and the power of ruling dynastic elites (e.g., Schele 1976; Schele and Miller 1986:78; Taube 1992). In carved monuments, Maya rulers were depicted as taking the K'awiil scepter during accession rites, and K'awiil effigies were prominently displayed in works of art depicting important royal events (Figures 4.7 and 4.8). In portraying themselves in the company of K'awiil, elite rulers thus explicitly associated themselves with powerful lightning storms that bring the lifegiving rains necessary for human sustenance as well as essential creation accounts concerning the regeneration of the maize god.

figure 4.7
Bird Jaguar dances with a K'awiil scepter while his wife dances with a bundle. Lintel 53, Yaxchilan, Mexico. (Drawing by Linda Schele, © David Schele, courtesy of the Foundation for the Advancement of Mesoamerica Studies, Inc. [Freidel et al. 1993:fig.613:b; Schele 2000].)

figure 4.8
Bird Jaguar holds a K'awiil scepter over three bound captives during a rite reaffirming his legitimacy and entitlement to the throne. Stela 11, Yaxchilan, Mexico. (Drawing by Linda Schele, © David Schele, courtesy of the Foundation for the Advancement of Mesoamerica Studies, Inc. [Schele 2000; Schele and Freidel 1990:figs. 2:12, 7:8; Schele and Miller 1986:fig. V.5a].)

The prominent role of lightning in protection among contemporary ritual practitioners suggests an additional role for the multifaceted Classic-period K'awiil. Perhaps K'awiil—in his form as a personified lightning weapon—was a central agent in combating disease-causing underworld creatures of the past. In the political theater of the ancient Maya state, the conspicuous display of K'awiil scepters and personified axes in monumental art may have communicated the ruler's prowess in protecting his or her kingdom against attacks by such denizens from the dark.[13]

In considering the extensive linkages of lightning and its material correlates with protection against underworld beings and curses today, I cannot help but wonder about Classic-period chipped-stone objects bearing K'awiil symbolism—especially those fashioned from dark chert or obsidian—and the roles that these compelling objects played in the past. Might some of these lightning-empowered blades have been the ancient Maya equivalent of a silver bullet—a magical armament that killed dangerous animal-human hybrids, such as werewolves or disease-carrying underworld creatures, on the spot? If so, then perhaps the bundle of black eccentric flint blades that was cached in a doorway during a termination rite (Agurcia Fasquelle and Valdés 1994) for Rosalila—an Early Classic temple that entombed earlier building foundations housing the physical remains of the Copan lineage founders—constituted an impressive security measure by the ancient Maya. Embodied with the power of lightning weapons, K'awiil and the cadre of royal ancestors personified in this bundle would have protected its sacred cargo from evil intrusions in perpetuity.

Female Figurine as Old Grandmother—The First Midwife

Interviewees identified the female figurine in the ritual toolkit as Old Grandmother in her role as the First Midwife (see Figure 4.3). The First Midwife is one of the most primordial of all nawales and strongly associated with female fertility, children, medicine, divination, and sorcery, among other things. Today, she is known by various names, but she is envisioned as an old white-haired grandmother bent over from years of experience in the affairs of women and as the nawal who "opens the path for children" (Mendelson 1958:123).

The First Midwife takes many forms in contemporary Santiago Atitlán. One of her most powerful material manifestations is that of a communal sacred bundle often referred to quite simply as "Maria" (Mendelson 1958:123). The Maria bundle physically embodies the essence of female generative powers and, as such, is also known as *ruk'u'x alaniem* (literally, "root of placenta"), the original placenta for every "fruit, human, animal or vegetable, found on the face of the earth" (Stanzione 2003:287). The bundle is physically housed in the Cofradía San Juan, where it is stored in a wooden box suspended by chains from the rafters. The box hangs in such a manner that it can be rocked like a cradle, leading to another of the bundle's names—Maria K'uxul or Maria Cradle (Stanzione 2003:280). As the power of all fertility itself, this female bundle—cradle of the past, present, and future—can only be handled by those with enough spiritual command to have intimate contact with its primal energies. These individuals include midwives, by whom this bundle is particularly revered (Mendelson 1958:123).

Traditionally, Tz'utujil midwifery was a sacred profession (e.g., Carlsen 1997; Douglas 1969; Paul 1974, 1975; Paul and Paul 1975; Prechtel and Carlsen 1988; Rogoff 2011).[14] As with other ritual specialists, women were called to become midwives via divine election. Today, some spiritually called midwives still bring women to the Maria bundle for special ceremonies related to conception, pregnancy, and birth. Midwives and other healers also bring sick children to the bundle as it has special pediatric medicinal powers. On several occasions, I have seen ritual practitioners place a gravely ill infant inside of the box, gently positioning the child in direct contact with the bundle, while the suspended box is rocked back and forth during prayers for a cure.

As a primordial nawal, the First Midwife is part of a broad symbolic complex linking her to many other things; here, I focus on two. First is her association with the most ancient of Tz'utujil pre-sunrise beings, the original Father-Mother, who was a deified tree. Second is her association with weaving on a backstrap loom. I would note that my understanding of the centrality of the Original Tree in contemporary Atiteco deep-time narratives, as well as its linkage to weaving on a backstrap loom, owes much to several superb articles written by Carlsen and Prechtel (1991; Prechtel and Carlsen 1988). These articles inform the following discussion.

Turning first to the Original Tree, an important Tz'utujil creation account called Kotsej Juyu Ruchiliew, or Flowering Mountain Earth, makes explicit the central role of this deity in the genesis of all life:

> Before there was a world (what we would call the "universe"), a solitary deified tree was at the centre of all that was. As the world's creation approached,

this deity became pregnant with potential life; its branches grew one of all the things in the form of fruit. Not only were there gross physical objects like rocks, maize and deer hanging from the branches, there were also such elements as types of lightning, and even individual segments of time. Eventually this abundance became too much for the tree to support, and the fruit fell. Smashing open, the fruit scattered their seeds; and soon there were numerous seedlings at the foot of the old tree. The great tree provided shelter for the young "plants", nurturing them, until finally it was crowded out by the new. Since then, this tree has existed as a stump at the centre of the world. This stump is what remains of the original "Father/Mother" (Ti Tie Ti Tixel), the source and endpoint of life. (Carlsen and Prechtel 1991:27)

The focus of Atiteco religion is oriented backward into the past—to this Original Tree—which, "if properly maintained, renews and regenerates the world" (Carlsen and Prechtel 1991:27).

The close relationship between potent deities from the beginning of time, such as the Original Tree, and midwifery is evident in the practice of weaving on a backstrap loom.[15] As Prechtel and Carlsen (1988:123) note, "Things related to weaving seem to curve stealthily back to something that is symbolically, perhaps archetypally, female." The loom itself is understood to be a collection of female pre-sunrise beings, and each of its sticks embodies one of the thirteen Ixoq Ahawa ("Female Nawales"; O'Brien 1975:154; Prechtel and Carlsen 1988:124; Tarn 1997:298). During births, midwives call upon the female powers of the animate loom, and they will physically manipulate its parts for certain obstetrical procedures, such as turning a breech baby.[16] In fact, the entire act of creating cloth on a backstrap loom is fused with metaphors of human gestation and birth, something evident in the understanding that cloth is not "made" but "born."[17]

Tz'utujil nomenclature for various components of the loom as well as the weaving process itself further illustrates the conflation of weaving and birth (Figure 4.9). The cloth on the loom is referred to as a child. The rope that attaches the loom to the support post is called the umbilical cord. And the tree, post, or hook that secures the rope and uppermost portion of the loom is always called Mother Tree, an aspect of the Original Deity (Prechtel and Carlsen 1988:126). Thus, when a woman sits down to weave she is literally bound to—via her woven progeny—the past and the ancestral deified tree at the center of existence, from which all creation was born.

The theme of an ancestral tree associated with birth, regeneration, and renewal has historical depth in the Maya area (see Martin 2006 for an excellent review). One of the more famous examples is portrayed in the carved sarcophagus lid of Lord K'inich Janaab-Pakal I, who ruled the site of Palenque, Mexico, from 615 to 683 CE (Robertson

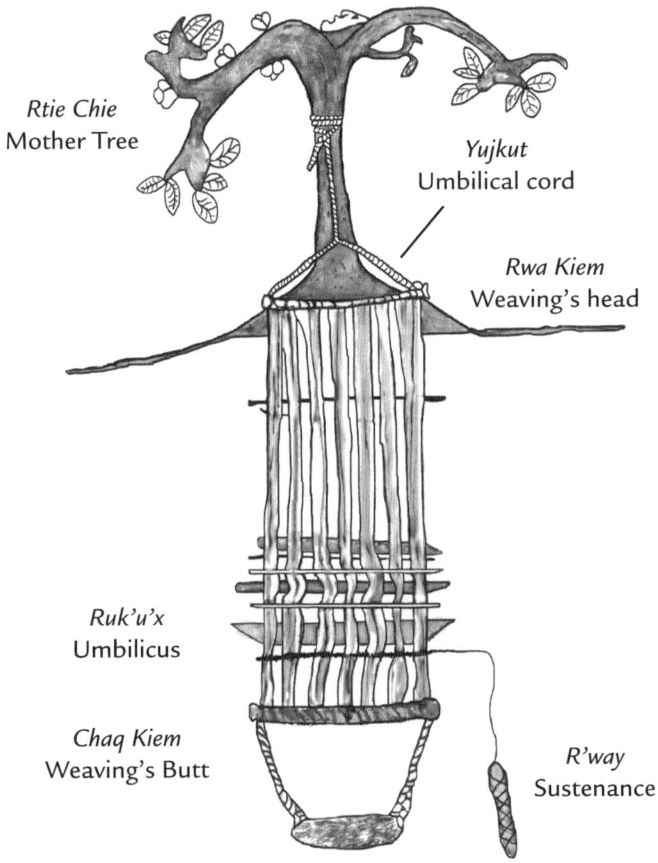

figure 4.9
Impressionistic rendition of a backstrap loom showing the relationship of the loom to the rope and tree, and the Tz'utujil names indicative of the birthing/weaving connection. (Illustration modified by author, after Prechtel and Carlsen 1988:fig. 2.)

1983; Ruz Lhuillier 1973). The cover of the lid depicts the deceased king ascending out from the skeletal pincers of an underworld centipede along the path of a deified tree (Grube and Nahm 1994; Taube 2003; Figure 4.10). We can be sure that Pakal's avenue of renewal is a tree as the ancient sculptor labeled its trunk and branches with the glyphic sign *te* or "tree." Additionally, the surface of the deified tree is marked with a mirror sign that is used to denote precious objects whose surfaces are "bright, shiny or wet" (Stone and Zender 2011:71), an important point to which I shall return shortly. Pakal is depicted as resting in an offering brazier at the base of this tree. He is recumbent in fetal-like position, a gesture that signals both his death and his imminent rebirth. The theme of trees linked with ancestral renewal and regeneration continues along the sides of the sarcophagus, where ten carvings of Pakal's predecessors are rendered as personified fruit trees spouting forth from fertile soil to form an "orchard of ancestral dead" (Schele and Freidel 1990:221).

figure 4.10
Detail from sarcophagus lid showing Pakal ascending from the underworld along the path of a deified tree, Palenque, Mexico. (Drawing by Linda Schele, © David Schele, courtesy of the Foundation for the Advancement of Mesoamerican Studies, Inc. [Schele 2000; Schele and Mathews 1998:fig.3.16.E1].)

But why would a deified tree be the pathway for human regeneration and renewal for the ancient Maya? How might we understand the resonance of this narrative among contemporary Tz'utujil *costumbristas* who maintain a vital, ongoing umbilical relationship with a deified tree? I propose that this most ancient nawal—like other pre-sunrise beings—continues to emerge in the present in a distinctly material form. It is shepherded into this world in the knowing hands of First Midwife, who guides her human descendants each time a woman gives birth. Specifically, the ancestral tree appears in the placenta.

The human placenta is a flat organ that connects a fetus to the inner wall of the maternal uterus via the umbilical cord. The umbilical cord, containing two arteries and one vein braided together like a rope, extends from the fetal navel area into the placenta, where it splits into a system of smaller blood vessels. The pattern formed by these small branching vessels distinctly resembles tree roots or tree limbs, depending upon one's perspective, a fact known by contemporary midwives, who refer to it in English as the "tree of life" (Figure 4.11).[18] The Tz'utujil name for the placenta also employs an arboreal theme in using the phrase *ruk'u'x ak'ala*, literally, "root of child" (Tarn 1997:380).[19]

During fetal development and at the moment of birth each human being is connected to the Original Tree via the umbilical cord. Like tender new sprouts, we visually germinate from its ancient roots and branches. Here, I am reminded of what one Atiteco conveyed to the anthropologist

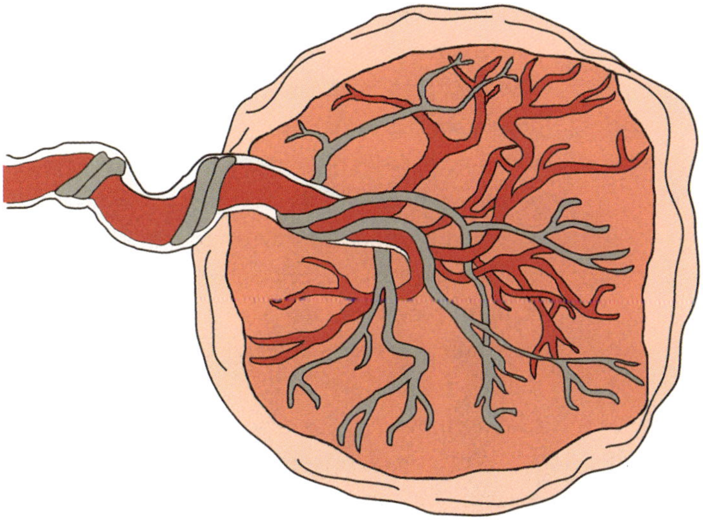

figure 4.11
Drawing of the fetal surface of a placenta at term showing the umbilical cord and treelike vessel pattern. (Illustration redrawn by author, after Bennett and Brown 1999:fig.51:3B.)

Michael Mendelson when he said that "each object in the world had a different father, but an umbilical linked it to this single large womb-root, which he also called a tree root, apparently in the sense of a lineage tree" (Tarn 1997:176).[20] This womb-root or tree root is indeed ancient. It is quite literally the source of human regeneration and ancestral renewal.

Returning to Palenque, the deified tree on the lid of Pakal's sarcophagus was marked with mirror signs denoting a bright, shiny, or wet surface, something that may reference the placenta as well. Notably, the entire fetal surface of the placenta, the side bearing the treelike vessel pattern, is covered with a layer of amnion. Amnion is a thin, glistening protective membrane.[21] The amnion covering the placenta gives its "tree" a very distinctive shiny, reflective, and wet appearance (Bennett and Brown 1999:977). Certainly, ancient Maya midwives would have been aware of the placental tree, its unique surface appearance, and its role in birth. I propose that Classic-period depictions of deified trees marked with mirror signs—especially those portrayed as pathways for renewal and rebirth—derived from this potent biological source.

Returning to the connection between the female figurine as First Midwife and handwoven cloth, we have seen that birth is inherently connected with weaving on a backstrap loom. But here I believe we can extend the biological linkage further. I suggest that the reason weaving "curves stealthily back to something that is symbolically, perhaps archetypally, female" (Prechtel and Carlsen 1988:123) is because it quite literally references anatomy and physiology during gestation and birth. When a woman sits down to weave, the spatial configuration of the weaver, loom, rope, and anchor replicates vital prenatal associations: the developing fetus (unfinished cloth on the loom), which extends outward from its mother's (weaver's) abdomen, is connected by the umbilical cord (the rope) to the placenta (Mother Tree; see Figure 4.9). To be born, the finished cloth, like a baby, must separate from its mother and be cut from its prebirth connection to the Original Tree. The act of weaving, then, is a visible enactment of the critical maternal and fetal relationship.[22]

The connection between the potent deified tree and prenatal anatomy is also evident in the following:

> Like the Mother Tree of the loom, the navel is associated symbolically with the tree at the center of the world. It offers testimony of our connection to our mother, from whom we received sustenance, and of her similar relationship with her mother, and she with her mother, and on and on. The navel presents a direct link to the ancestors, and in Maya belief it is from the ancestors that the living are derived. (Prechtel and Carlsen 1988:131)

The navel presents a direct link with our human predecessors as well the deified Original Tree, the potent pre-sunrise being, and the shimmering path of ancestral renewal that continues to emerge from the past and materialize in the present at each birth.

Conclusion

As we have seen throughout this essay, the Western ontological disjuncture that severs spirit from matter and segregates past from present dissolves in relationships between Tz'utujil ritual practitioners and puro-nawales. Objects embodied with nawal essence are not inert passive things but, instead, active participants in the co-creation, remaking, and preservation of local indigenous knowledge about the deep past. As Carolyn Dean (2010:40) noted for the animate Andean stone *wakas* she referred to as "remembered rocks,"[23] "people from conventionally literate societies tend to think that text substitutes for oral communication; it might be assumed, by analogy, that objects—like remembered rocks—substitute for orality as well" (Dean 2010:37–38). The same analogy holds true for the animate puro-nawal objects curated by Tz'utujil ritual practitioners today. The stories that travel with these embodied objects cannot be read like texts existing external to the self. Instead, they must unfold within the intimate spaces of a shared relationship. Ancestral knowledge and deep-time nawales continue to emerge in the present through such connections. Yet Western intellectual traditions understand the world through separations: present/past, subject/object, mind/matter, animate/inanimate, human/nonhuman, and so forth.[24] These dualisms slice apart relational and indigenous perspectives and, in the process, devalue the lived experiences of others while, at same time, leaving us to ponder the contradiction of why more people feel alone in an increasingly "connected" world.

While puro-nawal objects are participants in deep-time narratives about the past, they also create the present. As I have discussed, these potent materials serve a generative role in producing a particular type of person—a ritual practitioner—who has access to distinct types of knowledge and power. At the same time, the ritual practitioner serves a generative role in producing another type of person—a puro-nawal—who has access to knowledge and powers from the deep past. Thus, nawal objects and ritual practitioners are totally enmeshed or, in the parlance of today, entangled in social networks and practices where each co-creates the other. Neither type of person can exist outside of such relationships.

With the increasing dominance of Christian religions in Santiago Atitlán, puro-nawal objects take on an additional role as accomplices in the production of counterhegemonic epistemologies about how the past can be known. The ancestral past is still dreamed into the present via engagement with these powerful animate materials. For those who continue to walk in the footpath of the ancient nawales, the "binding of all times into one Time" (Mendelson 1957:497) remains extant today. Yet in the aftermath of significant social turmoil and religious change over the past half a century, this experience of time is no longer a "strong and permanent current of thinking and feeling in Atitlán" (Mendelson 1957:497).

In this new era, the cellular phone tower near the plaza has supplanted the Original Tree as the center of Santiago Atitlán (Carlsen 2003:154). And rather than obsidian lightning warriors combating curses, local "social cleansing" groups arbitrate witchcraft accusations through extra-judicial lynchings. At the end of this thirteenth *bak'tun*, the Tz'utujil are navigating new ways to know the present and the past. Only time will make clear whether the pre-sunrise beings embodied in dream bundles continue to counsel their post-sunrise human descendants, reaching across the eras, refusing to be silenced.

Acknowledgments

I gratefully acknowledge the support of the following institutions: the Wenner-Gren Foundation for Anthropological Research for fieldwork,

Dumbarton Oaks Pre-Columbian Studies for a fellowship allowing analysis and writing, and the National Science Foundation Short Courses on Research Methods in Cultural Anthropology for providing training in cognitive anthropological interview methods.

In Guatemala, I would like to thank Dolores Ratzán, David Glanville, Martin and Amanda Chif, and Nancy Hoffman for their friendship and support as well as Manuel Jamínez Tambríz, who assisted with interviews outside of the Atitlán area. This essay owes much to previous anthropological research conducted in Santiago Atitlán. In particular, I acknowledge my debt to Robert Carlsen, Vinny Stanzione, and Michael Mendelson. Many thanks to my supportive friends and colleagues: Barbara Arroyo, Harriet "Rae" Beaubien, Inga Calvin, Stephen Houston, Connie Koshewa, Colin McEwan, Megan O'Neil, Joanne Pillsbury, Karl Taube, Barbara Voorhies, and David Webster as well as the late Virginia Fields, who is greatly missed. I would like to thank Tony Aveni for organizing this symposium and extending a warm welcome to me. Very special thanks go to all the ritual practitioners who so generously shared their time and deep-time knowledge.

NOTES

1 The word *Atitecos* refers to the people who live in the highland Maya town of Santiago Atitlán, Guatemala.

2 The term *costumbres*, literally the "old ways," refers to a set of ritual practices that consciously maintains a connection to the way things were done in the past. Carlsen (2003:141) defines costumbres as "old inherited ways of knowing and doing." The word *costumbrista* refers to a person who engages in costumbres.

3 As Carlsen and Prechtel (1994:90) note, the word *aj'kun* and its plural form *aj'kuna* derive from the Tz'utujil verb *kumxic* (to go out and find). Thus, aj'kun can be glossed as "hunter." The aj'kun's hunting territory is the dream world. "Not only do aj'kuna diagnosis diseases through dream interpretation, but they often venture into the dream world to remedy a patient's affliction" (Carlsen and Prechtel 1994:97).

4 Rilaj Mam (Old Grandfather), commonly known as Maximón, is an important local deity in Santiago Atitlán. The position of telinel is held by an aj'kun for the duration of a single year. During this time, the most important task of the telinel is to carry the Mam on his shoulder in a series of ritual events occurring during Holy Week.

5 All of the names of interviewees are pseudonyms to protect the privacy of project participants.

6 In Santiago Atitlán, the term *nawal* or *rajawal* refers to a collection of potent beings (ancestors, dance figures, santos, earth and rain deities, and historical Maya kings, among others) who continue to empower the present with ancestral knowledge from the past. In other Maya areas, nawal can refer to an individual's animal co-essence or transformation. Atitecos use the word *ajelbal* for this phenomenon.

7 The "*aj'kun's* art" that is referred to in this quote concerns an individual's special relationships with embodied puro-nawal objects.

8 Interviewees noted that if the q'aaq'aal leaves the body at night but either fails to return or suffers fright during its nocturnal wanderings, then the individual will become sick and can die.

9 Mesoamerican peoples commonly associated chert with lightning (e.g., Miller and Taube 1993:106). As chert is not naturally found in the volcanic highland area of Santiago Atitlán, it was not included in the interview toolkit.

10 Interviewees also linked the greenstone axes in the toolkit with lightning. As with obsidian, ritual practitioners strongly associated greenstone axes with combating sorcery and witchcraft. But these items also had other significant ritual roles in rainmaking, agricultural fertility, and divination. Although an in-depth discussion of these associations is beyond the scope of the current essay,

ritual practitioners told me that greenstone axes were also known as *suu'baan* (literally, "wrapped") and, as such, were linked to personal and community sacred bundles, tamales, and swaddled infants. This latter association is notable in light of Tate's (2012:36) interpretation of some of the carved images in Olmec greenstone celts as representing human fetuses.

11 Redfield and Villa Rojas (1934:113–114) noted that the Yucatek Maya of Chan Kom also regarded found bits of obsidian and flint as protective weapons thrown by the Balams, spirits of the maize fields and towns. When these materials were found, a ritual practitioner often buried them in a communal purification ceremony known as Loh; "to protect the village from the evil winds, he buries toks (as these pieces are called) at each of the four entrances of the village" (Redfield and Villa Rojas 1934:113–114). I thank Karl Taube for bringing this reference to my attention.

12 In Santiago Atitlán, there is a specialized category of aj'kun called the nabeysil. Among his ritual duties, the nabeysil is the caretaker of the communal Martín bundle, related to rain and agricultural productivity. Traditionally, the job of the nabeysil was held by someone with a shriveled or deformed leg or some other condition that affected motion in the limbs, perhaps a reference to the snake leg of the lightning god K'awiil (Mendelson 1958:124).

13 Interestingly, Stuart (2005:166) suggests that during the Classic period, Maya state-sponsored "witchcraft—and not so much shamanism in its more general and vaguely defined sense—was an important component of rulership and its overarching ideology." Specifically, he proposes that some Maya *wahy* figures, previously interpreted as representing animal co-essences associated with elite nobles (Houston and Stuart 1989), likely depicted animate diseases related to ancient magical attacks. While this intriguing interpretation remains speculative, certainly the ancient Maya would have needed some powerful protection against diseasing-causing death gods and terrifying underworld beings. In a recent talk at the Dallas Museum of Art, Karl Taube (2011) also suggested that eccentric blades were likely potent symbolic weapons.

14 Christian recruitment efforts from the mid-twentieth century to the present focused on midwives (among others) following costumbres in an attempt to change beliefs and practices surrounding birth. Today, many midwives have received some formal instruction, often church sponsored, in Western medical delivery procedures. But the degree to which these recruitment efforts have fully changed traditional belief systems is debatable. In interviews, one midwife told me that many of the Christian midwives also have their collected sacred objects, but they will not show these materials to anyone for fear of retribution from within the church.

15 The backstrap loom consists of a series of separate wooden sticks that only takes the shape of a loom once the vertical wrap threads are in place. When in use, one end of the loom is anchored by a leather strap to the lower back of the weaver while the opposite end is attached to an elevated, fixed object.

16 To turn a breech baby, the midwife will place the appropriate loom members on the abdomen of "the expectant mother so as to coax the infant to turn, head down. Having done this, the midwife then physically turns the infant" (Prechtel and Carlsen 1988:124).

17 Prechtel and Carlsen (1988:123n2) note a gendered distinction between weaving on a backstrap loom versus a floor loom. Weaving on the larger floor loom, which is considered a machine, is a male activity. Thus, weavings from floor looms are considered "handmade." Only weavings from a backstrap loom are "born."

18 An Internet search of the terms *tree of life* and *placenta* will produce a number of images showing the treelike vessel pattern on the placenta as well as the practice of making imprints of the placenta tree as an artistic keepsake after birth. I thank Connie Koshewa for introducing me to the tree of life after a delivery in a small rural Texas birth clinic many years ago.

19 Tarn (1997:380) translates *ruk'u'x* as "root, stomach, heart, seed, core, center."

20 Michael Mendelson conducted anthropological research in Santiago Atitlán during the 1950s. He later changed career paths to pursue a profession in poetry and subsequently has written about the town under the nom de plume Nathaniel Tarn.

21 The shiny, wet, reflective amnion also makes up the amniotic sack holding the fetus.

22 For another birth–weaving connection, see Tate's (2012:43) argument that human embryos were the most common depiction carved on Olmec

greenstone "spoons." Follensbee (2008:98) argues that these "spoons" were very likely representations of weaving battens.

23 Dean (2010:21) uses the phrase "remembered rocks" to characterize those sacred rocks celebrated in Inca oral and performative cultures and "kept alive and potentially animate in the memories of those who encountered them." Remembered rocks, like contemporary Tz'utujil puro-nawal objects, connected a very particular material object with specific historic and cultural narratives.

24 In this regard, my research follows others in exploring the current "ontological turn" in anthropology and the rekindled interest in "animism" (e.g., Bird-David 1999; Descola 2013; Harvey 2006; Henare et al. 2007; Viveiros de Castro 2004).

REFERENCES CITED

Agurcia Fasquelle, Ricardo, and Juan Antonio Valdés

1994 *Secretos de dos ciudades mayas: Copán y Tikal / Secrets of Two Maya Cities: Copán and Tikal.* La Nación, S.A., San José.

Alcorn, Janis B.

1984 *Huastec Mayan Ethnobotany.* University of Texas Press, Austin.

Astor-Aguilera, Miguel Angel

2010 *The Maya World of Communicating Objects: Quadripartite Crosses, Trees, and Stones.* University of New Mexico Press, Albuquerque.

Aulie, Wilbur H., and Evelyn W. Aulie

1978 *Diccionario ch'ol–español, español–ch'ol.* Instituto Lingüístico de Verano, Mexico City.

Barrett, Deirdre

2001 *The Committee of Sleep: How Artists, Scientists and Athletes Use Dreams for Creative Problem-Solving—and How You Can Too.* Crown, New York.

2007 An Evolutionary Theory of Dreams and Problem Solving. In *The New Science of Dreaming,* edited by Deirdre Barrett and Patrick McNamara, vol. 3, pp. 133–153. Praeger, Westport, Conn.

2011 Answers in Your Dreams. *Scientific American Mind* (November/December):27–33.

Barrett, Deirdre, and Patrick McNamara (editors)

2007 *The New Science of Dreaming.* 3 vols. Praeger, Westport, Conn.

Bennett, V. Ruth, and Linda K. Brown (editors)

1999 *Myles Textbook for Midwives.* 13th ed. Churchill Livingstone, London.

Bird-David, Nurit

1999 "Animism" Revisited: Personhood, Environment, and Relational Epistemology. *Current Anthropology* 40(S1):67–91.

Borgatti, Stephen P.

1996 *ANTHROPAC 4.0.* Analytic Technolgies, Natick, Mass.

Brown, Linda A.

2000 From Discard to Divination: Demarcating the Sacred through the Collection and Curation of Discarded Objects. *Latin American Antiquity* 11(4):319–333.

Brown, Linda A., and Kitty F. Emery

2008 Negotiations with the Animate Forest: Hunting Shrines in the Guatemalan Highlands. *Journal of Archaeological Method and Theory* 15(4):300–337.

Bruce, Robert D.

1975 *Lacandon Dream Symbolism: Dream Symbolism and Interpretation among the Lacandon Maya of Chiapas, Mexico.* Ediciones Euroamericanas Klaus Thiele, Mexico City.

Bruchac, Margaret M.

2005 Earthshapers and Placemakers: Algonkian Indian Stories and the Landscape. In *Indigenous Archaeologies: Decolonizing Theory and Practice,* edited by Claire Smith and H. Martin Wobst, pp. 56–80. One World Archaeology 47. Routledge, New York.

Carlsen, Robert S.

1997 *War for the Heart and Soul of a Highland Maya Town.* University of Texas Press, Austin.

2003 Subversive Threads: An Interpretation of Maya Weaving and Culture. In *"With Their Hands and Their Eyes": Maya Textiles, Mirrors of a Worldview.* Etnografisch Museum Antwerpen, Antwerp.

2009 "Footpath of the Dawn, Footpath of the Sun": Maya Worldviews at Lake Atitlán. In *Maya Worldviews at Conquest*, edited by Leslie G. Cecil and Timothy W. Pugh, pp. 299–316. University of Colorado Press, Boulder.

Carlsen, Robert S., and Martin Prechtel

1994 Walking on Two Legs: Shamanism in Santiago Atitlán, Guatemala. In *Ancient Traditions: Shamanism in Central Asia and the Americas*, edited by Gary Seaman and Jane Day, pp. 77–111. University of Colorado Press, Niwot.

Carlsen, Robert S., and Martín Prechtel

1991 The Flowering of the Dead: An Interpretation of Highland Maya Culture. *Man* 26:23–42.

Christenson, Allen J.

2007 *Popol Vuh: The Sacred Book of the Maya.* University of Oklahoma Press, Norman.

Coe, Michael D.

1966 *An Early Stone Pectoral from Southeastern Mexico.* Studies in Pre Columbian Art and Archaeology 1. Dumbarton Oaks Research Library and Collection, Washington, D.C.

Coe, Michael D., and Richard A. Diehl

1980 *In the Land of the Olmec: The Archaeology of San Lorenzo Tenochtitlan.* 2 vols. University of Texas Press, Austin.

Coggins, Clemency C.

1988 The Manikin Scepter: Emblem of Lineage. *Estudios de cultura maya* 17:123–158.

Dean, Carolyn

2010 *A Culture of Stone: Inka Perspectives on Rock.* Duke University Press, Durham, N.C.

Descola, Philippe

2013 *Beyond Nature and Culture.* University of Chicago Press, Chicago.

Douglas, Bill G.

1969 Illness and Curing in Santiago Atitlán, a Tz'utujil-Maya Community in the Southwestern Highlands of Guatemala. PhD dissertation, Department of Anthropology, Stanford University, Calif.

Flannery, Regina, and Mary Elizabeth Chambers

1985 Each Man Has His Own Friends: The Role of Dream Visitors in Traditional East Cree Belief and Practice. *Arctic Anthropology* 22(1):1–22.

Follensbee, Billie J. A.

2008 Fiber Technology and Weaving in Formative-Period Gulf Coast Cultures. *Ancient Mesoamerica* 19(1):87–110.

Foster, George M.

1945 *Sierra Popluca Folklore and Beliefs.* University of California Press, Los Angeles.

Fought, John G.

1972 *Chorti (Mayan) Texts 1.* University of Pennsylvania Press, Philadelphia.

Freidel, David, Linda Schele, and Joy Parker

1993 *Maya Cosmos: Three-Thousand Years on the Shaman's Path.* Wiliam Morrow, New York.

Furst, Peter T.

2006 *Rock Crystals and Peyote Dreams: Explorations in the Huichol Universe.* University of Utah Press, Salt Lake City.

Gossen, Gary H.

1994 From Olmecs to Zapatistas: A Once and Future History of Souls. *American Anthropologist* 96(3):553–570.

1996 Maya Zapatistas Move to the Ancient Future. *American Anthropologist* 98(3):528–538.

Grube, Nikolai, and Werner Nahm

1994 A Census of Xibalba: A Complete Inventory of Way Characters on Maya Ceramics. In *The Maya Vase Book: A Corpus of Rollout Photographs of Maya Vases*, edited by Justin Kerr, vol. 4, pp. 686–715. Kerr Associates, New York.

Guiteras-Holmes, Calixta

1961 *Perils of the Soul: The World View of a Tzotzil Indian.* Free Press of Glencoe, New York.

Hallowell, A. Irving

1976 [1960] Ojibwa Ontology, Behavior, and World View. In *Contributions to Anthropology: Selected Papers of A. Irving Hallowell*, edited by Raymond D. Fogelson, pp. 357–390. University of Chicago Press, Chicago.

Hamann, Byron

2002 The Social Life of Pre-Sunrise Things. *Current Anthropology* 43(3):351–382.

Harvey, Graham

2006 *Animism: Respecting the Living World.* Columbia University Press, New York.

Henare, Amiria J. M., Martin Holbraad, and Sari Wastell (editors)

2007 *Thinking through Things: Theorising Artefacts Ethnographically.* Routledge, London.

Hendon, Julia A.

2012 Objects as Persons: Integrating Maya Beliefs and Anthropological Theory. In *Power and Identity in Archaeological Theory and Practice: Case Studies from Ancient Mesoamerica*, edited by Eleanor Harrison-Buck, pp. 82–89. University of Utah Press, Salt Lake City.

Hermitte, M. Esther

1964 Supernatural Power and Social Control in a Modern Maya Village. PhD dissertation, Department of Anthropology, University of Chicago, Chicago.

1970 *Poder sobrenatural y control social.* Instituto Indigenista Interamericano, Mexico City.

Holland, William R.

1962 Highland Maya Folk Medicine: A Study of Culture Change. PhD dissertation, Department of Anthropology, University of Arizona, Tucson.

Houston, Stephen, and David Stuart

1989 *The Way Glyph: Evidence for "Co-essences" among the Classic Maya.* Research Reports on Ancient Maya Writing 30. Center for Maya Research, Washington, D.C.

1998 The Ancient Maya Self: Personhood and Portraiture in the Classic Period. *RES: Anthropology and Aesthetics* 33:73–101.

Irwin, Lee

1994 Dreams, Theory, and Culture: The Plains Vision Quest Paradigm. *American Indian Quarterly* 18(2):229–245.

Looper, Matthew G.

2003 *Lightning Warrior: Maya Art at Quirigua.* Linda Schele Series in Maya and Pre-Columbian Studies. University of Texas Press, Austin.

Martin, Simon

2006 Cacao in Ancient Maya Religion: First Fruit from the Maize Tree and Other Tales from the Underworld. In *Chocolate in Mesoamerica: A Cultural History of Cacao*, edited by Cameron L. McNeil, pp. 154–183. University of Florida Press, Gainesville.

McAnany, Patricia

1998 Ancestors and the Classic Maya Built Environment. In *Function and Meaning in Classic Maya Architecture*, edited by Stephen D. Houston, pp. 271–298. Dumbarton Oaks Research Library and Collection, Washington, D.C.

Mendelson, E. Michael

1957 Religion and World View in Santiago Atitlán. PhD Dissertation, Department of Anthropology, University of Chicago, Chicago.

1958 A Guatemalan Sacred Bundle. *Man* 58:121–126.

Meskell, Lynn, and Rosemary A. Joyce

2003 *Embodied Lives: Figuring Ancient Maya and Egyptian Experience.* Routledge, London.

Miller, Arthur G.

1973 *The Mural Painting of Teotihuacan.* Dumbarton Oaks Research Library and Collection, Washington, D.C.

Miller, Mary Ellen

1986 *The Murals of Bonanpak.* Princeton University Press, Princeton, N.J.

Miller, Mary Ellen, and Karl A. Taube

1993 *The Gods and Symbols of Ancient Mexico and the Maya: An Illustrated Dictionary of Mesoamerican Religion.* Thames and Hudson, New York.

Miller, Walter S.

1956 *Cuentos Mixes.* Biblioteca de Folklore Indígena 2. Instituto Nacional Indigenista, Mexico City.

Monaghan, John

1995 *The Covenants with Earth and Rain: Exchange, Sacrifice, and Revelation in Mixtec Sociality.* University of Oklahoma Press, Norman.

1998 The Person, Destiny, and the Construction of Difference in Mesoamerica. *RES: Anthropology and Aesthetics* 33:137–146.

Montejo, Victor

1984 *El Kanil, Man of Lightning: A Legend of Jacaltenango.* Spanish-English ed. Signal Books, Carrboro, N.C.

Munch, Guido

1983 Cosmovisión y medicina tradicional entre los poplucas y nahuas del sur de Veracruz. In *Antropología y historia de los mixe-zoques y mayas,* edited by Lorenzo Ochoa and Thomas A. Lee, pp. 367–381. Universidad Nacional Autónoma de México, Mexico City.

Nash, June

1970 *In the Eyes of the Ancestors: Belief and Behavior in a Maya Community.* Yale University Press, New Haven, Conn.

1973 Época para cazar brujos: Estudio comparativo de algunos comunidades de Chiapas, México. *Estudio de cultura maya* IX:195–229.

O'Brien, Linda Lee

1975 Songs of the Face of the Earth: Ancestor Songs of the Tzutuhil Maya of Santiago Atitlán, Guatemala. PhD dissertation, Department of Music, University of California, Los Angeles.

O'Neil, Megan E.

2012 *Engaging Ancient Maya Sculpture at Piedras Negras, Guatemala.* University of Oklahoma Press, Norman.

Paul, Lois

1974 The Mastery of Work and the Mystery of Sex in a Guatemalan Village. In *Woman, Culture, and Society,* edited by Michelle Z. Rosaldo, Louise Lamphere, and Joan Bamberger, pp. 281–299. Stanford University Press, Stanford, Calif.

1975 Recruitment to a Ritual Role: The Midwife in a Maya Community. *Ethos* 3(3):449–467.

Paul, Lois, and Benjamin David Paul

1975 The Maya Midwife as Sacred Specialist: A Guatemalan Case. *American Ethnologist* 2(4):707–726.

Prechtel, Martín, and Robert S. Carlsen

1988 Weaving and Cosmos amongst the Tzutujil Maya of Guatemala. *RES: Anthropology and Aesthetics* 15:122–132.

Redfield, Robert, and Alfonso Villa Rojas

1934 *Chan Kom: A Maya Village.* Carnegie Institution of Washington, Washington, D.C.

Robertson, Merle Greene

1983 *The Sculpture of Palenque: The Temple of the Inscriptions 1.* Princeton University, Princeton, N.J.

Rogoff, Barbara

2011 *Developing Destinies: A Mayan Midwife and Town.* Oxford University Press, New York.

Ruz Lhuillier, Alberto

1973 *El Templo de las Inscripcíones, Palenque.* Colección científica arqueología 7. Instituto Nacional de Antropología e Historia, Mexico City.

Schele, Linda

1976 Accession Iconography of Chan-Bahlum in the Group of the Cross at Palenque. In *The Art, Iconography, and Dynastic History of Palenque, Part 3,* edited by Merle Greene Robertson, pp. 9–34. Robert Louis Stevenson School, Pebble Beach, Calif.

2000 The Linda Schele Drawing Collection. Electronic document, http://research.famsi.org/schele.html, accessed January 15, 2013.

Schele, Linda, and David Freidel

1990 *A Forest of Kings: The Untold Story of the Ancient Maya.* William Morrow, New York.

Schele, Linda, and Peter Matthews

1998 *The Code of Kings: The Language of Seven Sacred Maya Temples and Tombs.* Scribner, New York.

Schele, Linda, and Mary Ellen Miller

1986 *The Blood of Kings: Dynasty and Ritual in Maya Art.* Kimbell Art Museum, Fort Worth, Tex.

Smith, Michael E.

2003 What Did Mesoamerican Commoners Think about "Pre-Sunrise" Things? *Current Anthropology* 44(2):271–272.

Spero, Joanne M.

1987 Lightning Men and Water Serpents: A Comparison of Mayan and Mixe-Zoquean Beliefs. MA thesis, Department of Art, University of Texas, Austin.

Stanzione, Vincent James

2003 *Rituals of Sacrifice: Walking the Face of the Earth on the Sacred Path of the Sun; A Journey through the Tz'utujil Maya World of Santiago Atitlán.* University of New Mexico Press, Albuquerque.

Stone, Andrea, and Marc Zender

2011 *Reading Maya Art: A Hieroglyphic Guide to Ancient Maya Painting and Sculpture.* Thames and Hudson, New York.

Stuart, David

2005 The Way Beings. In *Glyphs on Pots: Decoding Classic Maya Ceramics,* edited by David Stuart, pp. 160–165. Sourcebook for the 29th Maya Meetings at Texas. University of Texas, Austin.

Tarn, Nathaniel

1997 *Scandals in the House of Birds: Shamans and Priests on Lake Atitlán.* Marsilio, New York.

Tate, Carolyn E.

2012 *Reconsidering Olmec Visual Culture: The Unborn, Women, and Creation.* University of Texas Press, Austin.

Taube, Karl

1992 *The Major Gods of Ancient Yucatan.* Studies in Pre-Columbian Art and Archaeology 32. Dumbarton Oaks Research Library and Collection, Washington, D.C.

2003 Maws of Heaven and Hell: The Symbolism of the Centipede and Serpent in Classic Maya Religion. In *Antropología de la eternidad: La muerte en la cultura maya,* edited by Andrés Cuidad Ruiz, Mario H. Ruz Sosa, and M. Josefa Igelsias Ponce de Leon, pp. 405–442. Sociedad Española de Estudios Mayas, Madrid.

2011 Stones of Lightning, Water, and Wind: The Symbolism of Flint and Jade in Ancient Mesoamerica. Talk given at the Dallas Museum of Art, Dallas.

Tedlock, Barbara

1981 Quiche Maya Dream Interpretation. *Ethos* 9(4):313–330.

1982 *Time and the Highland Maya.* University of New Mexico, Albuquerque.

1992 The Role of Dreams and Visionary Narratives in Mayan Cultural Survival. *Ethos* 20(4):453–476.

Tedlock, Dennis

1985 *Popol Vuh: The Mayan Book of the Dawn of Life.* Simon and Schuster, New York.

Thompson, J. Eric S.

1970 *Maya History and Religion.* University of Oklahoma Press, Norman.

Tozzer, Alfred M.

1907 *A Comparative Study of the Mayas and the Lacandones.* Macmillan, New York.

Umberger, Emily

1987 Antiques, Revivals, and References to the Past in Aztec Art. *RES: Anthropology and Aesthetics* 13:63–106.

Viveiros de Castro, Eduardo

2004 The Transformation of Objects into Subjects in Amerindian Ontologies. *Common Knowledge* 10(3):463–485.

Wilson, Richard

1990 Mountain Spirits and Maize: Catholic Conversation and Renovation of Traditions among the Q'eqchi' of Guatemala. PhD dissertation, Department of Social Anthropology, London School of Economics, London.

Wisdom, Charles

1940 *The Chorti Indians of Guatemala.* University of Chicago Press, Chicago.

"To Put in Order"

Classic Maya Concepts of Time and Space

MARKUS EBERL

> I know nothing of time. I am new every day. I am born when I wake up in the morning, I grow old during the day, and I die at night when I go to sleep.
>
> —Paul Auster, *City of Glass*

THE ANCIENT MAYA SHOULDERED THE burden of time. J. Eric S. Thompson (1950) opens his *Maya Hieroglyphic Writing: An Introduction* with a collage of god-numbers who carry time units with their *mecapal*, or head strap. This iconic image has defined our understanding of Classic Maya time as cyclical. For example, the Long Count works like the odometer of a modern car (Morley 1975 [1915]:60–63). It has five hierarchically arranged time periods, and each full cycle of a time period causes the next higher time period to advance by one position. When all five time periods reach completion or zero out, the cycle starts again with 1 on the next day. Linked to the Long Count are the 260-day *tzolk'iin* and the *haab*, the "vague year." The workings of both are often visualized as a mechanical clock with three gears. The 365 days of the haab form the cogs of the first gear that engage with the twenty day-signs of the tzolk'iin as the cogs of the second gear. Inside the latter runs a third gear with the thirteen coefficients of the tzolk'iin. The intersection of the three gears specifies the Calendar Round as the combination of tzolk'iin and haab.

The idealization of ancient Maya time as cyclical comes also from divinations in ethnohistoric documents. For example, the colonial Books of Chilam Balam present history in the form of the *k'atun* cycle (e.g., Edmonson 1982). Each k'atun is seventy-two hundred days, or slightly less than twenty years long, and named after the tzolk'iin date (always involving the day Ajaw) on which it ended. The entire cycle consists of thirteen k'atuns and covers 256 years. Alternating cities served as seats for each k'atun, thus expressing the cyclical nature of the count. The ups and downs of political and social history were recalled by k'atun. And each k'atun brought specific tidings: what had happened the last time a k'atun came to pass was thought to recur the next time it came around. Thus, the k'atun served to foretell the future.

The odometer-style nature of Maya calendars and their divinatory quality favor a cyclical interpretation of Maya time. The Maya carried, in Thompson's (1950) interpretation, the "burden of time." His work is cited not only to support cyclical versus linear time concepts but also to contrast the less advanced ancient with the presumably more sophisticated modern societies (Gell 1992:35; Howe 1981; Whitrow 1988:128, 154, 157). But the linear-versus-cyclical dichotomy overlooks the experiential basis of time. In the following, I shift the perspective from the "burden of time" to "burdened with time" in order to move from an abstract understanding of time to a study of its practice and bodily experience. I adapt Edmund Husserl's work to develop a cognitive model of time that I apply to Classic Maya culture.

Experiencing Time

Abstract time concepts downplay the personal experience of time. We are in time when we speak and when we act. Time forms an integral part of our existence. In his book *Zur Phänomenologie des inneren Zeitbewusstseins* (On the Phenomenology of Internal Time-Consciousness), Edmund Husserl (1966) develops a subjective perspective of time. He differentiates between the now moment, retention, and protention (Moran and Cohen 2012:280). When a musician plays a tone for several seconds, we perceive a continuous tone despite the fact that the initial tone is no longer audible and present. We experience the tone in the now—that is, as a cognitively concomitant moment that reaches from the present into the past and into the future (cf. specious present in James 1890:605–642). Unlike the knife-edge present, the now moment has duration.[1]

Franz Brentano (1874), Husserl's teacher, instead argues that we hear only the tone that the musician currently plays, but we constantly recall the initial tone and associate it inseparably with the present tone. Hearing is like viewing a filmstrip in which each frame represents an earlier aspect of the tone (Gell 1992:223); we string these frames or remembered now moments together into our experience of the now. Brentano's approach allows for multiple now moments and integrates the act of remembering into the act of experiencing the now moment (Mensch 2010:71–72).

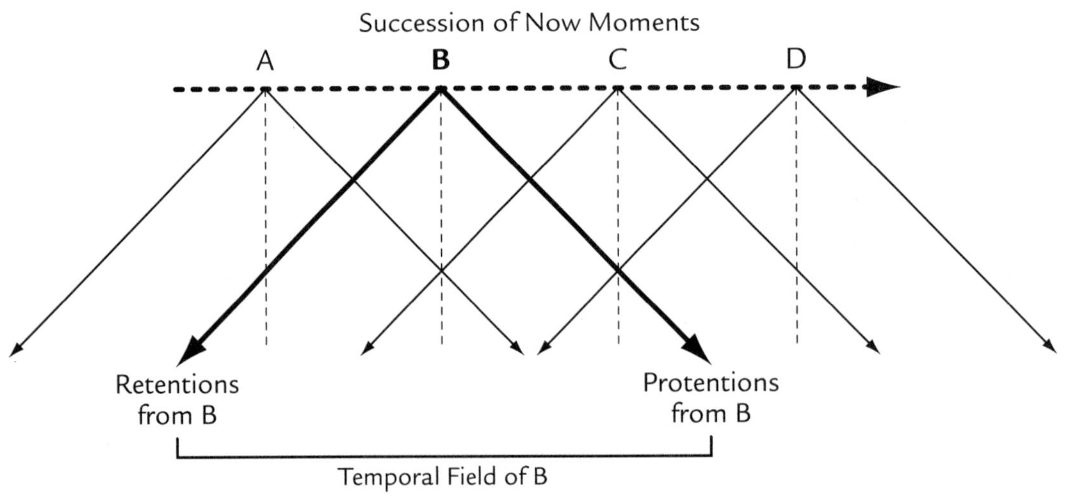

figure 5.1
Edmund Husserl's model of internal time-consciousness. (Drawing by Markus Eberl, after Gell 1992:225 and Husserl 1966:365.)

Husserl (1966) builds on Brentano's model and differentiates the reproduction from the retention or protention of experiences (Figure 5.1). Reproductions replay past experiences; we remember now moments and bring them from the past into the present. On the other hand, retentions and protentions are temporally removed aspects of a now moment. When we hear the musician play, we perceive the tone anew every second while remaining aware of only one tone. Upon hearing the tone for the first time, our mind creates a mental representation (Brentano's [1874:115] "in-existence"). As we continue to hear the tone, retentions direct our cognition to this mental image and create the experience of a lasting tone. The mental image is not stable, though, but changes with new perceptions. As the present moves on, retentions fall out of sync with perception and slowly sink below the temporal horizon that comprises the now moment. Modifications that come with new perceptions lead us to reevaluate past retentions. This process is not only backward looking but also future oriented. Protentions are mental perspectives that we form about forthcoming events (for example, we expect the musician to continue to play the tone) and that we continually modify as we come temporally closer to those events. Time is neither external nor abstract; instead, "subjective time arises as an inescapable feature of the perceptual process itself" (Gell 1992:231).

Reproducing Now Moments

Husserl's model of internal time consciousness differentiates between retention and reproduction. The former contributes to the cognitively ordered set of experiences that define the now moment. As the present advances, retentions sink into the past, and new perceptions continuously modify the arranged contents. These modified retentions then form the memory of the now moment. Husserl's concept of modification challenges the idea that the past can be reproduced "just the way it was."

A Classic Maya example illustrates the complexity of reproduction. K'inich Yax K'uk' Mo' founded the royal dynasty of Copan in 426 CE (Martin and Grube 2008:192–193). The Late Classic kings who dedicated the relevant hieroglyphic inscriptions recalled what had happened almost four centuries earlier (Figure 5.2). Yax Pasaj, the sixteenth ruler of Copan (763 to ca. 810 CE), dedicated Altar Q in 776 CE, and the altar's main text goes back to September 3, 426 CE, when the dynasty founder appears with his pre-accession name, K'uk' Mo' Ajaw, and grasps the K'awiil scepter at the *wi' te' naah*, or origin house (Stuart 2000, 2004a:235–238). Three days later, he leaves the origin house as K'inich Yax K'uk' Mo', his royal name. He embarks on a journey of approximately five months until he sets the K'awiil scepter down and arrives at Copan on February 6, 427 CE. The second inscription that recalls these events is Quirigua Zoomorph P. Quirigua was for a long time subordinate to Copan, and only its last rulers achieved independence. Among them was Sky Xul (785 to ca. 795 CE), the king who erected Zoomorph P (Martin and Grube 2008:223–224). According to his text, somebody leaves the origin house on September 3, 426 CE. Three days later, Tok Casper, the first king of Quirigua, plants a stone—that is, a carved monument—and ties the royal headband under the direct or indirect supervision of Copan's K'inich Yax K'uk' Mo' (called here K'uk' Mo' K'inich [?]).

The two inscriptions present the origins of rulership at Copan and Quirigua. They coincide in their dates, and both associate key events with the origin house.[2] On Altar Q, K'inich Yax K'uk' Mo' transforms into a divine king by changing his name and by grasping K'awiil, the royal scepter and god of rulership (Schele and Miller 1983). He then brings the K'awiil scepter to Copan and eventually hands it, as the row of sixteen kings on Altar Q's exterior shows, over to his successors. Zoomorph P structures the narrative around Tok Casper, Quirigua's dynasty founder. The person who comes from the origin house remains unspecified, but Tok Casper is implied because he initiates the following events. Unlike Yax K'uk' Mo', who grasps K'awiil and then leaves the origin house, Tok Casper leaves the origin house and then ties the royal headband. The K'awiil scepter and headband are both symbols of royal authority, but Maya rulers selected one or the

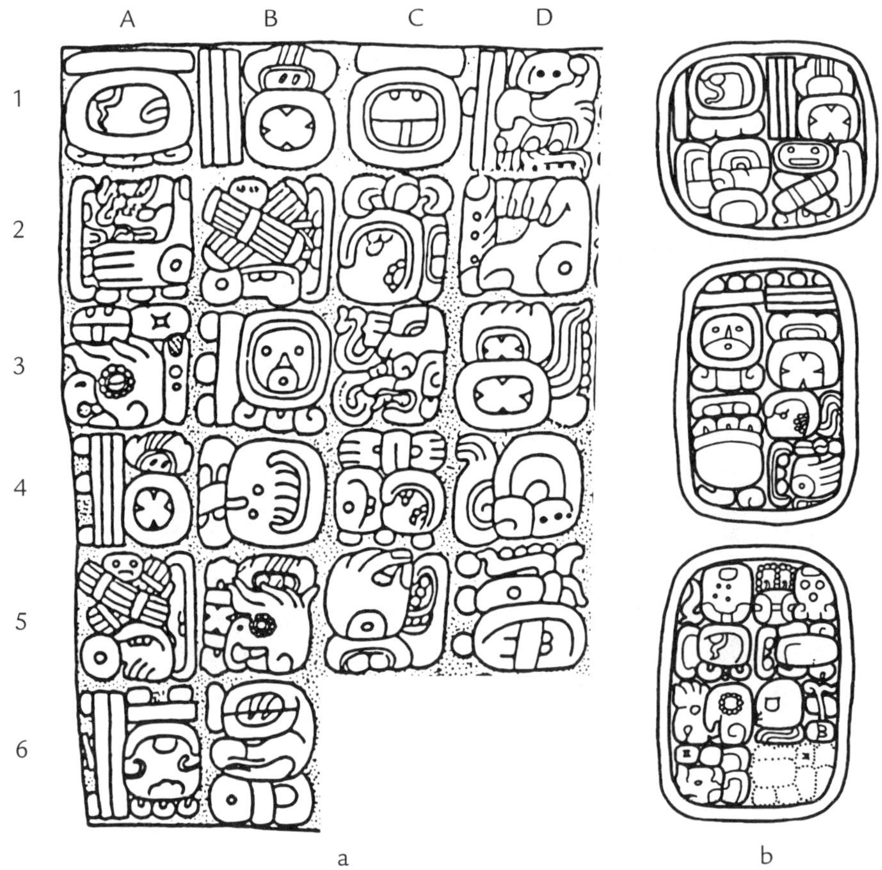

figure 5.2
Recalling the founding of royal dynasties at Copan and Quirigua: (a) glyphs A1–D5 on Copan Altar Q (after Schele and Freidel 1990:310); and (b) cartouches 5–7 on Quirigua Zoomorph P (after Looper 2003:38).

other during accession (Eberl and Graña-Behrens 2004:102–104). The chosen symbol alluded in part to their public identity. For example, Late Classic rulers from or affiliated with Calakmul's snake dynasty grasped the K'awiil scepter. By inverting the order and kind of foundational events, Copan Altar Q and Quirigua Zoomorph P highlight the unique charter of their royal dynasty.

The interpretation of Maya hieroglyphic texts has emphasized their reception. From this perspective, an ancient audience read Altar Q and Zoomorph P and identified how overlapping or diverging narratives mirrored historical events and ideology (Houston 1994; Marcus 1992a, 1992b, 1995; Martin and Grube 2008:127, 192–193; Stuart 1994). Husserl's phenomenology offers a complementary cognitive approach that emphasizes the author and his or her recollection of the past and in which modifications are an implicit aspect of understanding and representing time. The origin house serves as their spatial setting, and the dates overlap in both hieroglyphic texts. The foundational events share a spatiotemporal framework. The two texts also coincide in aspects that were graspable—for example, buildings identified as origin houses (Stuart 2000, 2004a)—and retellable even hundreds of years later. These physical manifestations frame perception and memory. At the same time, the texts highlight the unique royal charter of each dynasty. The places of the foundational events coincide but were charged with different temporal meanings and different kinds of past (Herzfeld 1991; Lass 1988; Lynch 1972; Munn 1992). Different perceptions thus create contestable cognitive images.

Creating Ordered Cognitive Images

In Husserl's model of time consciousness, intentions link perception with a mental representation. This representation changes dynamically through modifications and with new information from the physical senses. As we walk around a shopping mall, for example, we see the building from a different angle with every step. Instead of mistaking it for different malls, we correlate our changing visual impressions with the sense of our moving body and know that we remain in the same building. We navigate successfully from one store to the next by orienting ourselves through retentions and by making protentions or assumptions about how the mall will look around the next corner. As we move through the mall, retentions and protentions change continuously. This process allows us to refine our mental image of the building. We are not aware of what we perceive "in an abstractly empty way; any awareness has a constitutive structure" (McCumber 2011:140). The intentional object gives rise to an ordered set of experiences. Perception not only constitutes the perceived but is also constituted by it (Husserl 1966:91; Mensch 2010:115–117). As retentions and protentions bob to the present and then run off again, we become aware of this dynamic play and grasp the workings of our consciousness (Findlay 1975:18). Interpretations of Husserl's model highlight the ordered nature of the mental representation (Findlay 1975:11; McCumber 2011:139–140). Through an examination of *tz'ak*, the Classic Mayan word for "to put in order," I describe the ordering as an implicit outcome of moving in space and time (Figure 5.3; see glyph 573 in Thompson's [1962] catalog).

Classic Maya texts generally feature *tz'ak* in two contexts (see Riese 1984 for a full discussion). The glyph occurs before distance numbers—that is, the count of days between two events—and

figure 5.3
The glyph for *tz'ak* and its cognates: (a) Palenque Temple XIX platform, south face, glyph A8 (based on photoraph by Joel Skidmore); (b) Dos Pilas Stela 8, glyph A10 (after Houston 1993:111, fig. 4.14); (c) Yaxchilan Lintel 35, glyph A1 (after Graham 1979:79); (d) Palenque Temple XIX platform, west face, glyph D4 (based on photograph at www.mesoweb.com); and (e) Randall Stela, glyphs H4–J2 (after Miller and Martin 2004:190–191, pl. 107). (Drawings by Markus Eberl.)

has therefore been labeled as Distance Number Introductory Glyph or DNIG (Thompson 1950:160). The second context relates to the count of rulers (Grube 1988; Riese 1984; Schele 1992). Preceded by a number, *tz'ak* indicates the position of a Maya ruler with regard to the dynasty's founder (Riese 1984; *tal* [Figure 5.3c] is an alternative expression). The decipherment of T573 as **TZ'AK** originates with Berthold Riese (1984), who considered it alongside *hel* or *hal*. Riese preferred the latter readings because of similar words in K'iche', Pocom, and Yucatec Maya (also Thompson 1950:161–162). Recent discoveries support the **TZ'AK** reading. The Randall Stela mentions the eighteenth successor of a Maya lord and spells *successor* syllabically as "tz'a-ka-bu-li?" (Figure 5.3e; Stuart 2003). This substitution and *ka* phonetic complements—together with the recognition of a Ch'olan language as the language of the Classic period-hieroglyphic texts (Houston et al. 2000; Mora-Marín 2009)—the reading of T573 as **TZ'AK**.

Modern and colonial Ch'olan languages use *tz'ak* in contexts that are comparable to hieroglyphic texts (Table 5.1). *Tz'ak* and its cognates mean to order something that is incomplete and to make it complete again. In modern Ch'orti', the phrase "ne'n intz'aki ixin e ojroner twa' ayan uyokir alok'oy" translates as "I kept adding words so that it would make sense" (Hull 2005:109). "Niwijtz'in tz'akti'x chanmojy ujab'" means "My younger sister is [literally: added up to] already twenty years old" (see Pérez Martínez et al. 1996:228). In modern Ch'ol ritual language, *tz'äk* refers to "medicine and remedy," and the derived verb *tz'äk-an* means "to cure" (Josserand and Hopkins 2001:30). The Ch'ol of Tila identify soul loss as the main cause of illnesses because it results in an incomplete state of being. Curing returns the lost soul and makes the patient complete again. In Acalan Chontal, *-tz'ak* (written as *-dzac*) serves as numeral classifier for elements of a series (Smailus 1975:133). The early colonial Paxbolon-Maldonado papers list individual Acalan rulers: "tal-i u-chan-tz'ak ajaw u-k'aba Paxwa" or "there came the fourth ruler of the series named Paxwa" (translation based on Scholes and Roys 1948:384; Smailus 1975:28–29). *Tz'ak* identifies the position of each ruler in the dynasty. This meaning resonates in the modern Chontal *tz'äjca* for "one above the other, stacked" (Keller and Luciano G. 1997:261). The ordering completes a series.

Tz'ak precedes distance numbers in Maya hieroglyphic texts. It takes the form *u tz'ak-a* (it is the putting in order of) and is followed by the number of days and the calendar round of the next date. Thompson (1950:160) reasons that *u tz'ak-a* "has the function of announcing that a distance number is to follow." The label "distance number introductory glyph" is simplistic, though, as an example from Site Q or La Corona Panel 2 shows (Figure 5.4). The local ruler Chak Chan Nahb died on July 28, 667 (glyphs I7–K1), and the preceding passage links his death to two earlier events. Lady 6 Chan Ajaw had died 184 days before his death on January 25, 667 (glyphs I3–J3). The second distance number counts 3,456 days back to Chak Chan Nahb's accession on February 9, 658 (glyphs J5–J6). In glyphs I2 and I4, *u tz'ak-a* introduces both distance numbers and arranges the earlier events so that they line up and lead to Chak Chan Nahb's death.

The count of time and the count of rulers overlap conceptually, as the hieroglyphic bench from Copan Structure 21a shows (Figure 5.5). King Yax Pasaj had Structure 21a built for his accession, squeezing it between Structures 21 and 22, buildings of his famous predecessor Waxaklajuun Ubaah K'awiil (Martin and Grube 2008:204–205). The throne bench takes up the entire width of Structure 21a, and its hieroglyphic text runs, interrupted by large star signs, below its ledge. The text begins with "u tz'ak-a" (glyph A) and Yax Pasaj's name (glyphs B and C); it continues "i chum-wan," "then he is seated" (glyph D), "ti 6 Kaban," "in or on day 6 Kaban" (glyph E). "6 Kaban" refers to Yax Pasaj's accession date 9.16.12.5.17 6 Kaban 10 Mol or June 26, 763 CE. The bench text mimics a typical distance number (e.g., Figure 5.4): "u tz'ak-a [day count] i uht tzolk'iin" or "it is set in order these many days, [and] then it comes to pass tzolk'iin day so-and-so." On Structure 21a's bench, *Yax Pasaj* substitutes for the day count and his accession replaces *uht*. The accession statement further emphasizes the overlap between time and office. It normally reads "chum-wan ti ajawlel," "(s)he is seated in rulership," but here

table 5.1
Cognates of *tz'ak* in Ch'olan languages.

CH'OLAN LANGUAGE	DICTIONARY ENTRY	SOURCE
Proto-Cholan	*tz'äk, "complete, whole; enough; medicine; fish poison"	Kaufman and Norman 1984:134
Ch'olti'	tzactael, "cumplir, cabal" (to fulfill, to complete) tzactal, "cumplido" (fulfilled) tzactez, "cumplir" (to fulfill)	Robertson et al. 2010:305
Ch'orti'	t'saak, "good health, improvement in the body" t'saak-ar, "remedy"	Wisdom 1950:1123
Ch'orti'	tz'aki, "add, bring together"	Hull 2005:109
Ch'orti'	tz'aka, "conectar, añadir" (to connect, to add) tz'akar, "bastar, suficiente, añadir" (to satisfy, sufficient, to add) tz'akna, "curar" (to cure) tz'akpe'n, "curar, sanar, aliviar, añadir" (to cure, to heal, to ease, to add) tz'aksen, "ajustar, completar" (to adjust, to complete) tz'akta ujab', "cumplir años" (to celebrate one's birthday; literally: "to complete [one's] years")	Pérez Martínez et al. 1996:227–228
Ch'ol	ts'ac, "medicina, remedio" (medicine, remedy) (p. 121) ts'ajc, "pared de piedra, banqueta, cemento" (stone wall, sidewalk, cement) ts'ʌcal, "recaudo, medicina" (remedy, medicine) (p. 124) ts'ʌcan, "curar" (to cure) ts'ʌactesan, "completar" (to complete) ts'ʌactesʌntel, "cumplirse" (to be fulfilled)	Aulie and Aulie 1978:121, 124
Ch'ol	tz'ak, "medicina, remedio, curar" (medicine, remedy, to cure) tz'ök-an tz'ok-ö, "curar" (to cure) tz'ök-al, "color" (color) tz'ököl, "cabal, completo" (finished, complete)	Schumann G. 1973:97
Acalan Chontal	-dzac, "Clasificador numeral significando los elementos de una serie; es variante de -dzic" (Numeral classifier for the elements of a series; variant for -dzic) dzac-al, "mucho" (much)	Smailus 1975:133
Chontal	tz'ac, "medicina, remedio, curación" (medicine, remedy, cure) tz'äcä, "completo" (complete) tz'äcälin, "curar" (to cure) tz'äcchuye', "coser algún pliegue accidentalmente" (to sew a fold by accident) tz'äctan, "completarse, estar terminado" (to be completed, to be finished) tz'äjca, "uno sobre otro, apilados" (one above the other, stacked)	Keller and Luciano G. 1997:260–261

Yax Pasaj is seated "ti 6 Kaban," which translates as "on day 6 Kaban" or as "in day 6 Kaban"—in other words, in time. The Structure 21a text seems to say that Yax Pasaj was "put into the order" of Copan kings when he acceded to Copan's throne. As ruler, he embodied time and space.

In addition to ordering, *u tz'ak-a* also completes. The passage from the Site Q or La Corona Panel 2 links Chak Chan Nahb's accession and death (see Figure 5.4). Several scholars have also studied versions of the DNIG in which pairings substitute for *tz'ak* (Knowlton 2002; Riese 1984; Stuart 2005:99–100; also Schele 1992:139 for comparable substitutions in the count of rulers). Examples include wind and rain, night and day, star and moon, and green and yellow (see Figure

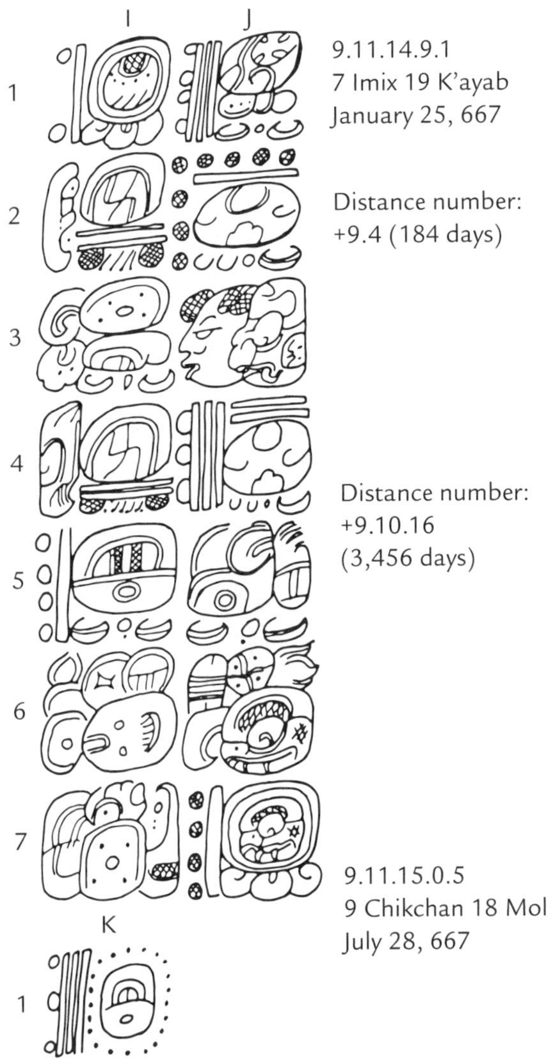

figure 5.4
Distance numbers on Panel 2 from Site Q or La Corona (glyphs I1–K1). (Drawing by Linda Schele [Mayer 1995:pl. 161].)

5.3d). Both elements complement each other, like night and day or green (unripe) and yellow (edible). They convey that *tz'ak* makes something whole and complete. A similar meaning applies to a Calendar Round that expresses the end of a haab month in a unique way (Houston and Mathews 1985:11; Figure 5.6a). The ancient Maya either counted the twenty days of a haab month from its seating (or *chum*) to the nineteenth day or from the first day to its completion on the twentieth day (Thompson 1950:119–121, fig. 19, 21–51). The Tonina fragment begins with the partially eroded tzolk'iin date 6 Eb? and continues with "huliy u tz'ak-te' Yaxk'iin," "it arrived the completion of [haab month] Yaxk'iin" or 20 Yaxk'iin. Completing one month also marks the transition to the next month. This applies also to the reign of rulers. Dos Pilas king Itzamnaaj K'awiil celebrated his first k'atun in office in 717 CE. But instead of calling it an anniversary, he looked forward and said that he arrived at "u 2 tz'ak u? K'atun [ti?] ajawle[l]," "the second completion of a K'atun in rulership" (Figure 5.6b).[3] The end of one cycle marks the beginning of the next cycle and thus the beginning of a series.

The count of rulers suggests a natural order. For example, Tikal ruler Animal Skull calls himself "u 22 tz'akbuil ajaw" or the twenty-second ruler of his royal dynasty (Figure 5.7a).[4] But while this and similar statements provide epigraphers with a convenient way to reconstruct royal dynasties, they look true and natural only in hindsight. Rulers used tz'ak counts not only to look back but also to stake a claim to the future. Quirigua king K'ahk' Tiliw

figure 5.5
The accession of Copan ruler Yax Pasaj (Hieroglyphic Bench from Copan Structure 10L-21a, glyphs A–E; the star sign between glyphs C and D is omitted). (Drawing by Markus Eberl, after photo and drawing in Morley 1920:pl. 31, with additional details from drawings in Jackson 2004:9 and Lounsbury 1989:84, fig. 6.11.)

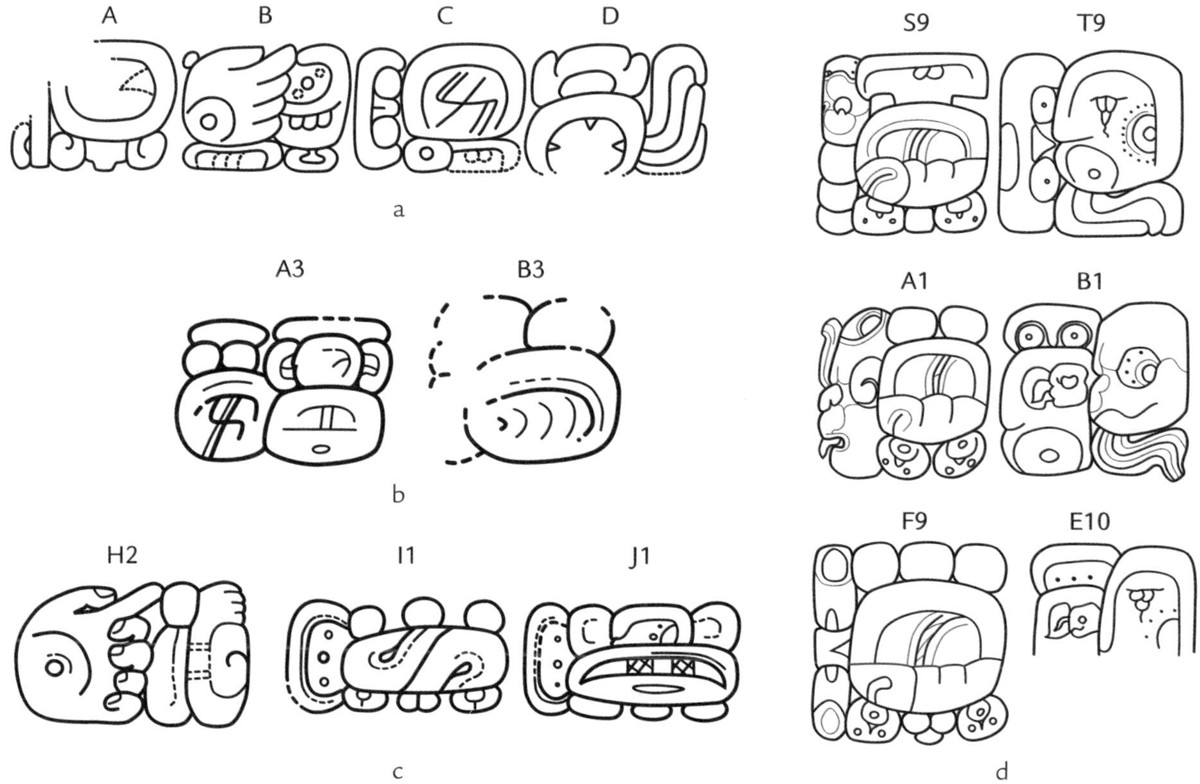

figure 5.6
The ordering of time and events: (a) Tonina Fragment 35, glyphs A–D (after Graham 2006:124); (b) Dos Pilas Hieroglyphic Stairway 1, Step III, glyphs A3–B3 (after Houston 1993:112); (c) Dos Pilas Hieroglyphic Stairway 4, Step I, glyphs H2–J1 (after Houston 1993:109); and (d) Palenque Temple of Inscriptions, East Tablet, glyphs S9–T9, and Center Tablet, glyphs A1–B1 and F9–E10 (shown top to bottom; based on drawings in Robertson 1983:figs. 95–96 and www.mesoweb.com photos). (Drawings by Markus Eberl.)

famously claimed to be the fourteenth successor of the founder of Copan after he defeated Waxaklajuun Ubaah K'awiil, Copan's thirteenth ruler (Riese 1984:276–279). Rulers thus imposed their subjective perspectives on the count of rulers. Yuknoom Ch'een II oversaw Calakmul's golden age in the seventh century CE. Later nicknamed Yuknoom the Great, his successors held him in such high esteem that they counted their place in the snake dynasty no longer from the dynasty founder but from him (Martin 2005:8). The order of rulers reflects strategic ordering, as Palenque's Temple XVIII exemplifies. Its now-destroyed stuccoed back showed how K'inich Janaab Pakal I arranged the succession order of his three sons (Figure 5.7b; Ringle 1996). The accompanying stucco glyphs praised his effort: "tihmaj a-wohl a-tz'akbuji," "you are pleased [that] you order them" (Stuart 2005:153). His sons heeded their father's order. K'inich Kan Bahlam II, the oldest son, succeeded his father, followed by K'inich K'an Joy Chitam, the middle son, and then K'inich Ahkal Mo' Nahb III, the son of Janaab Pakal's youngest son. K'inich K'an Joy Chitam, the middle son, reflected the new order by not calling himself the eleventh ruler of Palenque's dynasty but "the second *yok'iin?* of the west Kaloomte'," that is, he counted his dynastic position beginning with his father (Figure 5.7c).[5]

For the ancient Maya, the world needed to be ordered even to come into existence. 4 Ajaw 8 Kumk'u (September 6, 3114 BCE) marks the beginning of the Long Count and the creation

figure 5.7

The ordering of people and gods: (a) glyphs G–J from Tikal Burial 195 vessel 12U-38/27 (after Culbert 1993:50e); (b) stucco glyphs 539 from the rear wall of Palenque's Temple XVIII (drawing by Markus Eberl, after Schele and Mathews 1979:no. 539); (c) Palenque Palace Tablet, glyphs O13–P14 (drawing by Markus Eberl, based on drawing in Robertson 1985:fig. 258 and photographs at www.mesoweb.com); and (d) glyphs P1–R2 from the "Vessel of the Seven Gods" (K2796; only the first two gods are shown; drawing by Markus Eberl, after photograph by Justin Kerr at www.mayavase.com).

of the world. Vessels K2796 and K7750 show this momentous day when seven or eleven gods were seated in front of God L, the lord of the underworld. The accompanying hieroglyphic text explains that on 4 Ajaw 8 Kumk'u "it was put in order," followed by a location called *ik'-u-tahn*, "black is the center," and then a list of the gods seated before God L (Figure 5.7d). This text can be read in two ways. According to Linda Schele and Khristaan Villela (1996), the chaotic black universe was ordered by the council of gods and transformed into the present world. Alternatively, the gods, many of whom are elemental forces like the sky god and the earth god, were put into the appropriate order at the ik'-u-tahn. Since the *tz'ahk-a* verb is passive and can take only one argument, one has to assume that the scribe either omitted introducing the gods with *by* (e.g., using *u kabjiy*) in the case of the first interpretation or omitted a prepositional particle in front of *ik'-u-tahn* in the case of the second interpretation. Nonetheless, the ordering applies in both interpretations to space and time.

Finally, the ancient Maya employed *tz'ak* when they talked about the ordering of objects. Nim Li Punit Stela 15 mentions the celebration of an Early Classic period ending and explains that "u 6 tz'akal tuun," "the sixth completed stone-monument," was driven into the ground (Figure 5.8a). This passage suggests that Nim Li Punit's rulers tracked the order in which they dedicated each carved monument, which, by the end of the Late Classic, totaled thirty-two stelae. Maya rulers required a similar order when it came to the tribute payments owed to them, and many palace scenes show people bringing gifts and tribute. In the year 815 CE, Naranjo ruler Waxaklajuun Ubaah K'awiil received tribute

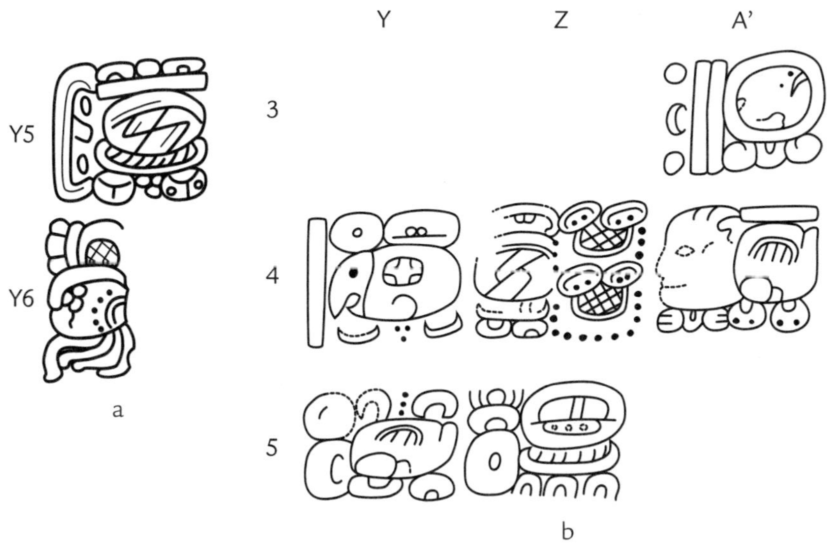

figure 5.8
The ordering of objects: (a) Nim Li Punit Stela 15, glyphs Y5–Y6a (after Robertson 1995:CD-ROM 11:T20163; Wanyerka 2004:72); and (b) Naranjo Stela 32, glyphs A'3–Z5 (after Graham 1978:86; Morley 1938:5:pl. 92d). (Drawings by Markus Eberl.)

(Le Fort and Wald 1995): "u tz'ak-aw k'uk' bahlam 5 k'al," "he piles up quetzal [and] jaguar, five times twenty" or, more freely, "he stacks one hundred feathers and pelts" (Figure 5.8b).[6] *Tz'ak* refers here to the piling of tribute in front of the ruler and to the completion of one's duty.

Distance numbers and counts of rulers are presented as natural orders in Maya hieroglyphic texts; I argue that they also represent a conscious effort of ordering. The expressions involving *tz'ak* position beings, events, and objects with regard to similar elements. The term implies a classification of the world into specific categories that then can be counted. Perception is not simply the passive input from our physical senses but an active cognitive process. The ancient Maya understood completion not as an end in itself but as necessary for a beginning: each end transitions into a beginning and initiates a series. Ordering is an essential aspect of the cognitive process and thus of being human and knowing how to be human.

Framing Now Moments

Now moments interweave physical senses, perception, and cognitive image. Time consciousness is, in Husserl's model, entirely internalized. French philosopher Maurice Merleau-Ponty (1962:137) criticized Husserl because "movement is not thought about movement, and bodily space is not space thought of or represented." Instead, perception relates to actual body movements. The position of our bodies in space defines what we perceive and what we can be conscious about. Voluntary body movements change our perspective and add new physical perceptions that modify now moments. The movements take place in space and determine how we relate to space (Goldstein 1923:163). Time consciousness is, therefore, also space consciousness.

The ancient Maya also brought space and time together in their celebrations of period endings. After defeating the neighboring kings of El Perú and Naranjo, Tikal's king Yik'in Chan K'awiil celebrated the end of the thirteenth *tuun* (9.15.13.0.0) on June 8, 744 (Martin and Grube 2008:49). The front of Tikal Stela 5 shows Yik'in Chan K'awiil with an incense pouch and a scepter or axe on top of the captured king of Naranjo. He carries a back rack that supports death god Ahkan, who is also known as Akan and God A' (Figure 5.9a; for reading the god's name as Ahkan, see Lacadena 2004:183n114). Ahkan's presence is a bad omen for the captive because he is associated with drinking, disease, death, and sacrifice (Grube 2004; Taube 1992:14–16; Zimmermann 1956:162–163). Yik'in Chan K'awiil's captive likely met a violent end during the period-ending celebrations.

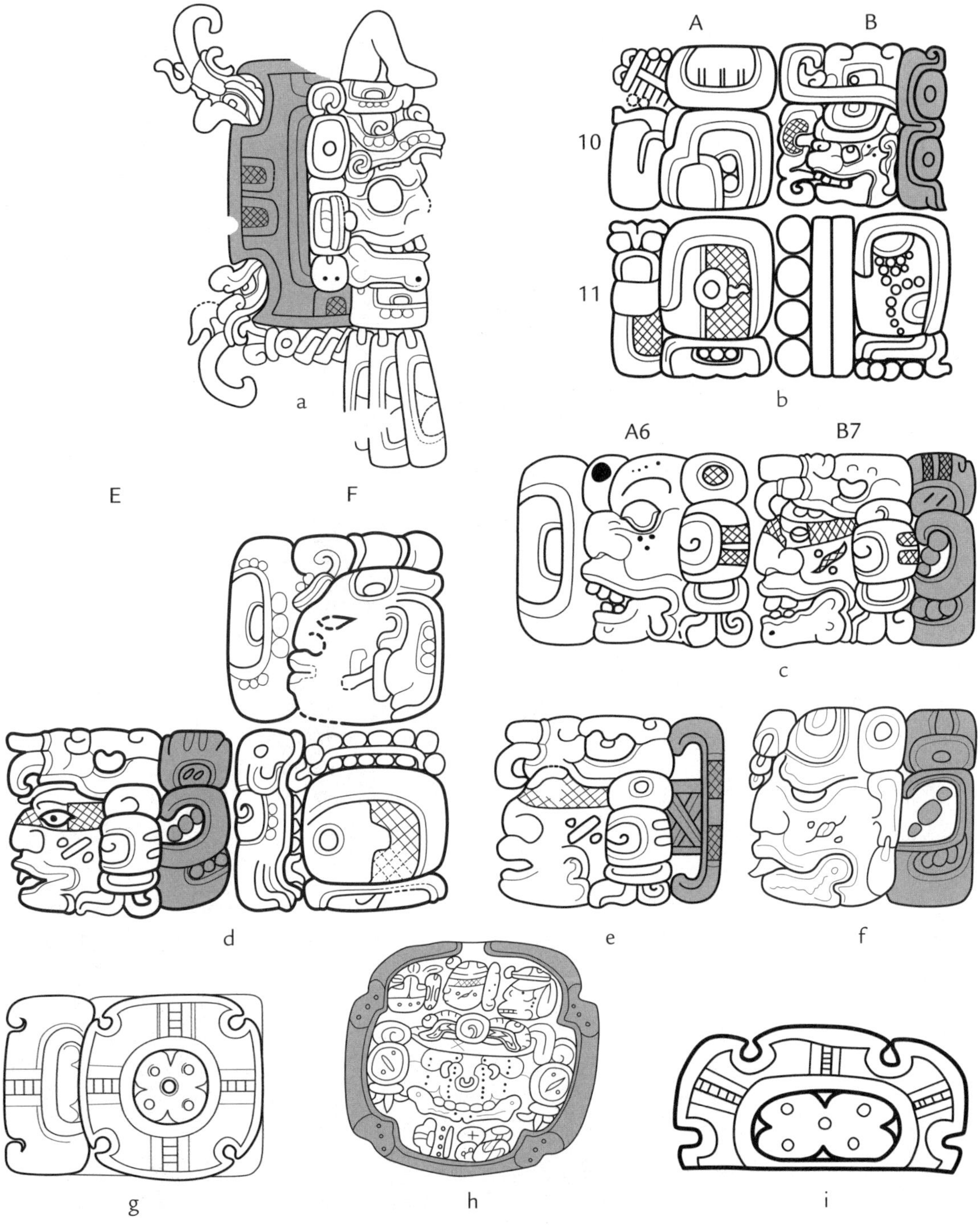

figure 5.9
Death god Ahkan in his otherworldly residence and portal (the latter are shaded in gray): (a) detail of the front of Tikal Stela 5 (based on Jones and Satterthwaite 1982:fig. 7 and photographs in the Tikal Project Image Database at www.famsi.org); (b) glyphs A10–B11 on Tikal Stela 5 (based on Jones and Satterthwaite 1982:fig. 7 and photographs in the Tikal Project Image Database at www.famsi.org); (c) glyphs A6–B7 from Tikal Temple IV Lintel 3 (after Jones and Satterthwaite 1982:fig. 74); (d) glyphs F5 through F6 from Tikal Temple IV Lintel 3 (after Jones and Satterthwaite 1982:fig. 74); (e) glyph G3 from Tikal Temple IV Lintel 3 (after Jones and Satterthwaite 1982:fig. 74); (f) glyph E13 from Palenque's Palace Tablet (after Robertson 1985:fig. 258); (g) glyph B9 on Copan Stela 9 (after Maudslay 1889–1902:1:pl. 110); (h) center of Naranjo Altar 1 (after Graham 1978:103); and (i) Yaxk'iin as haab patron on the back of Copan Stela 16 (after Morley 1920:pl. 10). (Drawings by Markus Eberl.)

figure 5.10
The paddler gods emerge from star signs, detail of Yaxchilan Stela 4. (Drawing by Markus Eberl, after Maler 1901–1903:pl. LXX.)

Death god Ahkan situates himself spatially. His name ends with *naah*, "building," the star sign, or two glyphs that have been read as **YAX:ja** (Grube 2004:67; Figure 5.9b–f). What is the nature of Ahkan's house? The star sign substitutes for *building* because the Classic Maya saw celestial objects (including the sun, the moon, and stars) as places where supernatural beings like Ahkan reside. For example, the paddler gods emerge or are born from star signs (Figure 5.10).[7] The center of Naranjo Altar 1 shows Ahkan (at least his name) inside or behind the *way* maws (see Figure 5.9h; Taube 2003). And similar to the doorways of houses, these celestial objects are portals through which gods travel and descend to earth (Freidel et al. 1993:269; Stuart 2005:73; Taube 2003).[8] Still unexplained is the *'yax-ja'* compound in some versions of Ahkan's name. On the front of Tikal Stela 5, Ahkan emerges from the *yax* glyph with its characteristic trilobe shape, the rounded interior, and the multiple parallel strokes (see Figure 5.9a).[9] The yax glyph shares these characteristics with the frame of the solar disk (see Figure 5.9g and 5.9i). In Early Classic versions, it surrounds the glyph for *k'iin*, "sun, day." Yax is, therefore, not only linked to the sun and other stars but also serves as a frame through which Ahkan emerges on Tikal Stela 5.[10] In other words, I suggest that in this context the assumed yax glyph is—or at least alludes to—a portal.

Ahkan's house is a conduit for the death god, and it manifests as building, star, and portal. *Yax* translates as "green-blue" and serves as the color of the center (Houston et al. 2009:40–42). It invokes the four-directional cosmogram shared by the ancient Maya and other Mesoamerican cultures (Freidel and Schele 1988). Particular colors, deities, animals, and plants are linked with the center and each cardinal direction. The cosmogram also invokes time through its association of east and west with the passage of the sun and its alignment with the calendar system. For example, the 260 days of the tzolk'iin line up as a four-sided cosmogram on pages 75 and 76 of the Madrid Codex. The substitutions at the end of Ahkan's name locate the death god in an otherworldy and centrally located house. Fittingly, Ahkan is called "ohl-is k'uh," "heart/center god," on Palenque's Palace Tablet (glyph block E14). The temporal associations reverberate in the hieroglyphic text of Tikal Stela 5 on which Ahkan is situated "tu ch'een 14 tuun," "in the cave of the fourteenth tuun" (see Figure 5.9b).

Classic Maya rulers saw themselves as the *axis mundi* in the center of the cosmogram, with time and space revolving around them (Freidel et al. 1993:123–172). To celebrate the period ending, Yik'in Chan K'awiil connected earth and otherworld and conjured Ahkan on earth and onto his back rack. His ritual performance completed a circular logic in which he created an interpretation of time and space while also portraying it as already existing (Keane 1997:18). The yax-shaped portal recalls Yik'in Chan K'awiil's position in the center of the universe. Tikal's divine king materialized space and time in concrete places and actions.

Manipulating Spacetime

Concretization applies not only to the present but also to the past (see also Brown, this volume; López Austin, this volume). The count of rulers indicates the positions of those rulers within their royal dynasty and sometimes names the dynasty founder as the starting point for the count (Riese 1984; see Figure 5.7a). Several inscriptions replace the dynasty founder with the *wi te' naah* or origin house (Figure 5.11). For example, Copan's Waxaklajuun Ubaah K'awiil identifies himself as the thirteenth successor of the origin house (Figure 5.11a). The latter was not an abstract place of memory; instead, buildings like Copan's Structure 10L-16 and Rio Amarillo's Structure 5 have been identified as origin houses (Stuart 2004a; Taube 2004b). By substituting a building for the dynasty founder, the ancient Maya created concrete places for ancestor worship and therefore materialized the past (cf. Barrett 1994:23, 1999:255–256).

In addition, the Classic Maya re-created mythical landscapes. The Five-Flower Mountain figures prominently in hieroglyphic texts (Taube 2004a). Its identification goes back to the work of David Stuart and Stephen Houston (1994) on Classic Maya place-names. They showed how the Classic Maya named natural and cultural features of the landscape,

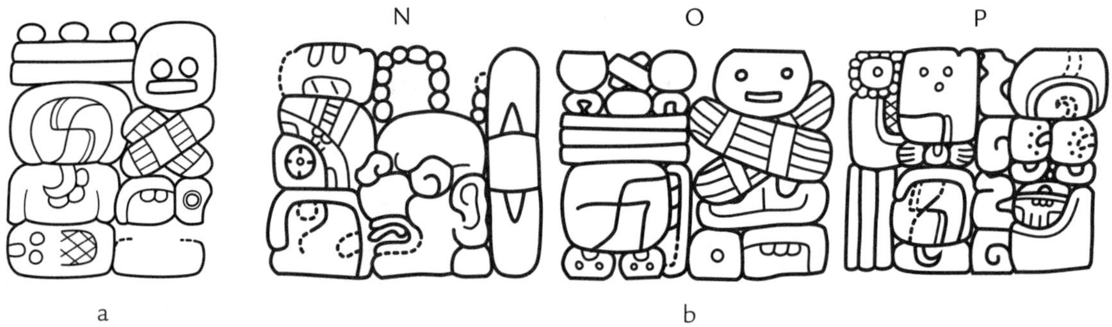

figure 5.11
The origin house in the count of rulers: (a) Copan Stela B, glyph B11 (after Maudslay 1889–1902:1:pls. 36B and 37B); and (b) Copan Reviewing Stand, right side, glyphs N–P. (Drawings by Markus Eberl.)

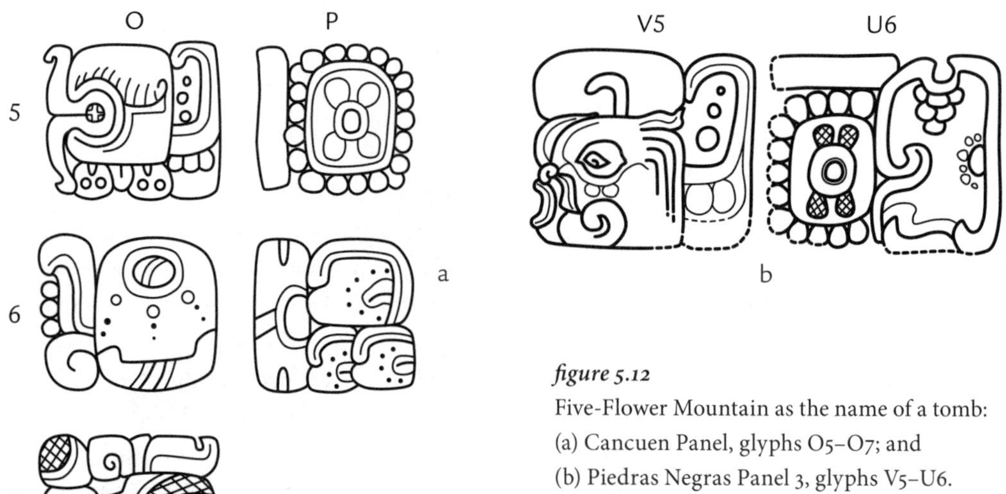

figure 5.12
Five-Flower Mountain as the name of a tomb: (a) Cancuen Panel, glyphs O5–O7; and (b) Piedras Negras Panel 3, glyphs V5–U6. (Drawings by Markus Eberl, based on photograph by Jorge Pérez de Lara at www.mesoweb.com.)

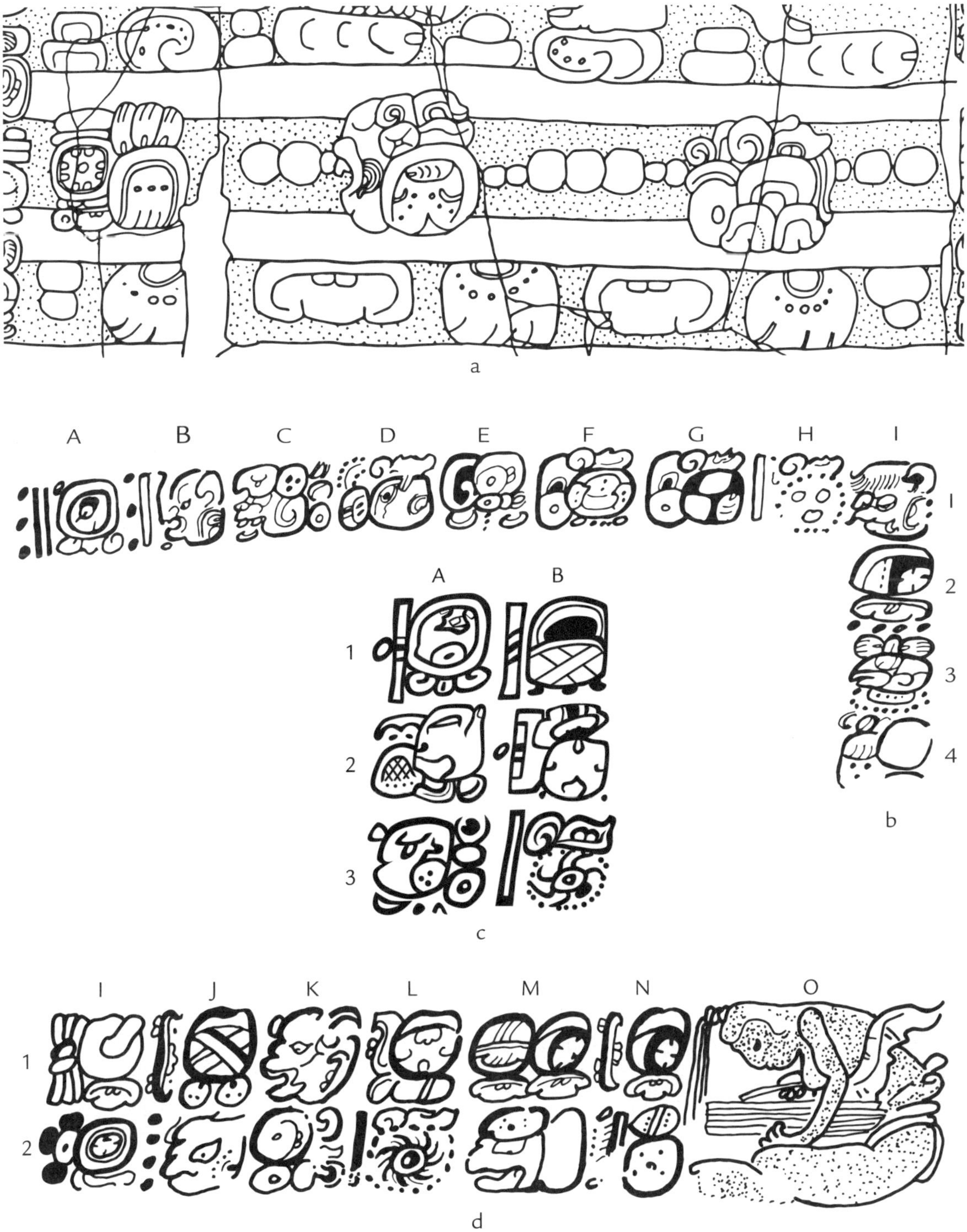

figure 5.13
Five-Flower Mountain in mythological and otherworldly contexts: (a) lower register of Palenque's Temple XIV panel (drawing number 150 by Linda Schele, © David Schele, courtesy of the Foundation for the Advancement of Mesoamerican Studies, Inc.); (b) glyphs A1–I4 of the "cosmic plate" (polychrome vessel K1609; detail of drawing number 7321 by Linda Schele, © David Schele, courtesy of the Foundation for the Advancement of Mesoamerican Studies, Inc.); (c) text from vessel K8457 (drawing by Markus Eberl, based on rollout photo by Justin Kerr); and (d) glyphs I1–O from polychrome vessel K6020 (drawing by Markus Eberl, after Kerr 1997:823).

including polities, cities, ceremonial centers, individual buildings, and tombs (Figure 5.12). After Piedras Negras Ruler 4 died on November 26, 757, he was buried in the Five-Flower Mountain (Figure 5.12b), and archaeologists assume that they have located his tomb in the form of Burial 13 within Structure O-13 (Houston et al. 1998). The Five-Flower Mountain designates a tomb in this case and others, but it also invokes ancient Maya beliefs about death and the afterlife (Figure 5.13). For example, Palenque's king K'inich Kan Bahlam II enters the otherworld as he dances on top of a water band that is marked with a five-flower location and two more toponyms (Figure 5.13a). The ancient Maya conceptualized the otherworld in a similarly detailed way on the "cosmic plate," on which a mythological event took place at the Black-Maw Place, the Black-Lake Place, and the Five-Flower Place (Figure 5.13b).[11] Ancient Maya creation myths also mention the Five-Flower Mountain. For example, Piedras Negras Altar 1 narrates the creation of the present era on 4 Ajaw 8 Kumk'u or September 8, 3114 BCE, and describes a creation event at the First Three-Mountain Place. The eroded text seems to be an abbreviated version of the elaborate account on Quirigua Stela C in which three throne-stones are driven into the ground on 4 Ajaw 8 Kumk'u. Piedras Negras Altar 1 adds that a Piedras Negras king witnessed the manifestation of the three stones from the Five-Flower Place. The Cancuen Panel juxtaposes the Five-Flower Mountain with the Red Three-Stones, likely a local variant of the Green-Blue or First Three-Stones at Quirigua (Figure 5.12a). Both locations form part of the landscape during the creation of time and space.

The Five-Flower Mountain challenges the Western separation of myth and history. Maya hieroglyphic texts invoke it as part of the landscape of creation, and they identify it as the name of burial pyramids and tombs. Scholars have identified actual Five-Flower Mountains at Piedras Negras (Structure O-13; Houston et al. 1998) and Quirigua (Structure 1B-2; Looper 2003:64–72). The urban landscape of these ancient Maya cities provided a physical frame that concretized mythic-historical pasts (Halbwachs 1992; Lass 1988; Leenhardt 1979:159; Munn 1992:113; Parmentier 1987). The landscape itself does not embody the past (contra Knapp and Ashmore 1999:13–14). Its significance emerges instead out of the knowledgeable and intentional interaction of humans with it (Boivin 2008; Gillespie 2011; Ingold 1995; Miller 2005; Thomas 2001). Through citation, prior actions motivate and structure subsequent ones (Gillespie 2011:31; Jones 2001; Joyce and Lopiparo 2005). Ordering time and space is a symbolic process (Munn 1990:1), and instead of embedding their practices in an immutable spatiotemporal context, the ancient Maya used distant objects and events as meaning horizons. By rebuilding the past, they created and manifested the sociotemporal framework that guided present and future actions.

Registered and Experienced Time

The elaborate calendar system of the ancient Maya has been compared to a mechanical clock with interlocking gears (Whitrow 1988:92–96), but this metaphor westernizes the Maya calendar. It makes the calendar understandable for Western readers but places it firmly within Western concepts of time and ideas about progress (Whitrow 1988:154, 157). For example, G. J. Whitrow (1988:96) remarks how the Maya "had the calendar but not the clock." While time is universal, cultures understand it differently, and it is critical that we recognize their particular perspectives and inherent biases (Gell 1992; Munn 1992). Previous interpretations contrasted and, therefore, implicitly aligned the Maya calendar system with the Western notion of abstract time.

Thompson (1950:frontispiece, 1954:145) describes the ancient Maya as bearing the burden of time, but I favor a more nuanced approach: the ancient Maya shouldered time; they were burdened *with* time. Time markers are depicted by god-numbers who carry time units on their heads. Rather than focusing on the weight of the abstract concept, I point to the physicality of the lived experience. Time must and can be carried forward. Agricultural people like the ancient Maya plant and harvest according to the season. Their schedule repeats year in and year out. For farmers to say, "We plant now as we did twelve months ago," requires a fixed reference point from which cyclical events can be said to recur. Linear and cyclical time concepts complement each other. The calendar wheels turn without obliterating an understanding of history as a linear and nonrepetitive process of events. Time is both ephemeral and concrete, both ongoing and finite, both cyclical and linear.

This emphasis on the physicality of time represents a shift from registered to experienced time. Husserl's phenomenology of internal time consciousness provides a basis to study how the ancient Maya perceived time. The position and movements of the human body define and constrain how we locate, perceive, and understand the world (Tilley 1999:34), and Classic Maya rulers embodied time. Sometimes they called themselves *u baah* or "the image of" time periods (Stuart 1996). On Calakmul Stela 9, a lady introduces herself as the image of the Calendar Round 11 Ajaw 18 Ch'en, which corresponds to the period ending of 9.11.10.0.0 or 662 CE (Figure 5.14). Less literally, Copan king Yax Pasaj acceded not only to the throne, but "on or in day 6 Kaban" (see Figure 5.5). Playing with the multiple meanings of preposition *ti*, the Structure 21a text provides the date of his accession while implying that he succeeded in the order of time expressed by Copan's dynasty of rulers. Hieroglyphic stairways with lists of rulers allowed individuals to experience the dynasty by walking up and down (Houston 1998:356–357). Tikal's Yik'in Chan K'awiil saw himself at the center of the four-sided cosmogram that fuses time and space. Merleau-Ponty (1962:139) observes that "we must therefore avoid saying that our body is *in* space, or *in* time. It *inhabits* space and time" (emphasis in the original). Moving in time means moving in space.

Sociocultural practices do not simply go in or through time and space, as Nancy Munn (1983:280) has written in her criticism of structuration theory, but actors produce their own spacetime (Munn 1986:11, 1992:106; Smith 2003:15; Urry 1991:160). Perception creates a mental representation of the perceived. The ordering is evident in what is included and omitted from the representation. But modifications that result from new perceptions change the representation and require new ordering. *Tz'ak*, "to put in order," captures this unending cognitive process. The end is never complete, but is only the transition to another beginning.

figure 5.14
A Calakmul lady impersonates the calendar round 11 Ajaw 18 Ch'en on Calakmul Stela 9, glyphs D1–E3 (the name phrase continues in glyph blocks not shown here). (Drawing by Markus Eberl, after Ruppert and Denison 1943:pl. 48c.)

Acknowledgments

Several people improved this essay with their comments, criticisms, and suggestions. I am particularly indebted to Anthony Aveni, Victoria and Harvey Bricker, John Carlson, Oswaldo Chinchilla, John Justeson, Danny Law, Alfredo López Austin, Leonardo López Luján, Simon Martin, Maria Mayo, Mary Miller, Christian Prager, Joel Skidmore, Mark Van Stone, and Marc Zender. The comments of two anonymous reviewers were also very helpful. Maria Mayo and Sarah Soliz corrected and edited my English.

1 Ancient Maya iconography arguably shows the experiential now moment in the form of scrolls, beads, and flowers that emerge from the mouths of humans and gods. They indicate smells—nobles imagined themselves as exuding a flowery aroma—and breath (Houston and Taube 2000:265–273). Dying people exhaled their wind soul with their last breath. Breath scrolls can be understood as a visual representation of the now moment (Mary Miller, personal communication 2012).

2 The initial chronologies of Copan Altar Q and Quirigua Zoomorph P coincide. The first date is 8.19.10.10.17 5 Kaban 15 Yaxk'iin (September 3, 426; this date and the following ones are based on the Julian calendar). Three days later is 8.19.10.11.0 8 Ajaw 18 Yaxk'iin (September 6, 426). Copan's Altar Q then jumps 153 days (7.13) forward to 8.19.11.0.13 5 Ben 11 Muwan (February 6, 427).

3 Bajlaj Chan K'awiil, the founder of Dos Pilas's dynasty, makes a similarly forward-looking statement on Dos Pilas Hieroglyphic Stairway 2 west, Step II, glyph blocks E1 and F2 (Houston and Mathews 1985:11, fig. 5.7). The statements with *tz'ak* look similar to statements involving *tal*. Dos Pilas's founder, Bajlaj Chan K'awiil, commemorates his sixtieth birthday as the finishing of "3 tal u k'atun," "his third k'atun" (see Figure 5.6c). The tablets in Palenque's Temple of Inscriptions detail the period-ending ceremonies of K'inich Janaab Pakal I and list them as the first, second, and third stone-seating in the series (see Figure 5.6d). Step VII of Yaxchilan Hieroglyphic Stairway 2 recounts three events in the deep past and differentiates between the first *tal ahil*, second tal ahil, and three *ahil* (glyphs A3, D2, and E5). These tal statements indicate the order of events, but they lack the idea of completion as beginning.

4 The Dallas Altar recounts how three noblewomen arrived at the site of La Corona over the course of the Late Classic period (Martin 2008). Its hieroglyphic text calls them the first, second, and third noblewoman from the snake dynasty. Hieroglyphic steps from Dzibanche list the wars of Early Classic snake rulers (Nalda 2004). Several of them add an ordinal number and *tal* after the name of a captive and presumably indicate in which order he was taken (Monument 8b: "u 2 tal"; Monument 10b: "8 tal"; Monument 11: "u 16 tal"; Monument 18: "u 5 tal"; and Monument 20: "[illegible number] tal"). On vessel K2914, the ruler mentions "tz'ak-baj kele[m]," "the youth[s] are put in order," seemingly talking about the four males lined up behind (Marc Zender, personal communication 2012).

5 *Yok'iin* remains an enigmatic title. K'inich Kan Bahlam II, the older brother of K'inich K'an Joy Chitam II, calls himself the tenth *yokte'* since a mythological king on the west jamb of the Temple of the Cross. Yok'iin appears as twelfth *tal yok'iin?* in the hieroglyphic text from Tikal Structure 5D-141 and as 6 yok'iin on the Puerto Barrios Altar.

6 *Tz'ak* appears on Naranjo Stela 32 in a transitive verb construction; the associated long count date is 9.19.5.9.12 12 Eb 5 K'ayab'. Similar instances of *u tz'ak-aw*, "(s)he stacks something up," appear on Naranjo Stela 23 (glyph F21) and Piedras Negras Stela 12 (glyph block D2) in the context of wars.

7 Down-pointing star signs interrupt the hieroglyphic text of Copan Structure 21a's bench (see Figure 5.5). When Yax Pasaj seated himself on the bench during his accession, he mirrored the paddler gods on Yaxchilan Stela 4, albeit upside down (see Figure 5.10). His accession transformed Yax Pasaj into a divine being, and he emerged from the otherworld as Copan's new ruler. One can speculate that this transformation took place in Structure 21a and that Yax Pasaj was revealed or unwrapped as ruler by opening the curtains that were likely draped over the building's doorway. Structure 21a may represent the otherworld, similar to the temples or "sweathouses" in Palenque's Cross Group (Houston 1996). Sometimes, the glyph for plaza or public space is similar to the full version of the star sign (e.g., Dos Pilas Stela 11, glyph block B1, and Stela 15, glyph block B7). The quatrefoil shape of Machaquila's main plaza makes it a giant portal glyph (Stuart and Houston 1994:33). During their ceremonies in the main plaza, the kings of Machaquila linked earthly and otherworldy spheres.

8 Hieroglyphic texts associate Ahkan not only with stars but also with *ch'een*, or "cave," and call him a cave god (see Figure 5.9b, glyph block A11, and Figure 5.9d, glyph block F6). Ahkan's simultaneous association with heavenly and subterranean features points

to a contrast between earth and otherworld and not between earth, sky, and underworld. I, therefore, call Ahkan an inhabitant of the otherworld.

9 The suffix *ja* (T181) could be a phonetic complement, the logograph for the moon, or a morphological suffix. The first possibility, *ja* as a phonetic complement, seems unlikely because the yax glyph seems to represent a portal in this context, and known Maya words for portal, like *ohl* and *way*, end in a consonant that is different from *j*. The second possibility, T181 as logogram for the moon, finds iconographic support in paired ancestor cartouches. For example, the top register of the temple side of Yaxchilan Stela 4 shows ancestors seated in a moon and a sun cartouche. The compound *yax-ja* could then form a couplet for "portal." Third, *ja* could be a morphological suffix, possibly the absolutive suffix *-aj* that marks unpossessed noun stems (Houston et al. 2001:42–46; Houston and Stuart 1998:76; Zender 2004). For example, *tup-aj*, "earspool," changes to *u tup*, "his or her earspool," when possessed. Different absolutive suffixes refine the semantic domain. Body parts are marked with *-is* (e.g., *baah-is*, "head") and personal property with *-aj* (*baah-aj*, "image"). If the *yax* in Ahkan's name reads here *ohl*, Maya scribes may have differentiated between *ohl-is*, "heart" (Zender 2004:201, fig. 8.2e), and *ohl-aj*, "portal."

10 The "halving" of the portal on Tikal Stela 5 is not unusual. Webster and colleagues (1998:331) discuss a comparable case for T-shaped iconographic elements on the facade of Structure 66S in Copan's Group 8N-11. Based on Bassie-Sweet 1991, they interpret the T shape as half of the cruciform glyph for *ohl*, "portal, heart," and suggest that it is a cave mouth on the horizon. Celestial objects emerge from and sink into it on their daily or nightly journeys across the sky.

11 The "cosmic plate" calls each of the three mythological places *chan-ch'een*, "sky-cave" (glyphs I1 and 2 in Figure 5.13b). Polychrome vessel K6020 shows a mythological palace scene in which several young women tend to an elderly god and a scribe kneels in front of them (see Figure 5.13c). The lengthy text ends with the statement that the preceding events "took place at the five-flower place sky-cave." Knowlton (2002) describes chan-ch'een, "sky-cave," and similar expressions as couplets in which two words form a new symbolic meaning. He compares *kab-ch'een*, "earth-cave," to Nahuatl *altepetl*, "town" (from *atl*, "water," and *tepetl*, "hill"; León-Portilla 1963:102, 1985:19), and suggests that it refers to "residence, settlement" (Knowlton 2002:11). Other scholars prefer to translate this term as "community" or "universe" (Martin 2004:106–109; Stuart 2004a:219, 2004b:3, 2005:90–91). The identification of the Five-Flower Place as chan-ch'een or "sky-cave" links it to Maya concepts of place and community and combines literal with metaphorical aspects.

REFERENCES CITED

Aulie, H. Wilbur, and Evelyn W. de Aulie

1978 *Diccionario ch'ol–español, español–ch'ol.* Instituto Lingüístico de Verano, Mexico City.

Barrett, John C.

1994 *Fragments from Antiquity: An Archaeology of Social Life in Britain, 2900–1200 BC.* Blackwell, Oxford.

1999 The Mythical Landscapes of the British Iron Age. In *Archaeologies of Landscape: Contemporary Perspectives*, edited by Wendy Ashmore and Arthur Bernard Knapp, pp. 253–265. Blackwell, Oxford.

Bassie-Sweet, Karen

1991 *From the Mouth of the Dark Cave: Commemorative Sculpture of the Late Classic Maya.* University of Oklahoma, Norman.

Boivin, Nicole

2008 *Material Cultures, Material Minds: The Impact of Things on Human Thought, Society, and Evolution.* Cambridge University Press, Cambridge.

Brentano, Franz C.

1874 *Psychologie vom empirischen Standpunkte.* Verlag von Duncker & Humblot, Leipzig.

Culbert, T. Patrick

1993 *The Ceramics of Tikal: Vessels from the Burials, Caches, and Problematical Deposits.* Tikal Report 25, pt. A. University Museum, University of Pennsylvania, Philadelphia.

Eberl, Markus, and Daniel Graña-Behrens

2004 Proper Names and Throne Names: On the Naming Practice of Classic Maya Rulers. In *Continuity and Change: Maya Religious Practices in Temporal Perspective; Fifth European Maya Conference at the University of Bonn, December 2000,* edited by Daniel Graña-Behrens, Nikolai K. Grube, Christian M. Prager, Frauke Sachse, Stefanie Teufel, and Elisabeth Wagner, pp. 101–120. Acta Mesoamericana 14. Verlag Anton Saurwein, Markt Schwaben.

Edmonson, Munro S.

1982 *The Ancient Future of the Itza: The Book of Chilam Balam of Tizimin.* University of Texas Press, Austin.

Findlay, J. N.

1975 Husserl's Analysis of the Inner Time-Consciousness. *Monist* 59(1):3–20.

Freidel, David A., and Linda Schele

1988 Symbol and Power: A History of the Lowland Maya Cosmogram. In *Maya Iconography,* edited by Elizabeth P. Benson and Gillett G. Griffin, pp. 44–93. Princeton University Press, Princeton, N.J.

Freidel, David A., Linda Schele, and Joy Parker

1993 *Maya Cosmos: Three Thousand Years on the Shaman's Path.* William Morrow, New York.

Gell, Alfred

1992 *The Anthropology of Time: Cultural Constructions of Temporal Maps and Images.* Berg, Oxford.

Gillespie, Susan D.

2011 Archaeological Drawings as Re-Presentations: The Maps of Complex A, La Venta, Mexico. *Latin American Antiquity* 22(1):3–36.

Goldstein, Kurt

1923 Über die Abhängigkeit der Bewegungen von optischen Vorgängen. Bewegungsstörungen bei Seelenblinden. *Monatsschrift für Psychiatrie und Neurologie* 54:141–194.

Graham, Ian

1978 *Naranjo, Chunhuitz, Xunantunich.* Corpus of Maya Hieroglyphic Inscriptions 2, pt. 2. Peabody Museum of Archaeology and Ethnology, Harvard University, Cambridge, Mass.

1979 *Yaxchilan.* Corpus of Maya Hieroglyphic Inscriptions 3, pt. 2. Peabody Museum of Archaeology and Ethnology, Harvard University, Cambridge, Mass.

2006 *Tonina.* Corpus of Maya Hieroglyphic Inscriptions 9, pt. 2. Peabody Museum of Archaeology and Ethnology, Harvard University, Cambridge, Mass.

Grube, Nikolai K.

1988 Städtegründer und "Erste Herrscher" in Hieroglyphentexten der klassischen Mayakultur. *Archiv für Völkerkunde* 42:69–90.

2004 Akan—The God of Drinking, Disease, and Death. In *Continuity and Change: Maya Religious Practices in Temporal Perspective; 5th European Maya Conference, University of Bonn, December 2000,* edited by Daniel Graña-Behrens, Nikolai K. Grube, Christian M. Prager, Frauke Sachse, Stefanie Teufel, and Elisabeth Wagner, pp. 59–76. Acta Mesoamericana 14. Verlag Anton Saurwein, Munich.

Halbwachs, Maurice

1992 *On Collective Memory.* Translated by Lewis A. Coser. University of Chicago Press, Chicago.

Herzfeld, Michael

1991 *A Place in History: Social and Monumental Time in a Cretan Town.* Princeton University Press, Princeton, N.J.

Houston, Stephen D.

1993 *Hieroglyphs and History at Dos Pilas: Dynastic Politics of the Classic Maya.* University of Texas Press, Austin.

1994 Review: Writing in New World Civilizations. *American Anthropologist,* n.s., 96(3):716–718.

1996 Symbolic Sweatbaths of the Maya: Architectural Meaning in the Cross Group at Palenque, Mexico. *Latin American Antiquity* 7(2):132–151.

1998 Classic Maya Depictions of the Built Environment. In *Function and Meaning in Classic Maya Architecture*, edited by Stephen D. Houston, pp. 333–372. Dumbarton Oaks Research Library and Collection, Washington, D.C.

Houston, Stephen D., Claudia Brittenham, Cassandra Mesick, Alexandre Tokovinine, and Christina Warinner

2009 *Veiled Brightness: A History of Ancient Maya Color.* University of Texas Press, Austin.

Houston, Stephen D., Héctor L. Escobedo, Donald Forsyth, Perry Hardin, David L. Webster, and Lori Wright

1998 On the River of Ruins: Explorations at Piedras Negras, Guatemala, 1997. *Mexicon* 20(1):16–22.

Houston, Stephen D., and Peter L. Mathews

1985 *The Dynastic Sequence of Dos Pilas, Guatemala.* PARI Monograph 1. Pre-Columbian Art Research Institute, San Francisco.

Houston, Stephen D., John Robertson, and David Stuart

2000 The Language of Classic Maya Inscriptions. *Current Anthropology* 41(3):321–356.

2001 *Quality and Quantity in Glyphic Nouns and Adjectives.* Research Report on Ancient Maya Writing 47. Center for Maya Research, Washington, D.C.

Houston, Stephen D., and David Stuart

1998 The Ancient Maya Self: Personhood and Portraiture in the Classic Period. *RES: Anthropology and Aesthetics* 33:73–101.

Houston, Stephen D., and Karl A. Taube

2000 An Archaeology of the Senses: Perception and Cultural Expression in Ancient Mesoamerica. *Cambridge Archaeological Journal* 10(2):261–294.

Howe, Leopold E. A.

1981 The Social Determination of Knowledge: Maurice Bloch and Balinese Time. *Man* 16(2):220–234.

Hull, Kerry

2005 A Dictionary of Ch'orti' Maya, Guatemala. Foundation for the Advancement of Mesoamerican Studies, Gainesville, Fla. Electronic document, http://www.famsi.org/reports/03031/index.html, accessed September 9, 2012.

Husserl, Edmund

1966 *Zur Phänomenologie des inneren Zeitbewusstseins (1893–1917).* Husserliana 10. Edited by Rudolf Boehm. Martinus Nijhoff, The Hague.

Ingold, Timothy

1995 Building, Dwelling, Living: How Animals and People Make Themselves at Home in the World. In *Shifting Contexts: Transformations in Anthropological Knowledge*, edited by Marilyn Strathern, pp. 57–80. Routledge, London.

Jackson, Sarah

2004 Interpreting Copán's Structure 21a and Hieroglyphic Bench. *Symbols* (Spring):7–9.

James, William

1890 *The Principles of Psychology.* Henry Holt, New York.

Jones, Andrew

2001 Drawn from Memory: The Archaeology of Aesthetics and the Aesthetics of Archaeology in Earlier Bronze Age Britain and the Present. *World Archaeology* 33:334–356.

Jones, Christopher, and Linton Satterthwaite

1982 *The Monuments and Inscriptions of Tikal: The Carved Monuments.* Tikal Report 33, pt. A. University Museum, University of Pennsylvania, Philadelphia.

Josserand, J. Kathryn, and Nicholas A. Hopkins

2001 Chol Ritual Language. Foundation for the Advancement of Mesoamerican Studies. Electronic document, http://www.famsi.org/reports/94017/index.html, accessed September 6, 2012.

Joyce, Rosemary A., and Jeanne Lopiparo

2005 PostScript: Doing Agency in Archaeology. *Journal of Archaeological Method and Theory* 12(4):365–374.

Kaufman, Terrence S., and William M. Norman

1984 An Outline of Proto-Cholan Phonology, Morphology, and Vocabulary. In *Phoneticism in Mayan Hieroglyphic Writing*, edited by John S. Justeson and Lyle Campbell, pp. 77–166. Institute for Mesoamerican Studies Publication 9. Institute for Mesoamerican Studies, State University of New York, Albany.

Keane, Webb

1997 *Signs of Recognition: Powers and Hazards of Representation in an Indonesian Society*. University of California Press, Berkeley.

Keller, Kathryn C., and Plácido Luciano G.

1997 *Diccionario chontal de Tabasco (mayense)*. Summer Institute of Linguistics, Tucson, Ariz.

Kerr, Justin

1997 *The Maya Vase Book: A Corpus of Rollout Photographs of Maya Vessels*, vol. 5. Kerr Associates, New York.

Knapp, A. Bernard, and Wendy Ashmore

1999 Archaeological Landscapes: Constructed, Conceptualized, Ideational. In *Archaeologies of Landscape: Contemporary Perspectives*, edited by Wendy Ashmore and A. Bernard Knapp, pp. 1–30. Blackwell, Malden, Mass.

Knowlton, Timothy W.

2002 Diphrastic Kennings in Mayan Hieroglyphic Literature. *Mexicon* 24(1):9–14.

Lacadena, Alfonso

2004 Passive Voice in Classic Mayan Texts: CV-*h*-C-*aj* and -*n*-*aj* Constructions. In *The Linguistics of Maya Writing*, edited by Søren Wichmann, pp. 165–194. University of Utah Press, Salt Lake City.

Lass, Andrew

1988 Romantic Documents and Political Monuments: The Meaning-Fulfillment of History in 19th-Century Czech Nationalism. *American Ethnologist* 15(3):456–471.

Leenhardt, Maurice

1979 *Do Kamo: Person and Myth in the Melanesian World*. Translated by Basia Miller Gulati. University of Chicago Press, Chicago.

Le Fort, Geneviève, and Robert F. Wald

1995 Large Numbers on Naranjo Stela 32. *Mexicon* 17(6):112–114.

León-Portilla, Miguel

1963 *Aztec Thought and Culture: A Study of the Ancient Nahuatl Mind*. Translated by Jack Emory Davis. University of Oklahoma Press, Norman.

1985 Nahuatl Literature. In *Literatures*, edited by Munro S. Edmonson, pp. 7–43. Vol. 3 of *Supplement to the Handbook of Middle American Indians*, edited by Victoria R. Bricker. University of Texas Press, Austin.

Looper, Matthew G.

2003 *Lightning Warrior: Maya Art and Kingship at Quirigua*. University of Texas Press, Austin.

Lounsbury, Floyd G.

1989 The Name of a King: Hieroglyphic Variants as a Key to Decipherment. In *Word and Image in Maya Culture: Explorations in Language, Writing, and Representation*, edited by William F. Hanks and Don S. Rice, pp. 73–91. University of Utah Press, Salt Lake City.

Lynch, Kevin

1972 *What Time Is This Place?* MIT Press, Cambridge, Mass.

Maler, Teobert

1901–1903 *Researches in the Central Portion of the Usumatsintla Valley: Report of Explorations for the Museum, 1898–1900*. Memoirs of the Peabody Museum of American Archaeology and Ethnology 2. The Museum, Cambridge, Mass.

Marcus, Joyce

1992a *Mesoamerican Writing Systems: Propaganda, Myth, and History in Four Ancient Civilizations*. Princeton University Press, Princeton, N.J.

1992b Royal Families, Royal Texts: Examples from the Zapotec and Maya. In *Mesoamerican Elites: An Archaeological Assessment*, edited by Diane Z. Chase and Arlen F. Chase, pp. 221–241. University of Oklahoma Press, Norman.

1995 Maya Hieroglyphs: History or Propaganda? In *Research Frontiers in Anthropology*, edited by Carol R. Ember, Melvin Ember, and Peter N. Peregrine, pp. 2–24. Prentice Hall, Englewood Cliffs, N.J.

Martin, Simon

2004 Preguntas epigráficas acerca de los escalones de Dzibanché. In *Los cautivos de Dzibanché*, edited by Enrique Nalda, pp. 105–115. Instituto Nacional de Antropología e Historia, Mexico City.

2005 Of Snakes and Bats: Shifting Identities at Calakmul. *PARI Journal* 6(2):5–15.

2008 Wives and Daughters on the Dallas Altar. Electronic document, http://www.mesoweb.com/articles/Martin/Wives&Daughters.pdf, accessed September 15, 2012.

Martin, Simon, and Nikolai K. Grube

2008 *Chronicle of the Maya Kings and Queens: Deciphering the Dynasties of the Ancient Maya*. 2nd ed. Thames and Hudson, London.

Maudslay, Alfred P.

1889–1902 *Biologia Centrali-Americana, or, Contributions to the Knowledge of the Fauna and Flora of Mexico and Central America: Archæology* 55–59. R. H. Porter and Dulau, London.

Mayer, Karl Herbert

1995 *Maya Monuments: Sculptures of Unknown Provenance, Supplement 4*. Academic, Graz.

McCumber, John

2011 *Time and Philosophy: A History of Continental Thought*. McGill-Queen's University Press, Montreal.

Mensch, James R.

2010 *Husserl's Account of Our Consciousness of Time*. Marquette Studies in Philosophy 70. Marquette University Press, Milwaukee, Wis.

Merleau-Ponty, Maurice

1962 *Phenomenology of Perception*. Translated by Colin Smith. Routledge and K. Paul, London.

Miller, Daniel

2005 Materiality: An Introduction. In *Materiality*, edited by Daniel Miller, pp. 1–50. Duke University Press, Durham, N.C.

Miller, Mary E., and Simon Martin (editors)

2004 *Courtly Art of the Ancient Maya*. Thames and Hudson, New York.

Mora-Marín, David F.

2009 A Test and Falsification of the "Classic Ch'olti'an" Hypothesis: A Study of Three Proto-Ch'olan Markers. *International Journal of American Linguistics* 75(2):115–157.

Moran, Dermot, and Joseph Cohen

2012 *The Husserl Dictionary*. Continuum, London.

Morley, Sylvanus G.

1920 *The Inscriptions at Copan*. Carnegie Institution of Washington Publication 219. Carnegie Institution of Washington, Washington, D.C.

1938 *The Inscriptions of Peten*. Carnegie Institution of Washington Publication 437. Carnegie Institution of Washington, Washington, D.C.

1975 [1915] *An Introduction to the Study of the Maya Hieroglyphs*. Dover, New York.

Munn, Nancy D.

1983 Gawan Kula: Spatiotemporal Control and the Symbolism of Influence. In *The Kula: New Perspectives on Massim Exchange*, edited by Jerry W. Leach and Edmund R. Leach, pp. 277–308. Cambridge University Press, Cambridge.

1986 *The Fame of Gawa: A Symbolic Study of Value Transformation in a Massim (Papua New Guinea) Society*. Cambridge University Press, Cambridge.

1990 Constructing Regional Worlds in Experience: Kula Exchange, Witchcraft, and Gawan Local Events. *Man* 25(1):1–17.

1992 The Cultural Anthropology of Time: A Critical Essay. *Annual Review of Anthropology* 21:93–123.

Nalda, Enrique (editor)
2004 *Los cautivos de Dzibanché*. Instituto Nacional de Antropología e Historia, Mexico City.

Parmentier, Richard J.
1987 *The Sacred Remains: Myth, History, and Polity in Belau*. University of Chicago Press, Chicago.

Pérez Martínez, Vitalino, Federico García, Felipe Martínez Álvarez, and Jeremias López y López
1996 *Diccionario del idioma ch'orti'*. Proyecto Lingüístico Francisco Marroquín, Antigua, Guatemala.

Riese, Berthold
1984 Hel Hieroglyphs. In *Phoneticism in Mayan Hieroglyphic Writing*, edited by John S. Justeson and Lyle Campbell, pp. 263–286. Institute for Mesoamerican Studies Publication 9. Institute for Mesoamerican Studies, State University of New York, Albany.

Ringle, William M.
1996 Birds of a Feather: The Fallen Stucco Inscription of Temple XVIII; Palenque, Chiapas. In *Eighth Palenque Round Table, 1993*, edited by Martha Macri and Jan McHargue, pp. 45–61. Pre-Columbian Art Research Institute, San Francisco.

Robertson, John S., Danny A. Law, and Robbie A. Haertel
2010 *Colonial ch'olti': Seventeenth-Century Moran Manuscript*. University of Oklahoma Press, Norman.

Robertson, Merle Greene
1983 *The Sculpture of Palenque I: The Temple of Inscriptions*. Princeton University Press, Princeton, N.J.
1985 *The Sculpture of Palenque III: The Late Buildings of the Palace*. Princeton University Press, Princeton, N.J.
1995 *Merle Greene Robertson's Rubbings of Maya Sculpture*. Pre-Columbian Art Research Institute, San Francisco.

Ruppert, Karl, and John H. Denison Jr.
1943 *Archaeological Reconnaissance in Campeche, Quintana Roo, and Peten*. Carnegie Institution of Washington Publication 543. Carnegie Institution of Washington, Washington, D.C.

Schele, Linda
1992 The Founders of Lineages at Copan and Other Maya Sites. *Ancient Mesoamerica* 3(1):135–144.

Schele, Linda, and David A. Freidel
1990 *A Forest of Kings: The Untold Story of the Ancient Maya*. Morrow, New York.

Schele, Linda, and Peter L. Mathews
1979 *The Bodega of Palenque, Chiapas, Mexico*. Dumbarton Oaks Research Library and Collection, Washington, D.C.

Schele, Linda, and Jeffrey Miller
1983 *The Mirror, the Rabbit, and the Bundle: "Accession" Expressions from the Classic Maya Inscriptions*. Dumbarton Oaks Research Library and Collection, Washington, D.C.

Schele, Linda, and Khristaan D. Villela
1996 Creation, Cosmos, and the Imagery of Palenque and Copán. In *Eighth Palenque Round Table, 1993*, edited by Martha J. Macri and Jan McHargue, pp. 15–30. Pre-Columbian Art Research Institute, San Francisco.

Scholes, France V., and Ralph L. Roys
1948 *The Maya Chontal Indians of Acalan-Tixchel: A Contribution to the History and Ethnography of the Yucatan Peninsula*. Carnegie Institution of Washington Publication 560. Carnegie Institution of Washington, Washington, D.C.

Schumann G., Otto
1973 *La lengua chol de Tila (Chiapas)*. Cuaderno 8. Centro de Estudios Mayas, Universidad Nacional Autónoma de México, Mexico City.

Smailus, Ortwin
1975 *El Maya-Chontal de Acalan: Análisis lingüístico de un documento de los años 1610–12*. Universidad Nacional Autónoma de México, Coordinación de Humanidades, Mexico City.

Smith, Adam T.
2003 *The Political Landscape: Constellations of Authority in Early Complex Polities*. University of California Press, Berkeley.

Stuart, David

1994 Review of *Mesoamerican Writing Systems: Propaganda, Myth, and History in Four Ancient Civilizations* by Joyce Marcus. *MesoAmerica* 21:251–253.

1996 Kings of Stone: A Consideration of Stelae in Ancient Maya Ritual and Representation. *RES: Anthropology and Aesthetics* 29/30:149–171.

2000 "The Arrival of Strangers": Teotihuacan and Tollan in Classic Maya History. In *Mesoamerica's Classic Heritage: From Teotihuacan to the Aztecs*, edited by Davíd Carrasco, Lindsay Jones, and Scott Sessions, pp. 465–513. University of Colorado Press, Boulder.

2003 On the Paired Variants of TZ'AK. Electronic document, http://www.mesoweb.com/stuart/notes/tzak.pdf, accessed October 9, 2012.

2004a The Beginnings of the Copan Dynasty: A Review of the Hieroglyphic and Historical Evidence. In *Understanding Early Classic Copan*, edited by Ellen E. Bell, Marcello A. Canuto, and Robert J. Sharer, pp. 215–247. University of Pennsylvania Museum of Archaeology and Anthropology, Philadelphia.

2004b The Paw Stone: The Place Name of Piedras Negras, Guatemala. *PARI Journal* 4(3):1–6.

2005 *The Inscriptions from Temple XIX at Palenque: A Commentary*. Pre-Columbian Art Research Institute, San Francisco.

Stuart, David, and Stephen D. Houston

1994 *Classic Maya Place Names*. Dumbarton Oaks Research Library and Collection, Washington, D.C.

Taube, Karl A.

1992 *The Major Gods of Ancient Yucatan*. Dumbarton Oaks Research Library and Collection, Washington, D.C.

2003 Maws of Heaven and Hell: The Symbolism of the Centipede and Serpent in Classic Maya Religion. In *Antropología de la eternidad: La muerte en la cultura maya*, edited by Andrés Ciudad Ruiz, Mario Humberto Ruz Sosa, and Maria Josefa Iglesias Pince de Leon, pp. 405–442. Sociedad Española de Estudios Mayas, Madrid.

2004a Flower Mountain: Concepts of Life, Beauty, and Paradise among the Classic Maya. *RES: Anthropology and Aesthetics* 45:69–98.

2004b Structure 10L-16 and Its Early Classic Antecedents: Fire and the Evocation and Resurrection of K'inich Yax K'uk' Mo'. In *Understanding Early Classic Copan*, edited by Ellen E. Bell, Marcello A. Canuto, and Robert J. Sharer, pp. 265–295. University of Pennsylvania Museum of Archaeology and Anthropology, Philadelphia.

Thomas, Julian

2001 Archaeologies of Place and Landscape. In *Archaeological Theory Today*, edited by Ian Hodder, pp. 165–186. Polity Press, Malden, Mass.

Thompson, J. Eric S.

1950 *Maya Hieroglyphic Writing: An Introduction*. Carnegie Institution of Washington Publication 589. Carnegie Institution of Washington, Washington, D.C.

1954 *The Rise and Fall of Maya Civilization*. University of Oklahoma Press, Norman.

1962 *A Catalog of Maya Hieroglyphs*. University of Oklahoma, Norman.

Tilley, Christopher Y.

1999 *Metaphor and Material Culture*. Blackwell, Oxford.

Urry, J.

1991 Time and Space in Giddens' Social Theory. In *Giddens' Theory of Structuration: A Critical Appreciation*, edited by Christopher G. A. Bryant and David Jary, pp. 160–175. Routledge, London.

Wanyerka, Phillip J.

2004 The Southern Belize Epigraphic Project: The Hieroglyphic Inscriptions of Southern Belize. Electronic document, http://www.famsi.org/reports/00077/, accessed September 19, 2012.

Webster, David L., Barbara W. Fash, Randolph J. Widmer, and Scott Zeleznik

1998 The Skyband Group: Investigation of a Classic Maya Elite Residential Complex at Copán, Honduras. *Journal of Field Archaeology* 25(3):319–343.

Whitrow, G. J.

1988 *Time in History: The Evolution of Our General Awareness of Time and Temporal Perspective.* Oxford University Press, Oxford.

Wisdom, Charles

1950 *Materials on the Chorti Language.* Microfilm Collection of Manuscripts on Middle American Cultural Anthropology 28. University of Chicago, Chicago.

Zender, Marc

2004 On the Morphology of Intimate Possession in Mayan Languages and Classic Mayan Glyphic Nouns. In *The Linguistics of Maya Writing*, edited by Søren Wichmann, pp. 195–209. University of Utah Press, Salt Lake City.

Zimmermann, Günter

1956 *Die Hieroglyphen der Maya-Handschriften.* Abhandlungen aus dem Gebiet der Auslandskunde 62B. Universität Hamburg, Hamburg.

6

Understanding Time, Space, and Social Organization in the Inca Ceque System of Cuzco

ANTHONY F. AVENI

One can imagine how puzzled nineteenth-century proto-anthropologists were when Javanese rice farmers responded to their question, "What are the lengths of your *mangsa* [months]?" with the reply, "They range from twenty-three to forty-one days." These months of unequal length closely approximated those in the old Javanese calendar. They constitute an excellent example of the cultural domestication of time in a highly organized agrarian community. Long after these irregular time spans were set up, without the aid of technology, Javanese calendar keepers devised a *bencet*, or gnomon, to partition the year into twelve months. The measurement principle consisted of equal units of length traveled by the shadow cast by the tip of the bencet at noon. Now there is a very special condition related to the behavior of shadows in the 7° south latitude of central Java: when the sun stands on the meridian at noon to the north of the zenith on the June solstice, the shadow length projected to the south is exactly double the length of that cast when the sun at noon lies south of the zenith at the December solstice and the shadow is projected to the north. To approximate the sequence of unequal months, Javanese chronologists halved the shorter length segment and quartered the longer, thereby producing a standardized physical representation of the twelve-month calendar. But how does this unusual time-parsing instrument, unique to the celestial environment in which it functioned, tie in with the agrarian cycle? The nineteenth-century descriptions of the organized work activities of Indonesian rice farmers were keyed to natural events. For example, the farmer works especially hard at midyear, when months are longest, to facilitate the growth of the staple crop before all can be left to mature during the short-month segment of the seasonal cycle that follows. While the commencement of the various agrarian activities was once casually determined, this schedule was replaced by a more formal seasonal calendar dependent on measuring devices keyed to celestial movement. The starting times were determined by the priests of various villages, each of whom consulted a local bencet (for details, see Ammarell 2005; Cohen-Stuart 1898:29–31; Crawfurd 1967 [1820]:285–326; Maass 1924:103–123). Like our Greenwich meridian, the Javanese bencet created an island-wide "standard time" that local farmers could use to adjust their natural clocks.

I chose to begin this discussion of one of the most complex and unusual systems delineating the relationship of natural time to cultural needs, the *ceque* system of the Inca capital of Cuzco, by referring to the nineteenth-century Javanese rice-farming calendar to make three points. First, in our attempts to understand indigenous calendars, close attention must be paid to the local earth-sky environment; second, information derived from accounts given by those who used the calendar is of seminal importance; and third, Western predilections about the structure of time often need not apply. My goal in this essay is to explain what we know about the calendars embedded within the ceque system in the context of the study of world calendars.

Descriptions of the Ceque System

The ceque system was the hallmark of urban social organization in Cuzco. One can envision it as a giant cosmographic map, a mnemonic device built into Cuzco's natural and man-made topography, that served to unify ideas about religion, social organization, calendar, hydrology, and astronomy in the Inca empire. The Spanish Jesuit chronicler Bernabé Cobo (1956 [1653]; Rowe 1979; see also Hamilton 1990) has left us, in his *History of the New World*, a thorough, detailed description of the system. An earlier chronicler, Juan Polo de Ondegardo (1981 [1585]), is generally credited with having produced the original document in 1585 that, together with other early colonial tracts, informed Cobo. The original document was likely acquired from *quipucamayoc*, keepers of the knotted records called *khipu* (for historical details see Bauer 1998:ch. 2; Zuidema 2010:ch. 1).

Cobo's conception of the radial structure of the ceque system is implied in his introduction: "From the Temple of the Sun as from the center there went out certain lines which the Indians call ceques; they formed four parts corresponding to the four royal roads which went out from Cuzco. On each one of those ceques were arranged in order the guacas and shrines which there were in Cuzco and its district, like stations of holy places, the veneration of which was common to all" (Rowe 1979:15). Testifying to the extraordinary power of the organization of landscape and architecture in the Inca state, Cobo goes on to describe and locate in detail the course of these imaginary radial lines called *ceques* (*ceque* means *raya*, or sight line, in Quechua; see Bauer 1998:5 for a list of source definitions). Cobo indicates that they were grouped zonally according to their location within each of the four *suyus*, or zones divided by roads leading out of Cuzco. The ceques emanated in a radial pattern from the Coricancha, the "Golden Enclosure," which Cobo calls the Temple of the Sun, at the city's center.

Cobo's description implies that each ceque was traceable in the landscape by a series of huacas (guacas), or sacred places in the landscape. These included natural or man-made temples, intricately carved rock formations, bends in rivers, fields, springs or other natural wells called *puquios*, hills, and even impermanent objects such as trees. More than one-third of the huacas are related to water and sacrificial rites connected with the agricultural calendar; for example:

> The seventh [huaca of the eighth ceque of the Chinchaysuyu quadrant of the city] was called Sucanca. It was a hill by way of which the water channel from Chinchero comes. On it, there were two markers as an indication that when the sun arrived there, they had to begin to plant the maize. The sacrifice which was made there was directed to the Sun, asking him to arrive there at the time which would be appropriate for planting, and they sacrificed to him sheep, clothing, and miniature lambs [llamas] of gold and silver. (Rowe 1979:27)

Each ceque was assigned one of a set of three hierarchical groups representing increased genealogical distance from the ruler. These social classes were required to tend to their huacas in a particular order. These included (a) *collana ceques* said to be maintained and worshipped by the primary kin of the Inca ruler; (b) those that were worshipped by his subsidiary kin (*payan ceques*); and (c) *callao ceques* tended to by that segment of the population not related to the ruler.

Cobo also tells us that the servants and attendants of these huacas came offering the appropriate sacrifices at the proper times. The assignments on the hierarchy of triad-based worship rotated sequentially, that is, (a), (b), (c); (a), (b), (c), and so forth in Hanan and (c), (b), (a), and so on in Hurin Cuzco, proceeding from one ceque to the next, all the way around the horizon in a clockwise direction in the two northern suyus and counterclockwise in the two southern suyus, as shown in Figure 6.1.

The organization of communal work activities, agriculture, and irrigation was also prescribed by the ceque system; for example, representatives of each of forty families drawn from the suyus participated in the ritual plowing that took place annually, each delegate plowing a designated portion of land (Zuidema 2010:ch. 6.5.2). Rules for the servicing and maintenance of the irrigation canals were specified by the order of the ceques.

How physically real is the ceque system? Are there still traces of it? Can it be mapped? Does is appear exactly as Cobo described it? Based on Cobo's and other chroniclers' descriptions, investigators have attempted to answer these questions by exploring the Cuzco landscape. Using Cobo's rich descriptions as a guide, they have discovered that several ceques still can be traced by identifying the locations of the huacas that composed each of them. Following up on earlier fieldwork by R. Tom Zuidema (1981) and me (Aveni 1981a), Brian Bauer (1998) has undertaken the most thorough attempt to map the ceque system to date. His method consisted of archaeological field surveys coupled with interviews of local inhabitants and archival searches in Cuzco that yielded toponyms relevant for constructing the system. In some cases, Bauer was able to locate specific shrines at point locations, while

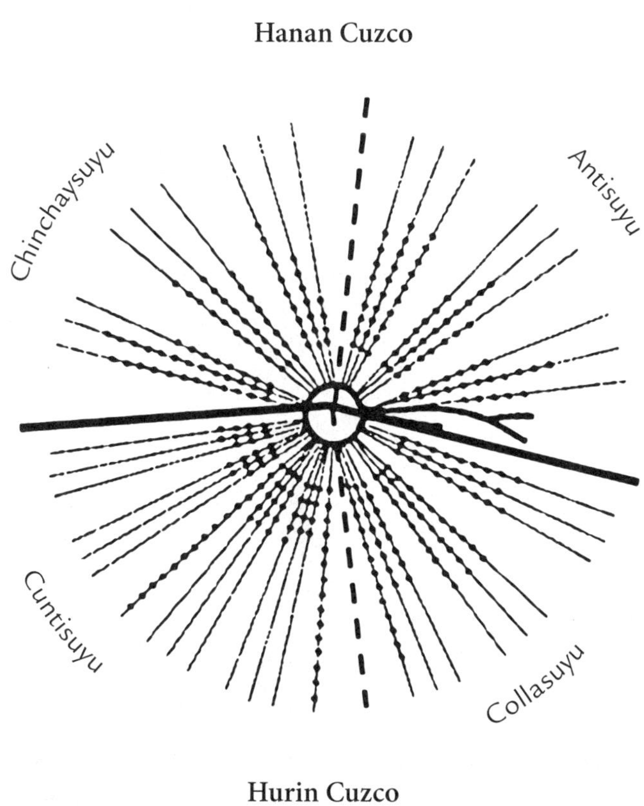

figure 6.1
This schematic diagram of the ceque system of Cuzco exhibits several principles of organization (Zuidema 2010:14). It illustrates the dualistic principle of halving (thick solid line) and then further dividing the city into quarters (thick dotted line), as well as the vertical splitting of it into upper and lower topographical sectors. Abstract straight visual lines called ceques (dotted in the diagram) further subdivided Cuzco into pie-like wedges. Ideally, ceques radiated from the center outward to the corners of the universe. The four suyus comprised the name of the empire, Tahuantinsuyu, or The Four Suyus.

in other instances he was able only to indicate the general area corresponding to a particular huaca. A major conclusion of Bauer's fieldwork is that the huacas of given ceques, when connected together in the sequences described by Cobo, do not form perfectly straight lines, as a strict reading of Cobo might imply (but compare, for example, Cobo 1956 [1653]:92:174, where he speaks of a huaca located apart from its ceque). This finding raises the interesting question of whether the ceque system was conceived as an organization of straight radial lines, as posited by Zuidema, that became nonlinear when they were physically traced over the bumpy landscape. Considerations of differences between the *ideal* and the *practical* nature of the system remain at issue. In this vein, Mesoamericanists are familiar with the competing methods and assumptions of mentalism versus materialism—for example, in the interpretation of the degree to which the Maya monumental inscriptions relate historical versus mythical events. In this essay, I will use Zuidema's interpretation of the ceque system as a template for laying out basic issues related to its interpretation.

Interpreting the Ceque System

Only a handful of researchers have contributed to the study of social and political models related to the ceque system, which may seem somewhat surprising given the importance Cobo attributes to it. The most detailed work in this area has been done by Zuidema, whose efforts have given special emphasis to the role of celestial observation, calendrics, and ritual in relation to the organization of the Inca capital. For a full exposition of his ideas, see his recently published capstone volume (Zuidema 2010), which contains a bibliography of his nearly one hundred relevant publications spread out over more than fifty years.[1] I begin with a brief account of Zuidema's version of the counting of the ceques and huacas, followed by his interpretation of the way they are ordered in the landscape.

Cobo lists nine ceques associated with three of the suyus: the northwest (centered around the district named Chinchaysuyu), the northeast (Antisuyu), and the southeast (Collasuyu). Fourteen more ceques were associated with the southwest (Cuntisuyu) quadrant, thus making a total of 9 + 9 + 9 + 14, or 41 radial lines.[2] Hierarchical ordering of the ceques proceeded from upriver (the Huatanay River) or Hanan Cuzco, the superior polity, to the inferior downriver, or Hurin Cuzco, polity in the moiety structure of the capital. These are labeled in Figure 6.1. Zuidema (2010:ch. 9) gives a more detailed exposition of the layout.

Zuidema's interpretation of ceques as physically straight lines of sight has led to the misconception that he treats all ceques as astronomical sight lines.[3] For example, Susan Niles (1987:178) notes that "it is difficult to apply any reading to Cobo, Molina, or Albornoz that would permit a reasonable attribution of any astronomical orientation for *all* the shrines or for *all* the *ceques*." Actually, only one ceque serves that function in Zuidema's theory. If ceques functioned as sight lines, in the sense of the last visible huaca at the horizon of a ceque having been observed in relation to a sky event as seen from the Coricancha, then the straightness of the line connecting all the huacas of that ceque would not have been a factor in evaluating any argument concerning astronomical alignments—that is, only the imaginary straight line linking the Coricancha and the huaca at the horizon point in question would need be taken into account. Finally, it is worth noting that the concept of straightness is not a purely Euclidean idea rooted in the minds of modern Western researchers, but instead is likely a genuinely indigenous concept that can be likened to our monthly calendar being based ideally on the phases of the moon (Aveni 1996): not a single one of the twelve months that make up the Christian year has a length equal to the actual measured phase cycle.

Employing the ethnohistoric record, Zuidema links the idea of ceque lines radiating from a center to chronicler Cristóbal de Molina's account of a *capacocha* (*capac hucha*) ritual. The term derives from *capac* (royal) and *hucha* (sin), thus, fault or obligation with respect to sinning. This ritual was an event in which a sacrificial victim was brought into Cuzco, sanctified, and then walked to a mountain and there sacrificed and buried along with precious objects. A capacocha ritual was accomplished "without going by the royal road directly, but without turning anywhere, traversing the gorges and mountains that they found before them until each one reached the part and place to receive said sacrifices" (Molina 1989 [1573]:126–127). Specifically, such a capacocha ceque might stretch more than one hundred kilometers along the Vilcanota River leading southeast out of Cuzco.[4] This example, and the chronicler's statements about the visual root of the word *ceque* as "raya" or "line" (Bauer 1992, 1998), might lead one to conceptualize ceques as straight lines. Bauer's study, on the other hand, accords primacy to the huacas as defining the courses of the ceque lines.

In sum, one can view the ceque system as an instrument that enjoined both lived and representational space, the former especially via the association of a large portion of ceques and huacas with water and irrigation, the latter by way of reference of these sacred places to myth and history. As Jeanette Sherbondy (1987) concludes in her examination of the relationship between Inca social organization and water rights, water-related ceques often served as boundary markers that refer to arable land among *ayllus*, or social groupings. Another set of associations in the system connects the *panacas*, or royalty-derived ceques, to the history of the Inca and commemorates achievements in the lives of these rulers.[5]

Other ceque-like systems exist in Andean cultures, though not necessarily related historically to the ceque system of Cuzco. Bauer (1998:146–154) offers a concise summary of evidence for the existence of other ancient ceques and systems of shrines, including those mentioned by Cristóbal de Albornoz in Anta and those at Huánuco Pampa, that exhibit a clear radial pattern. Contemporary Andean ritual systems that include radiating lines exist in the Sajama region of Bolivia. Still more recently, María de las Mercedes del Río (2005:85–93) finds the concept of radiality in texts related to the Aymara-speaking Sura federation of the altiplano region near modern Oruro. These involve the propagation of ceque-based concepts of social identity connected with solar worship descended from Inca times and the association of the descendants of the Inca Sucssu panaca with the great solar huaca of Potosí. She finds that two former Inca administrative centers, one near Oruro (Paria), the other near Cochabamba (Soras), consist of radial lines that lead to boundary markers (cf. especially Río 2005:87, Mapa 1.7). Río interprets the Soras system as an adaptation of ceques to local conditions centered on relations between the Inca rulership and local authorities. Additionally, Gary Urton (1990a, 1990b) has explored other representations of the organization of Andean public social space around ayllus and moieties via division into straight lines in contemporary Pacariqtambo as well as in Pre-Columbian Nasca.

Having set out a sense of its basic layout and relationship to social organization, I turn, in the next two sections, to a discussion of the astronomical and calendrical principles incorporated into the ceque system.

Astronomical Observation and the Ceque System

Cobo's description of the ceque system includes at least four direct references to astronomical observation. The nature of these observations involves a dialectical principle common in Andean philosophy, whereby the definition of one of a pair of polar opposites is derived from the view of the other, a process Billie Jean Isbell (1982:354) terms reversible dualism.

Contrary to the rules that underlie the structure of Western taxonomies, reversible dualism stresses the interdependence of polar opposites. Isbell (1982:357) argues that reversible dualism characterizes the structure of epistemologies throughout tropical cultures and especially that it is replicated in architectural spaces and the surrounding landscape, the environment of social space being "a mirror reflection of the dialectic elements in the heavens." Thus, rising and setting observation points are perceived as dialectical pairings out of which significant calendrical dates are derived.

These relevant observations and dates are as follows:

1ab. Sunrise/sunset on the two solstices, December 21 and June 21

2a. Sunrise on the two dates of annual solar zenith passages (October 30 and February 13 in the 13°30' S latitude of Cuzco), a phenomenon exclusive to tropical latitudes (cf. Aveni 1981b for a discussion of this principle)

2b. Sunset on the two dates opposite in the year to 2a, or the passages of the sun through the antizenith (April 26 and August 18; Zuidema 2010:34)

The observations related to the first opposing pair of alignments are the easiest to understand from a Western perspective. These include alignments connecting huacas belonging to ceques in Hanan Cuzco that mark the setting point of the sun on June 21 (Figure 6.2, alignment 1a) and its opposite, sunrise on December 21 and involving huacas of ceques in Hurin Cuzco (Figure 6.2, alignment 1b). Using specific places mentioned in the chronicles, one can tie these opposing directions, each going out from the capital, to stories of migration relating to mythic founders said to have passed into the center of Cuzco from the periphery (cf. Nair, this volume).

The empirical astronomical foundation of the pivotal zenith-antizenith axis (the alignments described in 2a and 2b) is based on a set of horizon features. The most well known among them involves a set of *sucanca*, or pillars, located on Cerro Picchu, also called Cerro Sucanca, a prominent hill overlooking the center of Cuzco from the west. In addition to Cobo's chronicle, that of an anonymous chronicler offers this detailed account of the well-known Cuzco sun pillars:

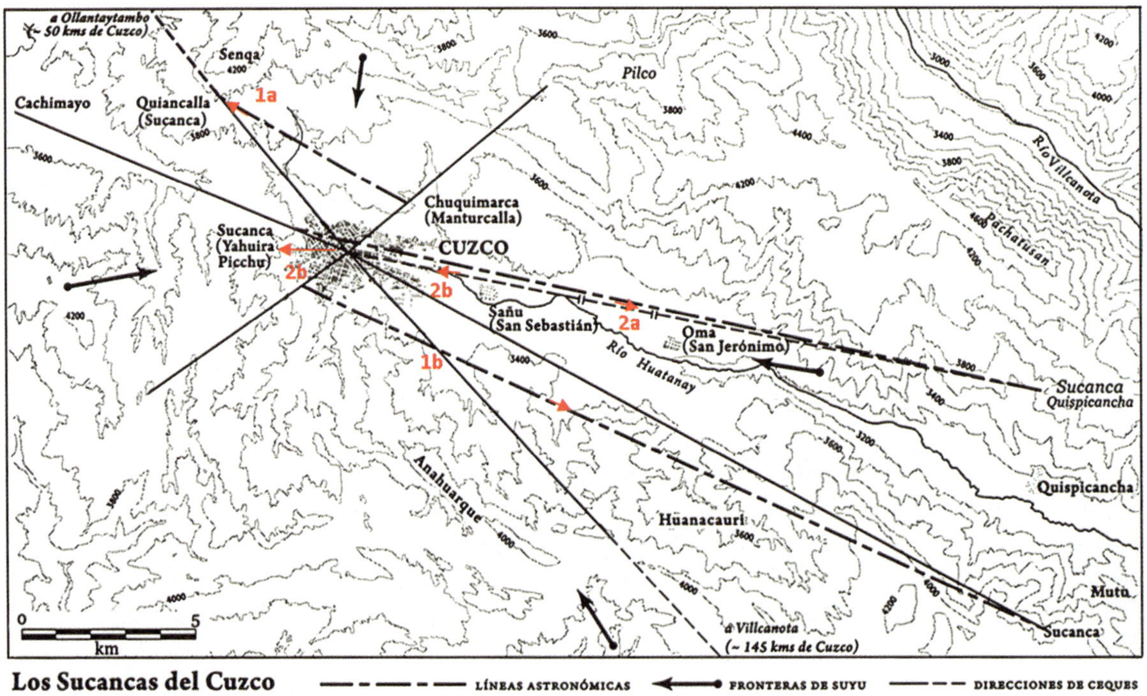

figure 6.2

The *sucancas* and ritual movements according to Zuidema's (2010:165) spatial model of the ceque system are associated with a number of alignments. Astronomical observations among huacas of the ceque system follow the principle of reversible dualism: (1a) sunset, June solstice; (1b) sunrise, December solstice; (2a) sunrise, date of solar zenith passage; and (2b) sunset, solar antizenith passage; offset alignments in common sets are due to variations in horizon elevation.

> In the highest mountains to the west as seen from Cuzco, they made four pillars in the form of small towers that can be distinguished from 2 or 3 leagues, with a distance from first to last of 200 *pasos*, and between the middle ones of 50 *pasos*, and the separation of these two cylinders was proper for their intended use; with the result that when they became ready for sowing in general and they started to sow in the higher places as the plants are later here, and when the Sun entered the central pillars it was time for sowing in Cuzco, and this was always the month of August. The place for taking the Sun to enter the central two pillars was another pillar in the middle of the plaza, a pillar of well dressed stone, one *estado* high, in the middle of the plaza, in a place for that purpose was singled out, that they call *Osno*, and they measured the Sun between the two pillars, and when it was exactly there, it was the time for sowing in the valleys of Cuzco and around. (Zuidema 1981:320)[6]

The time of the antizenith solar passage was incorporated in the ceque system via the reversal of an observation taken along a baseline to the rising sun on the day it passed the zenith (Figure 6.2, alignment 2a). Zuidema has employed the ethnohistoric record to connect the zenith-antizenith axis via an alignment with the sucanca at the place called Quispicancha, located east of the Coricancha (Figure 6.2, far right side).[7]

The discovery that these important agricultural dates happened, conveniently, to coincide with one of the prominent celestial phenomena in the environment of Cuzco may have led the Inca to regard the act of planting and the sun-under-the-world as symbolically going together. During this season of the year, the earth was at the peak of fertility and could be penetrated by man (with his plow) and the sun (with his rays). Viewed in this way, the sight line across the landscape connecting the horizon sun sightings on the days of zenith and antizenith passage (Figure 6.2, alignments 2a and 2b) becomes a reflection, in horizontal space, of the organizing principle of reversible dualism embedded in Andean thought. While some may disagree about the details, few will argue today against the notion that astronomical alignments had something to do with the foundation of the ceque system.[8]

The Ceque System as a Calendar

Time is motion, and the most accurate calendars are based on observations of the movement of the two brightest objects that dominate the sky: the sun and the moon. An age-old problem in the cultural mastery of astronomically based time, especially in sedentary agrarian societies, relates to the adaptation of moon time to solar, or seasonal, time. Nature has rendered this accommodation a difficult task because the length of the year measured by the annual movement of the sun is not precisely divisible by a whole number of lunar phase cycles, or synodic months (29.5 days).

In Western culture, a precise solar-lunar accommodation did not occur until the time of the Roman empire, when the lunar calendar was lengthened from approximately ten synodic months to 365 days divided into twelve unequal months with lengths ranging between 28 and 31 days. It is worth noting that many calendars, like that of early Rome, which had been based on the gestation period of cattle, simply did not count seasonal time periods during which little or nothing of importance transpired. The same was true of the Inca calendar.[9]

Urton (1997:189–190) notes that the lengths of the associated time intervals and their placement in the ceque system as a calendar were based on a variety of factors, among them astronomical periods observed from different parts of the city. Other factors included periods that happened to be of interest to the administration at particular times of the year, the status of groups in the sociopolitical organization, and mythical characteristics attributed to particular groups that might have rendered them more suited to conduct ritual activities at one time of the year rather than another. He regards the process of so organizing the Inca calendar as a "mathematics of rectification" (Urton 1997:190) designed to maintain the status quo based on values deemed appropriate by the administration for organizing relations among its people.

The total number of huacas that comprised the system (328) is a whole multiple (12) of the sidereal lunar period (27.3 days), or the time between two successive passages of the moon through the same stellar background, regardless of its phase. Also, in 328 days the moon passes the same star or star group four times; thus, $4 \times 3 \times 27.3 = 327.6$. Furthermore, this number is also 37 days short of the solar or tropical year, a period from the disappearance at dusk and the reappearence at dawn of the Pleiades as seen from the latitude of Cuzco—thus, $37 + 328 = 365$ (Zuidema 1977).[10] The 37 uncounted days have become the source of a major misconception that the Western mind harbors about the nature of seasonal timekeeping. That 328 is a whole multiple of the Inca eight-day week (i.e., $8 \times 41 = 328$), a period likely of nonastronomical origin, is often passed over, as is the fact that this number also approximates the gestation period of the llama. These facts offer very real parallels to the development of our own early Roman calendar, in which a market-week cycle was commensurate with the duration of the lunar calendar, which fell considerably short of the length of the seasonal (solar) year.

That the Inca calendar possessed a lunar base becomes even more evident when one groups in pairs the number of huacas summed along tripartite (a, b, c) sets of ceques. Considered in this way, the tripartite sets closely approximate double-month periods. For example, as depicted in the fourth set of entries from the center in Figure 6.3, the first three ceques following the thirty-seven-day period, when grouped together, contain twenty-six, thirty, and twenty-nine huacas.

Like the Javanese officials, Inca administrators slightly adjusted these months in order to enable them to correspond to periods of plowing and planting, the appearance and disappearance of water, and harvesting. The chronicles indicate that there was much concern that certain ritual activities be celebrated during a period commenced by a named new moon, the rituals becoming most important after the next full moon. Parallel examples from our own calendar include the tradition of commencing the harvest during the month initiated by the appearance of the harvest moon, or the first full moon following the autumnal equinox, and the opening of hunting season by the observation of the second, the hunter's moon.

The months of the year were divided among the population of Cuzco and its vicinity in such a way that each suyu was assigned the responsibility of tending to the ceremonies and sacrifices that were performed at the huacas associated with their particular months of the year in the day-counting scheme. Such a process would have helped promote social integration by giving each family a special role to play at a designated time. To give one example: the thirty-three-day month between November 21 and December 23 falls in the Chinchaysuyu ceque group, the first ceque of which bears the name Capac. This ceque is related to the rituals of Capac Raymi, especially to a celebration of the cargo of the Capac (Royal) ayllu; it terminates with the December solstice, which was celebrated on December 24 (Zuidema 2010:756–757).

An important step in establishing relationships between lunar- and solar-based calendars consists of relating a specific moon in the lunar cycle to the starting point of the seasonal year, the latter temporal reference point being based on the observation of Pleiades. In the Christian version of the calendar, the setting of the celebration of Easter offers a good parallel: it falls on the first Sunday following the first full moon after the spring equinox. In the Andes, the association of planting, irrigation, and the Pleiades probably developed when Inca predecessors recognized the approximate equivalence between the period of absence from the sky (about thirty-seven days) of the prominent star group and the time between the end of the harvest and the beginning of the next planting season (cf. Zuidema 2010:131ff., 148ff.; summary on p. 303.) The year would have been counted from the lunation that encompassed the June solstice, that is, not earlier than May 25 in the modern Gregorian version of the calendar.

The starting point of the ritual year was June 9, with the month that included the winter solstice in the southern hemisphere, which is connected with

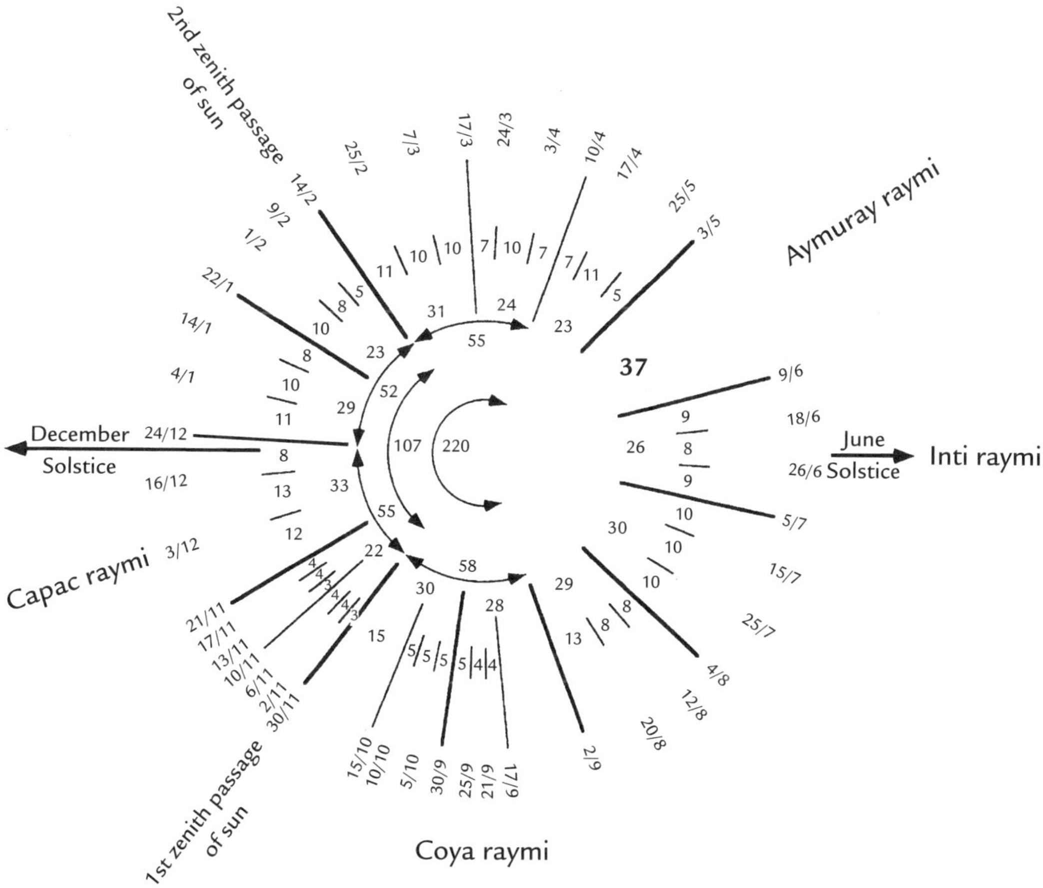

figure 6.3
Zuidema's model of the ceque calendar, showing various temporal divisions and dates (Ruggles and Urton 2006:fig. 8.1). Each huaca represents a day, each ceque a "week" of variable length, and each group of three ceques a "month" of variable length.

the Inti Raymi festival. This so-called feast of the sun terminates the least active period in the agricultural year. Sunrise at the June solstice would have occurred about three sun disks to the left (north) of the perpendicular to the west facade of the Coricancha (Aveni 1997; Zuidema 1981). This difference, translated into daily solar motion along the horizon, corresponds to a sunrise position about thirty days before or after the June solstice or to the date pair May 25 and July 20. The first date, which falls one synodic month before the solstice, is in excellent agreement with the timing of the start of the new year according to the chronicler Cristóbal de Molina (1943 [1573]:25), who begins his account of the calendar by stating that "they [the Incas] began counting the year in the middle of May, with the *first day of the moon*; which month of the beginning of the year they called *Haucay cusqui*, in which they celebrated the ceremonies called *Intiraymi*, which means 'Feast of the Sun', still celebrated today" (italics in original).

The "middle of May," which Molina refers to in the Julian calendar in use in Peru at that time, corresponds well with May 25 in the Gregorian. The "first day of the moon" probably meant the day of the moon's first reappearance as a thin crescent in the west after sunset. With such a definition of the "Feast of the Sun," it appears that the Inca preferred to schedule activities via a full moon on or around the June solstice.

The direction that the Coricancha faces, toward sunrise around May 25, coincides as well with that of a ceque leading radially outward from the Coricancha to a point on the northeast horizon where the Pleiades rose (Zuidema 1982). Its terminal huaca is called Cobo Susumarca, which agrees with one of the names of the Pleiades. The first annual appearance, or heliacal rise, of the Pleiades occurred about one-half lunar month before the June solstice, approximately June 9. Thus, if the Inca king sighted the first visible crescent moon falling after May 25, he would have found, on average, that the first full moon after that date would fall on the June solstice. In other words, *the first full moon after the first annual predawn rise of the Pleiades always will define the month that includes the June solstice.* Moreover, this particular moon would be observed not only close to the time of the reappearance of the Pleiades but also in the same area of the sky.

The thirteen-month seasonal year in the ceque calendar is depicted along the inner portion of Figure 6.3. These may be characterized as three "divisions" (see below). First, around the December solstice, it contains four double sidereal months of length 58, 55, 52, and 55 days, fitted to the ceques at observationally significant points. Second, two double synodic months, of length 60 and 59 days, plus an added 26-day sidereal month centered on the June solstice, complete the 365-day count. Three dual divisions of the solar year accommodate the two kinds of months as follows (these are labeled in Figure 6.3):

Division I:	258 and 107 days
Division II:	251 and 114 days
Division III:	220 and 145 days

For want of space, but at least to give an idea of the type of observational data the Inca likely employed, I shall allude only to the 107-day period in Division I and the 220-day period in Division III.[11] The 107-day period corresponds to the time between the two zenith passages; it consists of the 55- and 52-day double lunar sidereal months around the December solstice. The 220-day period includes the 58- and 55-day double sidereal months that precede and follow the middle pair.

An anonymous (1906) chronicler recognized that each of these four double months was shorter than two synodic months, "the moon of one month always reaching the other" (Zuidema 2006:275). The fact that information exists on lunar rituals applied exclusively to these four double months allows one to describe how the system likely worked in practice. Thus, the observation of the first visible crescent (close to new moon) after the December solstice commenced the third double month (of fifty-two days). At that time, recently initiated noble boys engaged in a ritual battle between their two moieties. During the night of the following full moon, all noble men and women danced through the streets and joined together at sunrise, encircling the king in the plaza. Four (Molina 1989 [1573]) or six (Cobo 1956 [1653]) days later, the most important concluding rituals took place. At night, runners followed the rushing waters of the river down to Ollantaytambo, stayed there for two days, and then returned during the daytime in a competitive race. Just before reaching Cuzco, they passed the sucanca where the June solstice had been observed. During the waning phase of that moon, the race finished just before the sun passed the zenith for the second time. So it was that a single moon within a two-month period "carried on time" between two "seats" (to use chronicler Guaman Poma's words) of the sun.[12]

While the ceque calendar may give the impression of imprecision and irregularity, as did the Javanese example discussed in the introduction, it is based on a sequence of regular observations that make sense in the temporal framework of indigenous cultural needs. These consist of three types: (a) solar observations, extended or abbreviated in particular months; (b) lunar- and stellar-related observations; and (c) the division of months into shorter periods, or "weeks," likely influenced by sociopolitical concerns (Zuidema 2010:758).[13]

Conclusions

We can think of the ceque system of Cuzco as a unique mnemonic scheme that brought together space, time, and social hierarchy. When we delve into how it worked, we ought not be surprised to find it alien to Western ways of mastering time, indeed, as alien as the early anthropologists viewed the structure of the Javanese rice farmers' calendar.

What stands out most about the role of precise observation in setting up the system is the way the rulership used the natural order they perceived in their own particular land- and skyscape as a means of structuring social order. The ceque system functioned like clockwork only when each particular class performed its assigned function in the proper place at the correct time of the seasonal year. Thus, the people of Hanan-Cuzco were in charge of the June observations and offerings based on a western-facing alignment, while those in Hurin-Cuzco took care of the December solstice, which was related to the opposing easterly direction. Between the moiety divisions ran the seminal zenith-antizenith axis, the pivotal alignment that partitioned the agricultural year. Applying the culturally embedded principle of reversible dualism, the Inca devised a brilliant scheme that helped them to know and interpret the vagaries of a harsh, difficult, and changeable agricultural environment. The ceque system served to map out and prescribe proper human action based upon residence and kinship in a radial, quadripartite, geographic framework.

Acknowledgments

I am indebted to Catherine Allen, Colin McEwan, Tristan Platt, Gary Urton, and Tom Zuidema for helpful discussions and criticisms of earlier drafts of this essay.

NOTES

1 Zuidema's approach to the study of cultures is strongly influenced by the Dutch structuralist approach (so designated after Lévi-Strauss's work became popular) in which he was schooled at Leiden during the 1940s. Benchmarks in the development of this form of structuralist anthropology and how it has influenced Zuidema's thinking about Andean social systems are explained by Urton (1996).

2 Some disagreement exists regarding the number of ceques and huacas. Polo, for example, says there were 340, but later he states that there were over 400. Albornoz (1989) reports on huacas not listed in Cobo's account, but many of these may not have figured in the system. Zuidema (2010:23) points out where Polo specifically states that 12 special huacas of the 340 he had tallied were counted twice. Other researchers, such as Bauer (1998), claim that 42 rather than 41 is the correct number of ceques; however, Zuidema (2010:23–28) points out that Cobo considered a pair of Cuntisuyu ceques, labeled part *collana* and part *cayao*, to be regarded as one. As we shall see, the issue of number is fundamental to Zuidema's interpretation of the ceque system as a calendar.

3 Much has been made of the issue of the straightness of the ceques. Recently, Zuidema (1990:632) has been more explicit in his explanation that the ceque system was an *abstract* system of coordinates, but critics (e.g., Bauer 1998:10–12) have noted that his original model had emphasized ceque directionality over huaca location.

4 There are other well-known Andean parallels that might lead one to conceptualize ceques as straight lines. These include the linearity, continuity, and juxtaposition of *chhiutas* and their possible relation to *chapas* (Urton 1984; Zuidema 1990:639–640) and the khipu (Zuidema 1977:220).

5 Zuidema argues that the rulers are mythic rather than historical figures. In his 2010 book, he presents his most detailed analysis of Inca myth (see especially chapter 5 and the summary in chapter 1, pp. 35–37).

6 Zuidema (1981) and I (Aveni 1981a, 1997) were able to trace the huacas of the ceque that contained the pillars. Once we located their approximate position, we found that as seen from the position of the *osno*, or *ushnu*, which we located in the Plaza de Armas (Zuidema 2010:807), sunset would have been visible

between the two central pillars on approximately August 18. This date corresponds with the solar antizenith passage (see Figure 6.2, alignment 2b). Of no importance in Western astronomy, because it is not an observable phenomenon in the latitudes where Western astronomies developed, the zenith (overhead) passage of the sun occurs twice annually between the tropics of Cancer and Capricorn (latitudes 23.5 degrees north and south, respectively); the dates are latitude dependent. The reciprocal dates in the year correspond to the time when the sun crosses the point underfoot, variously called the nadir or antizenith.

7 Thus: "A legendary account describes the approach to Cuzco, along the whole axis, of a newly crowned king. A myth describes the Incaic conquest of Cuzco along the last part of the axis. In April, a procession was repeated during the eight days that ended on the day of antizenith sunset. It took place along the axis in the valley east of Cuzco in the direction of the town and toward sunset on the mountain called Sucanca by antonomasia. In August, observations were made from the ushnu in the plaza following the month-long sequence of sunsets through the four pillars (erected on Mt. Sucanca). The multiple uses of this axis to observe antizenith sunsets—from the far-off other sucanca, from the valley, and from the ushnu—formed the core of the Inca calendar, as they defined the months used for planting (around August) and harvesting (around April/May)" (Zuidema 2006:273).

8 In a series of papers that appeared between 1987 and 1995, Dearborn and Schreiber, Bauer, and Bauer and Dearborn have leveled criticisms regarding the interpretation of both astronomical and ethnohistoric data, to which both Zuidema and I have offered responses. Too lengthy to relate here, the relevant publications are reproduced in their entirety in Aveni 2008:125–180, to which I refer interested readers. These critiques focus on what the authors term the absence of direct evidence and a lack of precision in Zuidema's interpretations. Dearborn and Schreiber (Aveni 2008:125) generally accept the solstitial alignments. While they admit that solar zenith passage is cited in the chronicles and find Zuidema's case for antizenith observations logical and strong, they contend that clear alignments attesting to its observation do not exist and, furthermore, that the August 18 date, based on our field measurements and huaca locations, may be at least ten days in error. They conclude that the idea that Inca sun watchers reversed the alignment (cf. Figure 6.2, alignments 2a and 2b) lacks physical evidence and must remain uncertain. Dearborn (Aveni 2008:162) further criticizes Zuidema's interpretation of the exactitude assigned to the antizenith passage date and alignment based on data taken from the chronicler Molina regarding the festivals associated with these dates. For a rebuttal of these arguments, see Zuidema 2010:chs. 4.4, 10.5–6, and Aveni 2008:158.

9 For a fuller discussion of the relationship between timekeeping and human activity across diverse cultures, see Aveni 1989.

10 See Zuidema 1982 for a discussion of the importance of the Pleiades in Andean cosmology.

11 For a full exposition, see Zuidema 2006:269ff. and 2010:ch. 11.

12 For a more thorough discussion of the other double month rites, consult Zuidema 2005, 2006:275ff., and 2010.

13 Of all the aspects of Zuidema's interpretation of the ceque system, the calendar has come under the heaviest criticism (compare, for example, Bauer 1992, 1998:160). Ziolkowski and Sadowski (1989:esp. 67, 131, 136, 146ff., 197) have dedicated practically an entire volume to an assessment of Zuidema's calendrical theory. They regard most of the historical evidence employed by Zuidema as "circumstantial," and they contend that his methodology consisted of laying out the divisions and subdivisions of the year and then looking for arguments to support his decoding (Ziolkowski and Sadowski 1989:202). In line with Bauer's complaint about the lack of direct physical evidence in the archaeological record, Ziolkowski (Ziolkowski and Sadowski 1989:208) summarizes his long-standing criticisms of Zuidema's calendrical model on the grounds that "there is no immediate ethnohistorical data on it."

Albornoz, Cristóbal de

1989 Instrucción para descubrir todos las guacas del Píru y sus camayos y haziendas. In *Crónicas de América*, edited by Pierre Duviols, vol. 48, pp. 135–199. Historia 16, Madrid.

Ammarell, Gene

2005 The Planetarium and the Plough: Interpreting Star Calendars of Rural Java. In *Songs from the Sky: Indigenous Astronomical and Cosmological Traditions of the World*, edited by Von Del Chamberlain, John Carlson, and Mary Jane Young, pp. 320–335. Center for Archaeoastronomy, College Park, Md.

Anonymous

1906 Discurso de la sucesión y gobierno de los yngas. In *Juício de límites entre el Perú y Bolivia: Prueba presentada al gobierno de república argentina*, vol. 8, edited by Victor Maurtua, pp. 149–165. Chunchos, Lima.

Aveni, Anthony F.

1981a Horizon Astronomy in Incaic Cuzco. In *Archaeoastronomy in the Americas*, edited by Ray A. Williamson, pp. 305–318. Ballena Press, Los Altos, Calif., and Center for Archaeoastronomy, College Park, Md.

1981b Tropical Archaeoastronomy. *Science* 213:161–171.

1989 *Empires of Time: Calendars, Clocks, and Cultures*. Basic, New York.

1996 Astronomy and the Ceque System. *Journal of the Steward Anthropological Society* 24(1/2):157–172.

1997 *Stairways to the Stars*. Wiley, New York.

Aveni, Anthony F. (editor)

2008 *Foundations of New World Cultural Astronomy: A Reader with Commentary*. University Press of Colorado, Boulder.

Bauer, Brian

1992 Ritual Pathways of the Inca: An Analysis of the Collasuyu Ceques in Cuzco. *Latin American Antiquity* 3(3):183–205.

1998 *The Sacred Landscape of the Inca: The Ceque System of Cuzco*. University of Texas Press, Austin.

Cobo, Bernabé

1956 [1653] *Historia del Nuevo Mundo. Obras del P. Bernabé Cobo de la Compañia de Jesus*. Edited by P. Francisco Mateos. Biblioteca de Autores Españoles, vol. 92, bks. 11–14. Editorial Atlas, Madrid.

Cohen-Stuart, A. B.

1898 *Javaansche Tijrekening*. Landsdrukkorij, Batavia.

Crawfurd, John

1967 [1820] *The History of the Indian Archipelago*, vol. 1. Cass, London.

Hamilton, Roland

1990 *Inca Religion and Customs by Father Bernabe Cobo*. University of Texas Press, Austin.

Isbell, Billie Jean

1982 Culture Confronts Nature in the Dialectical World of the Tropics. In *Ethnoastronomy and Archaeoastronomy in the American Tropics*, edited by Anthony F. Aveni and Gary Urton, pp. 353–363. Annals of the New York Academy of Sciences 385. New York Academy of Sciences, New York.

Maass, Alfred

1924 Sternkunde und Sterndeuterei im Malaiischen Archipel. *Tydschrift voor Indische Jaal-, Land en Volkenkunde* 64:103–173.

Molina, Cristóbal de

1943 [1573] *Relación de las fábulas y ritos de los incas*. D. Miranda, Lima.

1989 [1573] *Fábulas y mitos de los incas*. Edited by Henrique Urbano and Pierre Duviols, pp. 1–136. Historia 16, Madrid.

Niles, Susan

1987 *Callachaca: Style and Status in an Inca Community*. University of Iowa Press, Iowa City.

Polo de Ondegardo, Juan

1981 [1585] Los errores y supersticiónes de los indios sacadas del tratado y averiguacíon que hizo el Licenciado Polo. In *El catecismo del III Concilio Provincial de Lima y sus complementos pastorales* [1584–1585], edited by Juan Guillermo Durán, pp. 457–478. Publicaciones de la Facultad de Teología de la Universidad Católica Argentina, Buenos Aires.

Río, María de las Mercedes del
2005 Etnicidad, territorialidad y colonialismo en los Andes: Tradicíon y cambio entre los soras de los siglos XVI y XVII. Instituto Francés de Estudios Bolivianos (IEB), La Paz.

Rowe, John
1979 An Account of the Shrines of Ancient Cuzco. *Ñawpa pacha* 17:1–80.

Ruggles, Clive, and Gary Urton (editors)
2006 *Cultural Astronomy in New and Old World Cosmologies.* University Press of Colorado, Boulder.

Sherbondy, Jeanette
1987 Organizacíon hidráulica y poder en el Cuzco de los incas. *Revista española de antropología americana* 17:117–153.

Urton, Gary
1984 Chuta: El espacio de la practica social en Pacariqtambo, Perú. *Revista andina* 3:7–43.
1990a Andean Social Organization and the Maintenance of the Nazca Lines. In *The Lines of Nazca*, edited by Anthony F. Aveni, pp. 173–206. American Philosophical Society, Philadelphia.
1990b *The History of a Myth: Pacariqtambo and the Origin of the Inkas.* University of Texas Press, Austin.
1996 R. Tom Zuidema, Dutch Structuralism, and the Application of the "Leiden Orientation" to Andean Studies. *Journal of the Steward Anthropological Society* 24(1/2):1–36.
1997 *The Social Life of Numbers: A Quechua Ontology of Numbers and Philosophy of Arithmetic.* University of Texas Press, Austin.

Ziolkowski, Mariusz, and Robert Sadowski
1989 *Time and Calendars in the Inca Empire.* British Archaeological Reports, International Series 479. BAR, Oxford.

Zuidema, R. Tom
1977 The Inca Calendar. In *Native American Astronomy*, edited by Anthony F. Aveni, pp. 219–259. University of Texas Press, Austin.
1981 Inca Observations of the Solar and Lunar Passages through Zenith and Anti-Zenith at Cuzco. In *Archaeoastronomy in the Americas*, edited by Ray A. Williamson, pp. 319–342. Ballena Press, Los Altos, Calif., and Center for Archaeoastronomy, College Park, Md.
1982 Catachillay: The Role of the Pleiades and of the Southern Cross and α and β Centauri in the Calendar of the Incas. In *Ethnoastronomy and Archaeoastronomy in the American Tropics*, edited by Anthony F. Aveni and Gary Urton, pp. 203–229. Annals of the New York Academy of Sciences 385. New York Academy of Sciences, New York.
1990 Ceques and Chapas: An Andean Pattern of Land Partition in the Modern Valley of Cuzco. In *Festscrhift für Thomas S. Barthel*, edited by Bruno Ilius and Mathias Laubscher, pp. 627–643. Lang, Frankfurt.
2005 The Astronomical Significance of a Procession, a Pilgrimage and a Race in the Calendar of Cuzco. In *Current Studies in Archaeoastronomy: Conversations across Time and Space; Selected Papers from the Fifth Oxford International Conference at Santa Fe, 1996*, edited by John Fountain and Rolf Sinclair, pp. 353–370. Carolina Academic Press, Durham, N.C.
2006 Solar and Lunar Observations in the Inca Calendar. In *Cultural Astronomy in New and Old World Cosmologies*, edited by Clive Ruggles and Gary Urton, pp. 261–278. University Press of Colorado, Boulder.
2010 *El calendario inca, tiempo y espacio en la organización ritual del Cuzco: La idea del pasado.* Fondo Editorial del Congreso del Perú and Fondo Editorial de la Pontificia Universidad Católica del Perú, Lima.

7

Space and Time in the Architecture of Inca Royal Estates

STELLA NAIR

HOW CAN ROYAL ARCHITECTURE MARK OUT time? How does one construct references to an ancient time or memorialize specific moments in the past when the landscape is animate and dynamic? What does it mean to construct buildings from stones that do not sit passively but can move and be transformed? These are just a few of the questions that Inca architects had to answer with their designs, and in doing so, they addressed the entangled and complex relationship between time and space.

The Inca were brilliant engineers. They harvested, transported, carved, and laid high-quality stones to form impressive installations across the Andes (Ogburn 2004; Protzen 1983, 1985, 1986, 1993). They radically redirected and strictly controlled water movement, above and below the ground (Wright et al. 2006; Wright and Valencia Zegarra 2000; Wright, Witt et al. 1997; Wright, Wright et al. 1997). They created towering timber and reed roofs and built the largest covered space in the Americas until the arrival of Europeans (Lee 1988, 1992).

But they were not only consummate engineers; they were also skilled transformers of the transcendental, particularly as it related to space. From their use of sacred lithic materials in buildings to their careful reshaping of the sanctified landscape in the form of terraced hillsides and valleys, the Inca crafted their architecture to articulate pan-Andean notions of the sacred (Dean 2007, 2010b; Niles 1992, 2004). In the process, they radically reshaped space in the Andes (Bauer 1998; Gasparini and Margolies 1977; Hyslop 1990; Nair 2009; Niles 1980, 1989, 1993, 2004; Protzen 2000). They controlled access to objects and places; determined who could travel, where they could go, and what they could see; and they carefully distributed imperial Inca settlements across a vast land in order to make visible their domination.

But for the Inca, architecture reshaped far more than space; it also manipulated and commented upon time. In Quechua, *pacha* means "place" *and* "time" (Aguilar 1970; Gonzalez Holguin 1989 [1608]; Torres Rubio 1754).[1] It is a physical, temporal, and spatial descriptor (Salomon et al. 1991:14–16). When used as part of compound words or phrases, *pacha* helps to define special and often potent manifestations of space and time. Examples include the seasons, earthquakes, heaven, the underworld, the creator of the world, and the end of the world.[2]

As an expression of both time and space, architecture is an illustration of Inca understandings of pacha. In architecture, space only unfolds in relation to time. And just as space has implications for and beyond architecture, so does time. Both have ideological and conceptual roles, particularly in relation to power. Therefore, in this essay I will examine the sophisticated ways in which time was articulated, exposed, amplified, and manipulated in the Inca built environment to convey messages of imperial conquest and personal authority.

In order to move away from more generalized discussions of the Inca built environment, I have focused this essay on royal estates and the subtle yet highly effective ways in which time and space were addressed in these regal landscapes. In particular, I will examine royal estate architecture's relationship with time in the articulation of conquest, inheritance, motion, and sensory perception.

The Spaces of Conquest and Succession

One way that the Inca used architecture to manipulate space and time was through their use of materials that were deeply symbolic, for example, stones. In the high Andes, stone was often understood as a sacred material, and thus it is significant that most royal estates in the Inca heartland are built with stone masonry and integrated with striking outcrops (Figure 7.1). This type of construction was no accident but instead, an intentional design strategy to literally root imperial architecture in a vast, sanctified landscape (Dean 2007; Niles 1992). In

figure 7.1
Machu Picchu, estate of the ruler Pachacuti. (Photograph by Stella Nair.)

terms of space, this strategy allowed architecture to make visible the Inca relationship with sacredness and thus strengthen their imperial authority.

Carolyn Dean (2007, 2010b) has noted that these architectural gestures had a temporal component. By incorporating primeval lithic materials and places into their buildings, the Inca and their royal estates projected the sense of an ancient presence, thus belying their recent expansion into newly conquered territories. This reference to an age-old beginning would have been important to the colonizing desires of the Inca. As recent conquerors, the Inca had a vast array of subjects who questioned the imposition of Inca rule. In response, the Inca purposefully integrated their architecture with ancient materials and landscape features, creating a counternarrative of a long-standing Inca presence. Thus, the Inca used architecture to rewrite their temporal relationship to Andean space.

While Inca architecture made a generalized reference to an ancient period, it also marked specific moments in that distant past by denoting construction time.[3] For example, the Inca left bosses (an element of the construction process) on occasional stone blocks (Figure 7.2). These lithic projections were a record of the moment in the past when the Inca physically shaped the sanctified landscape (Dean 2007). By denoting facture (the process of making), the Inca made visible specific moments when they had the power to shape sacred materials, thus visually articulating their supreme authority.

In the Inca heartland, estates played an exceptional role in redefining space and denoting time. Here, however, they commented not only on a generalized Inca presence but also that of distinct individuals. The architecture of these estates reflected the royal patron and the specifics of his rule. For example, Susan Niles (2004) has shown how drastically different Pachacuti's royal estates were from that of his grandson Wayna Capac, revealing the distinct ways in which architecture and landscape were manipulated to address very different political climates and personal desires. As we shall see, the architecture of these estates reveals the specific ways in which temporal concerns became entangled with royal space.

Pachacuti's reign was marked by new conquests in the Inca heartland. He is credited with conquering the lands surrounding the Cuzco Valley in the early fifteenth century and expanding out across the south-central Andes. Thus, most of the architecture built under his rule would have been erected in

figure 7.2
Bosses on a wall at Ollantaytambo. (Photograph by Jean-Pierre Protzen.)

newly conquered lands, highly visible to local populations who were recently subjugated. In these areas, Pachacuti built royal estates for himself and his family, such as Pisac, Ollantaytambo, and Machu Picchu. For this *sapa Inca* (ruler), royal estates were critical and visible aspects of his conquests.

Set precariously on a narrow saddleback between two soaring mountain peaks, Machu Picchu is a masterful example of an architectural style that typified the patronage of Pachacuti. Buildings and terraces are integrated with natural elements, such as water sources, sloping hillsides, and dramatic cliff faces. For example, a series of baths are integrated with a cascading water channel and staircase. Nearby are a long staircase and wall that almost bisect the site in two. Throughout the site an extensive array of terraces reshape the folds and cliffs of the mountainside, while a collection of finely made Inca buildings are so deftly integrated with the local bedrock and outcrops that the structures seem to emerge seamlessly from the sacred rock of the mountainside (Figure 7.3).

But the "natural" setting of Machu Picchu's buildings was not at all natural; instead, it was carefully orchestrated on both the large and small scale (Wright and Valencia Zegarra 2000; Wright, Kelly et al. 1997; Wright, Witt et al. 1997; Wright, Wright et al. 1997). Stone outcrops that appear natural were actually subtly carved or framed, and buildings were meticulously designed to appear entangled with natural features (Dean 2006, 2007; Niles 2004; Figure 7.4). How one moved to and through the site, as well as what one could see both near and far, were precisely planned (Julien 1990). This carefully composed setting and experience served as a visible and tangible manifestation of the power of the patron, Pachacuti. Not only did this architecture provide a complex stage set from which he could enact his

figure 7.3
Machu Picchu, view of terraces and buildings. (Photograph by Stella Nair.)

figure 7.4
Structure currently referred to as the "Temple of the Sun," where the June solstice was tracked. (Photograph by Stella Nair.)

rule in the newly conquered region, it also visually articulated Pachacuti's message that his rule was supported by the sacred landscape.

While the design of royal estates reveals the Inca preoccupation with controlling people through space (physically and conceptually), the people who actually experienced royal space changed according to context, which had important temporal implications. This idea can be clearly seen in Chinchero, the royal estate of Pachacuti's son (Nair 2003, n.d.). Topa Inca built Chinchero at the end of his reign in the late fifteenth century (Figure 7.5). By this time, the Urubamba Valley had been stabilized. Under the direction of both Pachacuti and Topa Inca, the region had been spatially transformed with Inca roads, warehouses, shrines, and, of course, royal estates. Instead of being composed largely of rival ethnic groups, the Urubamba Valley during the time of Topa Inca's rule had a rapidly growing population of wealthy and powerful Inca nobles, as well as their servants.

Topa Inca located his estate up a narrow canyon, above the valley floor (Alcina Franch 1969, 1970, 1971, 1972, 1974a, 1974b, 1976; Alcina Franch, ed. 1976; Alcina Franch et al. 1976; Ballesteros Gaibrois 1976; Nair 2003, 2009, n.d.). The architecture he deployed has all the markers of the Inca style, yet its articulation of and relationship with sacred natural features is very different compared to Pachacuti's estates. The reason for this change in the design's emphasis lies with the specific political issues of his reign. Topa Inca built his estate not as a signal of the Inca presence in newly conquered lands, but as a lavish retreat to share with his favorite lady, Chequi Ocllo. Although it was customary for Inca rulers to have several wives and many mistresses, it was not common for them to build royal estates explicitly to share with one of their secondary ladies. Perhaps this is why Topa Inca's principal wife and sister, Mama Ocllo, was said to be jealous of the other wives and lovers of her husband (Figure 7.6).

Competition among a ruler's sexual partners was not merely about affection and status, it also involved issues of succession. In the case of Chinchero, it was about which woman's son would inherit the *mascaypacha*, or royal fringe of rule. When this estate was built, Topa Inca is reported to have been trying to elevate Capac Huari, the son he had with Chequi Ocllo, as his successor. Maria Rostworowski de Diez Canseco (1953) has described how the promotion of this son over the son of his principal wife would have been very controversial. Although the Inca did not have primogeniture, they did favor the children of the principal wife. The realization of such a change in succession would have depended largely on the support of leading Inca nobles, particularly those most closely related to the ruler and relevant wives. Topa Inca and Mama Ocllo's closest relatives belonged to Pachacuti's *panaca* (royal family group), which reaped the rewards of the rich lands of their father's royal estates in the Urubamba Valley. Thus, the

figure 7.5
Topa Inca Yupanqui, drawing 110 [110] of Felipe Guaman Poma de Ayala, *El primer nueva corónica y buen gobierno*, 1615. (Photograph courtesy of The Royal Library, Copenhagen.)

royal residents of the Urubamba Valley became the key players in determining who would succeed Topa Inca.

It is this audience that seems to have driven the design of Chinchero. By situating his royal estate at the top of a steep canyon, Topa Inca placed it above that of the lands of his primary wife and other Inca nobles, establishing a vertical hierarchy. By making the primary facade of Chinchero the one facing the Urubamba Valley, rather than Cuzco, the Inca capital, Topa Inca made clear who his intended audience was. About 90 percent of Chinchero's architecture is devoted to the elaborate reception area for those traveling from the Urubamba Valley (Figure 7.7). Besides the extensive terraces, this area includes intricately carved outcrops, impressive buildings, and water features. Thus, the vast majority of Chinchero's built environment was designed

figure 7.6
Mama Ocllo, drawing 138 [138] of Felipe Guaman Poma de Ayala, *El primer nueva corónica y buen gobierno*, 1615. (Photograph courtesy of The Royal Library, Copenhagen.)

to be seen and experienced by people coming from the Urubamba Valley.

Clearly, space and spatial practices were critical to the siting and design of Chinchero and Topa Inca's plans for succession. But time was also a vital factor because succession is fundamentally about time, specifically in terms of mortality and legacy. Chinchero was an estate that did not so much look back to an ancient past, like Pachacuti's estate, as it looked forward to a time when a new ruler would be chosen. Thus, for Topa Inca the future greatly influenced the design of Chinchero.

Both Pachacuti and Topa Inca used their royal estates to transform the Inca heartland spatially and emphasize their own power. Yet the specific events that drove these spatial practices were very different, and each estate alludes to distinct aspects of time. While Pachacuti used architecture at his

figure 7.7
Chinchero, north facade. Terraces are above the Inca road leading from the Urubamba Valley. (Photograph by Stella Nair.)

estates to mark out imperial conquest and the Inca relationship with the ancient natural world, Topa Inca created a built environment as a statement of authority directed at other Inca elites, in whose hands his son's future transition to power lay (Nair 2003, 2009, n.d.). Pachacuti, due to his conquest campaigns, emphasized ideas of a long-standing presence and the ancient past in his architecture, while Topa Inca, because of the importance of succession, used his architecture to highlight his vision of a powerful present and preferred future by signaling his chosen successor.

While these estates commented on aspects of time (past and future as they related to the present), they also conveyed the fact that the sapa Inca experienced time differently than his subjects. For the Inca, the ruler never died, but instead changed form, moving from that of a living body to that of a venerated mummy or *mallki* (Bauer 2004; Dean 2010a; Isbell 1997; Salomon 1995). In their new forms, Pachacuti and Topa Inca would have continued to live at their impressive royal estates as they had when they were made of flesh; plazas would have been filled with performers, and on special occasions the desiccated sapa Inca would have been sung to, fed, and consulted. Hence, Machu Picchu, Chinchero, and other royal estates would not have become quiet memorials to dead rulers, but instead, they would have remained critical royal centers of imperial control, their purpose continuing in the eternal present.

Motions of Time and the Inca State

The Inca did not limit their expression of time to the past, present, and future but also used architecture to locate themselves within astronomical time (Aveni 1982; Bauer 1992, 1998; Bauer and Dearborn 1995; Dearborn and Schreiber 1986; Dearborn and White 1982, 1983; Reinhard 2007; Urton 1981; Zuidema 1964). Anthony Aveni (this volume) has shown how the Inca used the built environment to track astronomical time in the case of the *ceque* system of Cuzco. In this case and in others, the Inca used aspects of the built environment such as stones to mark the solstices, the passage of Venus and the Pleiades, and the solar antizenith, as well as lunar cycles and the movement of the sun.

Examples of architecture that mark cosmic time can be seen at the royal estate of Machu Picchu, where a large outcrop was encircled and partially encased in an Inca building. This stone and building were used to mark the June solstice (Figure 7.8; Dearborn and White 1983). A similar one exists at Pisac, another of Pachacuti's estates, which may have been used for astronomical observations. Although no such stone remains at Chinchero today, there likely were architectural markers in the past that allowed Topa Inca and his staff to trace astronomical cycles, such as the arrival of the solstices.

The tracing of celestial cycles was important to the Inca and was intimately linked to agricultural and ritual time (Aveni, this volume). For example, most farming communities tracked the solstices because this information helped them to know when to plant and harvest specific foods—celestial time was intimately linked to agricultural time. Perhaps because of this relationship, the tracing of astronomical movements was linked to rituals. Thus, the outcrops, pillars, and structures that the Inca used to trace astronomical time were also tangible expressions of agricultural and ritual time.

For example, the tracking of the sun in Cuzco was an important ritual sponsored by the state. It involved large numbers of people and took place

figure 7.8
Carved stone in the "Torreon" at Machu Picchu (now referred to as the "Temple of the Sun"). (Photograph by Jean-Pierre Protzen.)

outside so that ceremonies could command a large audience. Brian Bauer and David Dearborn (1995:153) have argued that this emphasis on the sun in public festivals was an attempt to validate and promote Inca rule, as Inca rulers declared themselves descendants of the sun. Because of the large audiences involved in these rituals, specific spaces were required in which the festival could be held. Most Inca settlements, including royal estates, had impressive plazas that could house such ceremonies. When we consider the architectural landscape of astronomical and agricultural time, then, we should not confine ourselves to the physical markers that helped to measure and map celestial movements but also include the vast built environment that would have held the often politically tinged rituals associated with these critical temporal cycles.

This built environment includes not only the particular spaces for temporal observance or ritual celebration but also that used for proscribed processional movements across space. These kinds of movement became important acts of remembrance, locating both people and events in time and space, and royal palaces often played an important role. During his investiture, for example, the royal heir would take a pilgrimage to learn of the accomplishments of his ancestors and to be introduced to the powerful *wak'a*. His movement through and veneration of specific spaces, objects, and buildings connected him to a larger sacred history that reached back across time. This pilgrimage involved palace buildings and outcrops, among other places and objects. With each stop, songs were sung, recounting the events of each place and its patron. As a result, Inca rulers and their achievements were located in time within the ever-growing empire of Inca spaces (Niles 1999; Rostworowski de Diez Canseco 1992).

The story of royal pilgrimage is a reminder of how Inca buildings served as mnemonic devices for people and for events. In the Andes, architecture has functioned as a critical memory aid in the telling of oral histories, such as with the Inca (Cummins 1998; Harrison 1989). Hence, Inca buildings, such as those at royal estates, along with outcrops and other spaces of significance could together serve as a tangible memory palace or as remembered materials, allowing travelers or family members to hear about past rulers and the key events of their reign (Dean 2010b; for a comparative example, see Linda Brown's discussion of the Tz'utujil *nawales*, this volume). Inca buildings were not only objects defining space, but they were also objects recording time. Thus, we should not only discuss how stately buildings at Inca royal estates were used or how they impressed local populations and visiting nobles, but we should also consider the role some of these structures may have played in creating a living history of the great achievements of a ruler, as well as important events related to his family life.

The story of royal pilgrimage also highlights the issue of motion, in particular, its relationship to time. While astronomical, ritual, and agricultural time were all important to the Inca, time in the form of motion was their preoccupation (Cummins and Mannheim 2011; Salomon et al. 1991). The Inca understood the world as being in motion and animate. Not only could humans and animals move, but so could objects. Much of this motion was related to time in that it was constant, cyclical, and transformative. Stones did not sit passively in Inca buildings; they could shimmer, move, and be transformed. Tom Cummins and Bruce Mannheim (2011) have described this phenomenon as Inca kinetics, or motion that was linked to time, and argued that we must understand Inca architecture as an expression of this awareness of motion.

For an empire that considered control over others as paramount, such that the Inca state tried to control where subjects lived, what they produced, where they went, and what they wore, one can see how the view that everyone and everything in the Andes was in motion would be of great concern to the Inca. It also suggests why the Inca became so concerned with materially marking time, whether in terms of an ancient past, continuing present, or professed future. Whoever could control such a dynamic animate world, even for a short period, was one who possessed great power.

The fact that a state with great colonizing ambitions would be concerned with the power of motion is not surprising. Paul Virilio, who has written extensively about time, space, and power, has argued, "Whoever controls the territory possesses it. Possession of territory is not primarily about laws and contracts, but first and foremost a matter of *movement* and circulation" (Armitage 2000). Laws and contracts can be ignored or overlooked; real authority means being able to have quick and ready access to goods, peoples, and information so that subjects can be fed, rebellions put down, markets carried out, and the message controlled (Virilio 1986, 2002; Virilio et al. 1994). If the state cannot move and goods, people, and messages cannot circulate in a timely manner, all authority of the state is lost. To have real power, in other words, one must not only master space, but also the temporal experience of it. Although Virilio is concerned with time and the modern state, his description would have resonated with the Inca.

To understand Inca attempts to possess territories and control access in the Andes, therefore, we must appreciate the critical role of time as expressed in motion or movement. This argument may seem innocuous, but it has important implications for settlement studies. While scholars have noted the importance of the layout of Inca installations in their possession of space, we might consider those same installations as expressions of time and in particular of the Inca being able to access places, goods, and people in a timely manner, which Virilio sees as central to the possession of territories.

In addition, we can also see the manipulation of time in space as a critical way in which the Inca reminded his subjects of his power over their bodies. Thus, the Inca emphasized a phenomenological experience of time in space (see Baldwin 2007; Casey 2000; Eberl, this volume; Husserl 1962; Merleau-Ponty 1962; Ricoeur et al. 1996; Seamon 1993; Seamon and Mugerauer 1985; Tibbetts 1969; Tuan 1990). The Inca manipulated the real and perceptual experiences of time and space, thus making things rapidly accessible or seemingly forever out of reach, depending on the ways that they controlled

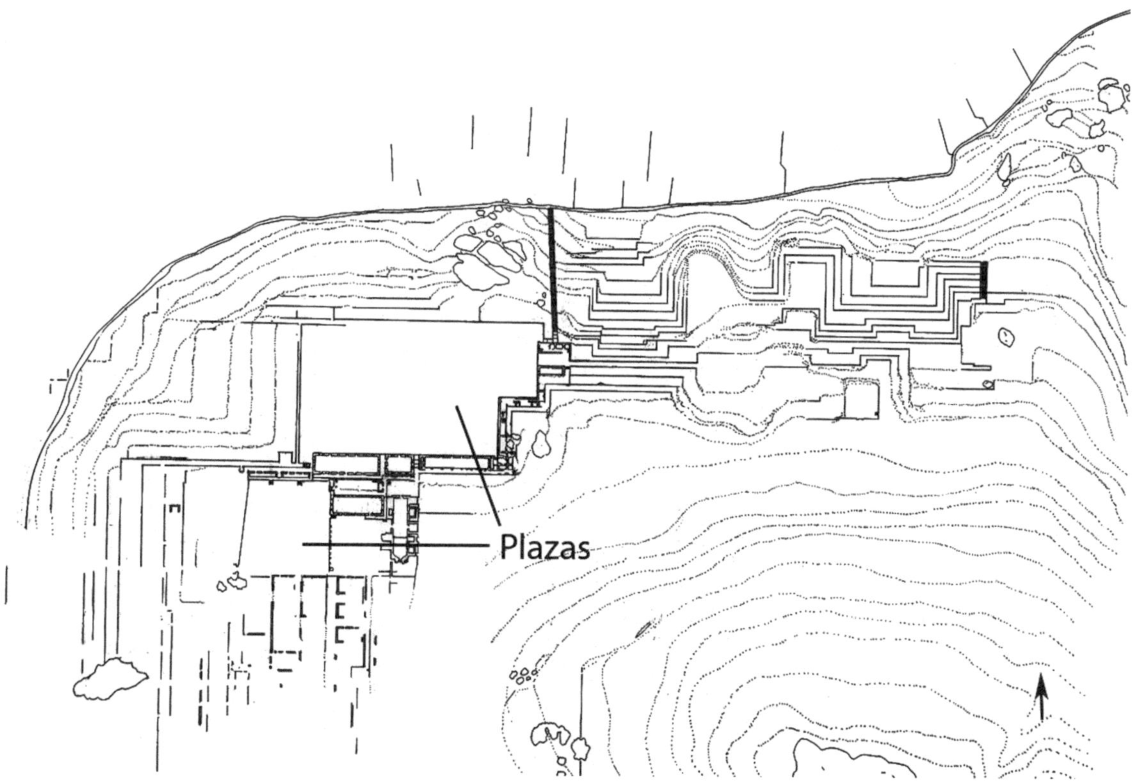

figure 7.9
Chinchero, map showing Inca remains and the two plazas (large and small). (Illustration by Stella Nair.)

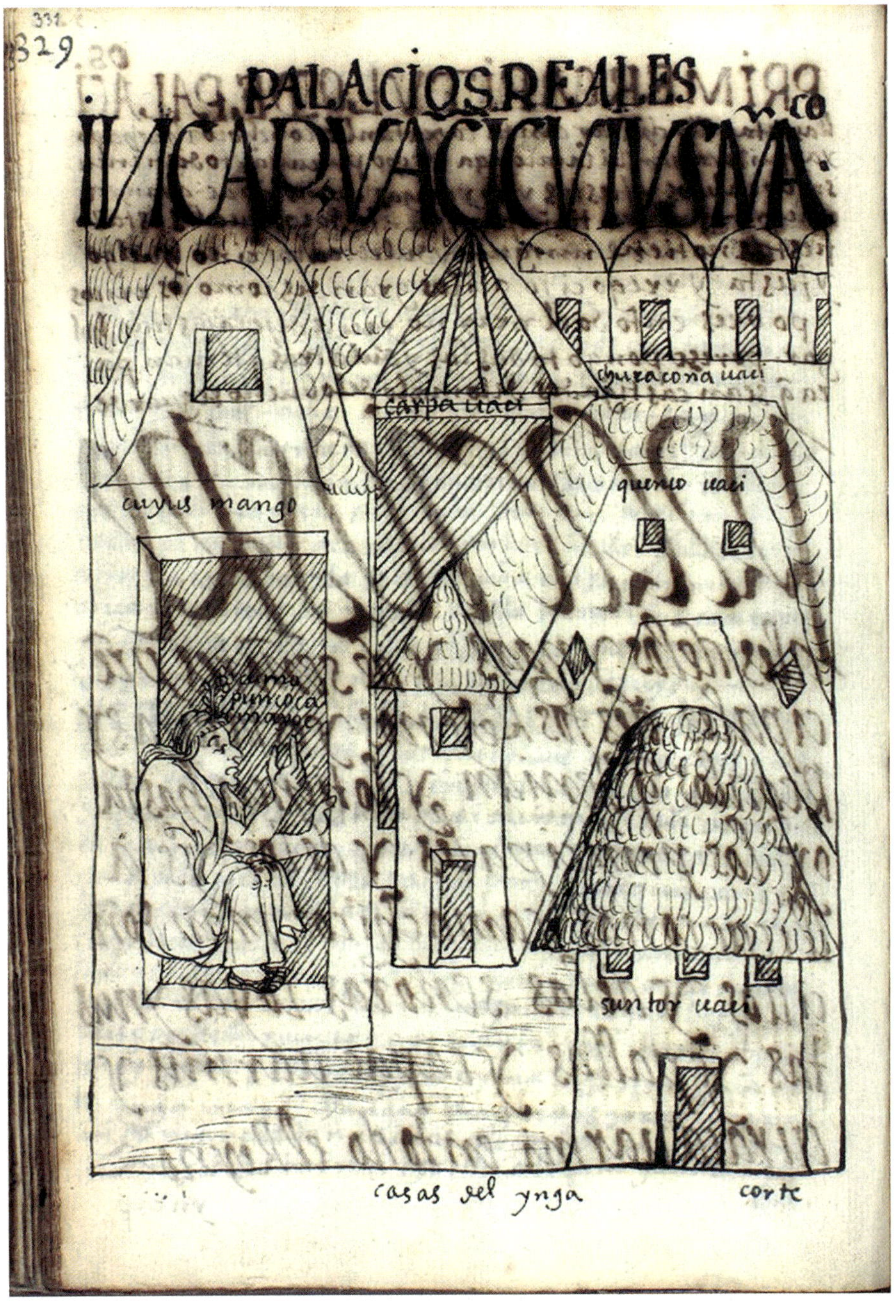

figure 7.10
Depiction of major building in an Inca royal palace, such as the *cuyusmanco* and the *carpawasi*, drawing 329 [331] of Felipe Guaman Poma de Ayala, *El primer nueva corónica y buen gobierno*, 1615. (Photograph courtesy of The Royal Library, Copenhagen.)

movement with their architecture. This compressing and stretching of time allowed for Inca power over their subjects to be far more than merely witnessed; instead, it was to be bodily experienced.

As an example, we can take a closer look at Topa Inca's royal estate as a temporal experience. If we look at the installations in and around Chinchero in terms of "speed" (rate of time through space) as an expression of power, it is clear that Topa Inca would have had all that he needed. Not only is the estate filled with elaborate terraces that could have grown a variety of crops, but the buildings associated with royal estates would have held goods to fulfill all his most urgent needs. The estate itself is divided into two sections, both centered on plazas (Figure 7.9). The largest plaza and its supporting buildings would have housed the more public functions, while the smaller, less accessible plaza

figure 7.11
Map showing some of the holdings associated with Topa Inca's royal estate. ©2013 Google.

and its buildings likely serviced the ruler's more private needs.

We are given some insight into the type of buildings that royal estates would have had by colonial writings and drawings (Figure 7.10). For example, Guaman Poma provides a list of building types within royal estates, and they encompass both public functions—such as the *cuyusmanco* (palace), *carpawasi* (auditorium), and *camachiconawasi* (place for government officials)—and private functions, such as a *punonawasi* (house for sleep and sex) and the *uachawasi* (house for "orphan" children or, more specifically, the house for the children the ruler had with lower-status women). In addition, royal estates had the infrastructure that facilitated easy access to ritual drinking (*aqhawasi*) and that served urgent needs, such as the *acawasi* (house of feces). Thus, royal estates provided quick means to fulfilling the desires and most pressing needs of a ruler, both professional and personal. For the ruler, time was contracted such that he had timely access to things and places he needed.

Other functions that were less urgent were laid farther out on the temporal map (Figure 7.11). Machu Colca, a large storage center, could be accessible by an hour's walk. The shrine at Cuper Bajo, about forty-five minutes away along the road to Cuzco, held a unique miniature wall made up of multicolored stones. This shrine overlooked Lake Piuray, a sacred lake that brought water to Cuzco during both the imperial Inca and Spanish colonial periods. About forty-five minutes away, on the road to the Urubamba Valley, was a *tambo*, or guard station, monitoring travelers. These travelers would have started their journey at Urquillos, where both Topa Inca and Mama Ocllo had lands. This road was finished while the other road from Cuzco was just begun. The construction stages are revealing, as they highlight the architectural priorities of the patron, which must have been even more pronounced given Topa Inca's advanced age. His own "personal time" therefore also influenced the architecture of Chinchero, and he is said to have died after living at the royal estate for only a few

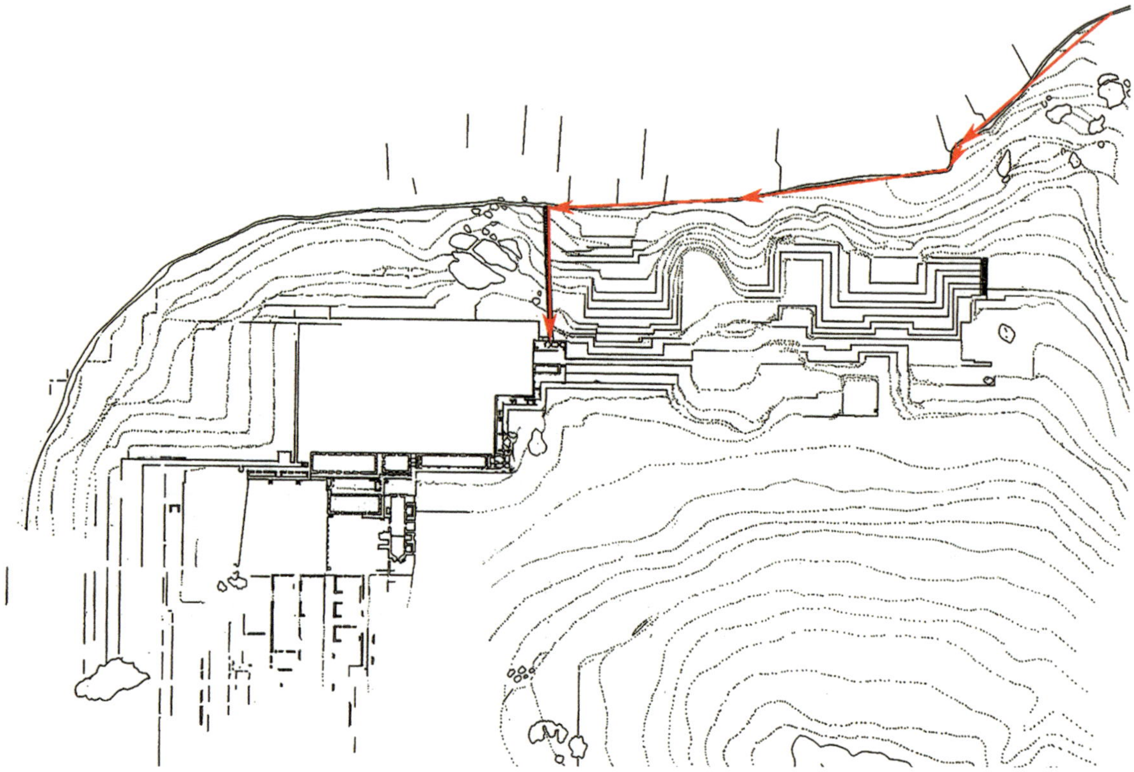

figure 7.12
Plan of Chinchero showing path to the main plaza. (Illustration by Stella Nair.)

short years. Arrayed around the royal estate was a series of places that that ruler would have needed access to, each laid out in the landscape at a temporal distance that allowed the ruler power of access.

But understanding the full meaning of temporal practices is not readily accomplished simply by looking at a map. That is because the Inca made use of the landscape to manipulate temporal experiences of distance. For example, locating Chinchero above the Urubamba Valley not only expressed spatial (and social) hierarchy but also allowed for the manipulation of time across space. As anyone who lives in Chinchero or Urquillos will tell you, the walk from the valley to Topa Inca's estate is a rather brutal one. The road starts with a gentle incline but then gets rapidly steeper. The result is not only that the body feels heavier but also that time becomes stretched. Neither sensation, of course, is real, as a person does not suddenly gain weight and time does not abruptly slow down, but light–heavy and slow–fast are how space and time are experienced on the road.

Relief does eventually come after several hours of hiking, when the road turns a final corner and the visitor finds himself standing at the base of Chinchero, its magnificent stone walls within easy reach. Thus, time is suddenly compressed. But this experience is an illusion. First, one must walk down a flat, well-built road with the terraces looming above. It is an easy walk, so time is normalized. But the visitor must then climb a long staircase. While this staircase may seem a bit long on a site plan, it does not appear nearly as long as it feels (Figure 7.12).

Nor is it adequately portrayed in the photograph shown in Figure 7.13. As anyone who has decided to walk straight up a mountain face can relate, with each step up time seems to stretch exponentially, making one more and more aware of the power of outside forces on the body. The Inca actively magnified this perceptual experience by

figure 7.13
The beginning of the long steep staircase that leads to the main plaza at Chinchero. Note that at this point, there is a significant drop-off to the left of the climber. (Photograph by Stella Nair.)

starting the staircase at a steep angle that becomes even steeper as it gets higher. There are no platforms to rest on, just a seemingly endless series of steps that disappear into the horizon. To give a sense of scale, in this image you see about four terraces of stairs, yet on the staircase you pass fourteen terraces. It is an exhausting climb in which the final goal seems increasingly out of reach, rather than closer.

This stretching and compressing of time occurs throughout one's experience of Chinchero, as it does in so many other royal estates, where all of the bodily senses are manipulated in relation to time. The visitor is reminded of the power of the Inca over space, time, and the body. In the case of Chinchero, visitors from the Urubamba Valley would have likely been the nobles from Pachacuti's panaca, the brothers and sisters of Topa Inca and his sister-wife, Mama Ocllo. One can only imagine what they thought as they walked this theatrical path to Chinchero. But one thing was most likely clear: the power of the ruler to manipulate both time and space.

This control of time in space on the local setting is not very different from what occurred on the larger landscape. The extensive Inca road network allowed Inca rulers, elites, troops, administrators, messengers, and servants quick access to all parts of Tawantinsuyu, but it also prevented everyone else from traveling. For those not given official permission to travel on the roads, the penalty was severe and even included death. In the Andes, where movement across landscape has a long and diverse history, the prohibition of travel for most people was devastating. For colonized people, or punished nobility, roads were not means of access but became visible symbols and embodied experiences of dispossession and displacement.

Conversely, these same roads allowed the swift movement of the Inca state, such that representatives could rapidly move across the Andean landscape to reach places, people, and goods that they needed access to. This brisk movement, which was facilitated by an infrastructure of way stations, storage centers, and administrative centers, allowed Inca elites and their sanctioned representatives to experience time in a way that was previously unknown. And this temporal access was crucial to Inca success. Inca sites, such as tambo (way or guard stations) and *colca* (storage centers), were mapped out on the landscape to provide travelers with ready access to places of rest and refreshment. These carefully placed sites allowed armies to quickly replenish themselves before marching on to squash revolt or conduct a surprise attack on new territories. It also enabled *chaski* (messengers) to carry news swiftly across long distances, bringing crucial information to imperial officials. It was the temporal aspects of the Inca built environment (mapped, measured, and experienced) as much as its spatial qualities that enabled the Inca state to be so successful.

Movement through the landscape was not merely a functional act (moving people, goods, and ideas from one place to another), it also was a meaningful act of remembrance that has long been practiced in the Andes (Bauer and Stanish 2001; Dulanto, this volume; Sallnow 1987). Movement through space simultaneously enacts memory and reiterates a relationship to the land. By restricting movement of populations across the landscape, the Inca with their network of roads could alter or suppress memories of dominated populations, especially those memories that tied people to specific places. The past times or histories of these populations were threatened with alteration and erasure. The Inca thus used movement as a way to mark time, memory, and conquest in space.

Conclusion

While space was of critical importance to the Inca, it was always intimately linked with time, particularly in regard to the built environment. For the Inca, references to permanence, inheritance, motion, and the senses were all important in denoting their relationship to the complex layers of the temporal world. To be bound up with time was to be linked to powerful forces.

But it was not enough for the Inca to be linked to time and space; they also had to show their power to control both. In space, the Inca designed key architectural gestures that manifested their presence while controlling movement of both subjects and elites. But it is in the control of time that they were perhaps the most proficient. When they referred to a primeval past with their architecture, for example, they were able to give an impression of their ancient (and accepted) presence in the sacred landscape. Thus, they used ideas about time to signal that they belonged in the Andes, like the sacred wak'a and *apu*.

To shape perceptions of time was a particularly great achievement for the Inca because time was not something stable and readily grasped but instead was powerful and elusive. To the Inca, everything was in motion, and thus, everything took part in a dynamic system of circulation and transformation. While the Inca could not control time itself, they did develop strategies to mark, influence, appease, or manipulate the world of circulating elements. The Inca used architectural gestures to record celestial movements, to situate their ritual and agricultural calendars, to give a sense of their enduring presence in the landscape, and as mnemonic devices that facilitated their remembrance of history. But they also used architecture to manipulate the experience of time. In doing so, the Inca demonstrated how the control of movement and circulation meant the true possession of territories and people.

Postscript

Ironically, it was the sophisticated manipulation of space and time by the Inca that later facilitated the swiftness of the European invasion. Specifically, the temporal and spatial designs that constituted the Inca's architecture of conquest also enabled the rapid advance and onslaught of the invading Iberians. The Spanish started off with a distinct advantage: by the time they arrived in the Andes, the diseases they brought with them had devastated the vast lands of Tawantinsuyu. Included in the body count were the ruler and his intended heir. Their sudden deaths triggered a brutal civil war between two of Huayna Capac's sons, Atahualpa and Huascar. But what happened next rested largely on the fact that the Spanish, like the Inca, recognized that the control of movement was a key component in their conquest of the Andes.

As Heidi Scott (2009) has so eloquently explained, the journey down the western rim of South America was a difficult one for many early sixteenth-century travelers. For the Spanish, however, it did not begin that way. At first they sailed rapidly, making stops along their journey so that they could rest and gather supplies. On the ships were many African slaves and Indian captives, who were forced by the Spanish to explore the unknown lands at each stop. Often, these non-Europeans were required to penetrate thick jungles that grew along the equatorial regions. Here, they gathered crucial supplies for the journey and made notes about what was encountered. The Spaniards, who tended to stay within the confines of their ships, took pride in their ability to survive so successfully (never acknowledging that their success was based on intermediaries who braved the elements). To the Spaniards, their rapid mobility across space was a sign of their power.

When their pace suddenly slowed, the Iberians blamed not their declining abilities, but instead a harsh and unwelcoming landscape. As the African and indigenous slaves and servants who were sent out into the jungles eventually died from exposure to the difficult terrain, the Spaniards were forced to move into the new lands themselves. Their inexperience in the landscape took a dramatic toll on the Spanish sailors, much as it had on the African and indigenous people before them. Iberians began getting sick and dying, which slowed the ships' progress on land and thus at sea. For the Spanish, their slow speed was not the fault of their own inabilities in the Americas but of the inhospitality if not outright hostility of Andean space.

But they did not credit Andean space when it enabled them to quicken their pace. When the surviving Spaniards landed on the edge of Tawantinsuyu, their temporal experience changed dramatically. Because the local population had been decimated by the onslaught of European diseases and the ensuing civil war, the Spanish could land on Andean shores with little trouble. The lack of large hostile populations aided their movement, but what really allowed the Spaniards to speed across the land was Inca architecture. The very architecture that empowered the Inca now facilitated the movement of the Spanish. Storage houses (colca), way stations (tambo), and highways (*capac ñan*) serviced the Spaniards in their desire to race to Cajamarca. The Spanish had wide paved Inca roads to travel upon, storage centers to provide them food and supplies, and way stations to give them shelter. Although it was critical to their speed, the Spanish never gave credit to this landscape that aided their progress but instead claimed that their rapid travel across the Andes was a sign of their superiority (and God's will). Nevertheless, the Inca built environment proved again to be a perfect example of Virilio's argument that time (space-time) and not law is the source of state authority. Without the Inca built environment and its ability to quicken time across space, the Spaniards would have remained marooned on the coast, vulnerable to threats both natural and man-made. And Andean history might have developed very differently.

NOTES

1 For example, Torres Rubio defines *pacha* as "tiempo, lugar" ("time, place"; 1754:93, recto), "*pacha*—lugar" (1754:131, verso). An anonymous author defines *pacha* as "tiempo, suelo, lugar" ("time, earth, place"; Aguilar 1970:157). Gonzalez Holguin (1989 [1608]:268) also defines *pacha* as "tiempo suelo lugar." For a nuanced discussion of the complex ways in which *pacha* was understood in the colonial period, see Salomon et al. 1991:14–16.

2 "*Pacha, allppa*—suelo, tierra" ("ground, earth"; Torres Rubio 1754:142, recto); "*hanacpacha*—el cielo, lugar alto" ("the heaven, high place"; Torres Rubio 1754:84, recto); "*pacha*—el vientro" ("womb"; Torres Rubio 1754:221, verso); *pacha*, "vientre" ("womb"; Torres Rubio 1754:230, verso); "*Pachacamac*—Dios" ("God"; Torres Rubio 1754:123, recto); "*Pachacamac*—el hacedor del mundo" ("the maker of the world"; Torres Rubio 1754:162, recto); "tiempo" ("time"; Torres Rubio 1754:143, recto); *pacha cuyuy*, "templar la tierra" ("shake the earth"; Torres Rubio 1754:143, recto); "*pachauti*—el fin del mundo" ("the end of the world"; Torres Rubio 1754:93, recto); "*ttipiy pacha*—tiempo de cosecha" ("harvest time"; Torres Rubio 1754:103, verso); "*rupay pacha*—verano" ("summer"; Torres Rubio 1754:145, recto); "*Paraymitta*, vl. *Pacha*—Tiempo de agues" ("Time of the omens [?]"; Torres Rubio 1754:96, recto); "*pachan, pachallan*—

el proprio enterro" ("one's own burial"; Torres Rubio 1754:162, recto); "*pachallanmi*—estarse entera una cosa" ("to be an entire thing"; Torres Rubio 1754:93, verso). Examples from the anonymous author include "*pachakamaj'*—se decía un idolo muy célebre que adoraban los indios. Creador del mundo" (a celebrated idol adored by the Indians); "*pacakuti, pachat'i'jraj'a*—fin del mundo" ("end of the world"; Aguilar 1970:167).

3 I thank Tom Cummins (personal communication 2012) for suggesting that we consider these markings as an expression of "construction time."

REFERENCES CITED

Aguilar, Rafael

1970 *Arte, y vocabulario en la lengua general del peru, llamada, quichua, y en la lengua española.* Universidad Nacional Mayor de San Marcos, Lima.

Alcina Franch, José

1969 Excavaciones en Chinchero (Cuzco): Informe preliminar. *Internationalen Amerikanischenkongresses* 39(XXXVIII/1):421–428.

1970 Excavaciones en Chinchero (Cuzco): Temporadas 1968 y 1969. *Revista española de antropolgia americana* 5:99–122.

1971 El sistema urbanistico de Chinchero. *Revista del museo nacional* 3:124–146.

1972 Arqueología, historia, y antropolgía. In *Homenaje al Professor Carriazo*, vol. 2, pp. 1–18. Facultad de filosofía y letras, Universidad de Sevilla, Seville.

1974a Chinchero: Urbanismo y ruralismo. *Mundo hispánico* 312:18–21.

1974b Chinchero, villa inca. *Archaeologia paris* 66:58–64.

1976 *Arqueología de Chinchero: 1 la arquitectura.* I. G. Magerit, S.A., Madrid.

Alcina Franch, José (editor)

1976 *Chinchero: Misíon científica española en el Perú, 1967–1973.* Catalog de la Exposición de los Trabajos, Dirección General de Relaciones Culturales, Madrid.

Alcina Franch, José, Miguel Rivera, Jesusu Galvan, M. Carmen Garcia Palacios, Mercedes Guines, Balbina Martinez-Caviro, L. J. Ramos, and T. Varela

1976 *Arqueología de Chinchero 2: Cerámica y otros materials.* Ministerio de Asuntos Exteriores, Madrid.

Armitage, John

2000 The Kosovo War Took Place in Orbital Space: Paul Virilio in Conversation. Electronic document, http://www.ctheory.net/articles.aspx?id=133, accessed February 16, 2014.

Aveni, Anthony F. (editor)

1982 *Archaeoastronomy in the New World.* Cambridge University Press, Cambridge.

Baldwin, Thomas

2007 *Reading Merleau-Ponty: On Phenomenology of Perception.* Routledge, London.

Ballesteros Gaibrois, Manuel

1976 Breve historia de Chinchero. In *Chinchero: Misión científica española en el Perú (1967–1973)*, edited by J. A. Franch, pp. 7–8. Catalog de le Exposición de los Trabajos, Dirección General de Relaciones Culturales.

Bauer, Brian S.

1992 Ritual Pathways of the Inca: An Analysis of the Collasuyu Ceques in Cuzco. *Latin American Antiquity* 3(3):183–205.

1998 *The Sacred Landscape of the Inca: The Cusco Ceque System.* University of Texas Press, Austin.

2004 The Mummies of the Royal Inca. In *Ancient Cuzco: Heartland of the Inca*, pp. 159–184. University of Texas Press, Austin.

Bauer, Brian S., and David S. P. Dearborn

1995 *Astronomy and Empire in the Ancient Andes.* University of Texas Press, Austin.

Bauer, Brian S., and Charles Stanish

2001 *Ritual and Pilgrimage in the Ancient Andes.* University of Texas Press, Austin.

Casey, Edward S.

2000 *Remembering: A Phenomenological Study.* 2nd ed. Studies in Continental Thought. Indiana University Press, Bloomington.

Cummins, Tom

1998 Let Me See! Reading Is for Them: Colonial Andean Images and Objects "Como Es Costumbre Tener los Caciques Señores." In *Native Traditions in the Postconquest World*, edited by Elizabeth Hill Boone and Tom Cummins, pp. 91–148. Dumbarton Oaks Research Library and Collection, Washington, D.C.

Cummins, Tom, and Bruce Mannheim

2011 The River around Us, the Stream within Us: The Traces of the Sun and the Inka Kinetics. *Res* 59/60:5–21.

Dean, Carolyn

2006 The Trouble with (the Term) Art. *Art Journal* 65(2):24–32.

2007 The Inka Married the Earth: Integrated Outcrops and the Making of Place. *Art Bulletin* 89(3):502–518.

2010a The After-Life of Inka Rulers: Andean Death Before and After Spanish Colonization. *Hispanic Issues On Line* 7(Fall):27–54. Electronic document, http://hispanicissues.umn.edu/assets/doc/03_DEAN.pdf, accessed October 30, 2014.

2010b *A Culture of Stone: Inka Perspectives on Rock.* Duke University Press, Durham, N.C.

Dearborn, David S. P., and Katharina J. Schreiber

1986 Here Comes the Sun: The Cuzco-Machu Picchu Connection. *Archaeoastronomy* 9(1–4):15–37.

Dearborn, David S. P., and Raymond. E. White (editors)

1982 *Archaeoastronomy at Machu Picchu.* Annals of the New York Academy of Sciences 385. New York Academy of Sciences, New York.

1983 The "Torreon" of Machu Picchu as an Observatory. *Journal for the History of Astronomy,* Archaeoastronomy supplement, 14(5):s37–s49.

Gasparini, Graziano, and Luise Margolies

1977 *Arquitectura inka.* Centro de Investigaciones Históricas y Estéticas, Facultad de Arquitectura y Urbanismo, Universidad Central de Venezuela, Caracas.

Gonzalez Holguin, Diego

1989 [1608] *Vocabvlario de la legva general de todo el Perv llamada legua qquichua o del Inca.* Facsimile edition of 1952. Universidad Nacional Mayor de San Marco, Editorial de la Universidad, Lima.

Harrison, Regina

1989 *Signs, Songs, and Memory in the Andes: Translating Quechua Language and Culture.* University of Texas Press, Austin.

Husserl, Edmund

1962 *Ideas: General Introduction to Pure Phenomenology.* Collier Books, New York.

Hyslop, John

1990 *Inka Settlement Planning.* University of Texas Press, Austin.

Isbell, William Harris

1997 *Mummies and Mortuary Monuments: A Postprocessual Prehistory of Central Andean Social Organization.* University of Texas Press, Austin.

Julien, Catherine

1990 La metafora de la montana. *Humbolt* 100:84–89.

Lee, Vincent R.

1988 *The Lost Half of Inca Architecture.* Sixpax Manco, Wilson, Wyo.

1992 Reconstructing the Great Hall at Inkallacta. Paper presented at the 32nd Annual Meeting of the Institute of Andean Studies, Berkeley.

Merleau-Ponty, Maurice

1962 *Phenomenology of Perception*. International Library of Philosophy and Scientific Method. Humanities Press, London, and Routledge and K. Paul, New York.

Nair, Stella

2003 Of Remembrance and Forgetting: The Architecture of Chinchero, Peru, from Thupa 'Inka to the Spanish Occupation. PhD dissertation, Department of Architecture, University of California, Berkeley.

2009 Inca Architecture and the Conquest of the Countryside. In *Architecture: Design Methods; Inca Structures*, edited by J. Dehlinger and H. Dehlinger, pp. 114–125. Kassel University Press, Kassel.

n.d. *At Home with the Sapa Inca: Architecture, Space, and Legacy at Chinchero*. University of Texas Press, Austin, in press.

Niles, Susan A.

1980 Civil and Social Engineers: Inca Planning in the Cuzco Region. PhD dissertation, Department of Anthropology, University of California, Berkeley.

1989 Looking for "Lost" Inca Palaces. *Expedition* 30(3):56–64.

1992 Inca Architecture and the Sacred Landscape. In *The Ancient Americas: Art from Sacred Landscapes*, edited by Richard F. Townsend, pp. 346–357. Art Institute of Chicago, Chicago.

1993 La arquitectura incaica y el paisaje sagrado. In *La antigua America: El arte de los parajes sagrados*, edited by Richard F. Townsend, pp. 346–357. Grupo Azabache, Mexico City.

1999 *The Shape of Inca History*. University of Iowa Press, Iowa City.

2004 The Nature of Inca Royal Estates. In *Machu Picchu: Unveiling the Mystery of the Incas*, edited by Richard L. Burger and Lucy C. Salazar, pp. 49–70. Yale University Press, New Haven, Conn.

Ogburn, Dennis E.

2004 Evidence for Long-Distance Transportation of Building Stones in the Inka Empire, from Cuzco, Peru to Saragura, Ecuador. *Latin American Antiquity* 15(4):419–439.

Protzen, Jean-Pierre

1983 Inca Quarrying and Stonecutting. *Ñawpa pacha* 21:183–214.

1985 Inca Quarrying and Stonecutting. *Journal of the Society of Architectural Historians* 44(2):161–182.

1986 Inca Stonemasonry. *Scientific American* 254(2):94–105.

1993 *Inca Architecture and Construction at Ollantaytambo*. Oxford University Press, New York.

2000 Inca Architecture. In *The Inca World: The Development of Pre-Columbian Peru, A.D. 1000–1534*, edited by Cecilia Bákula, Laura Laurencich Minelli, and Mireille Vautier, pp. 193–217. University of Oklahoma Press, Norman.

Reinhard, Johan

2007 *Machu Picchu: Exploring an Ancient Sacred Center*. World Heritage and Monument Series. Cotsen Institute of Archaeology, University of California, Los Angeles.

Ricoeur, Paul, Pol Vandevelde, and Edmund Husserl

1996 *A Key to Husserl's Ideas I*. Marquette Studies in Philosophy. Marquette University Press, Milwaukee, Wisc.

Rostworowski de Diez Canseco, Maria

1953 *Pachacutec Inca Yupanqui*. IEP, Lima.

1992 *Historia del Tahuantinsuyu*. 3rd ed. Perugraph Editores, Lima.

Sallnow, Michael J.

1987 *Pilgrims of the Andes: Regional Cults in Cusco*. Smithsonian Institution Press, Washington, D.C.

Salomon, Frank

1995 The Beautiful Grandparents: Andean Ancestor Shrines and Mortuary Ritual as Seen through Colonial Records. In *Tombs for the Living: Andean Mortuary Practices*, edited by Tom D. Dillehay, pp. 315–353. Dumbarton Oaks Research Library and Collection, Washington, D.C.

Salomon, Frank, Jorge Urioste, and Francisco de Avila

1991 *The Huarochirí Manuscript: A Testament of Ancient and Colonial Andean Religion.* University of Texas Press, Austin.

Scott, Heidi V.

2009 *Contested Territory: Mapping Peru in the Sixteenth and Seventeenth Centuries.* University of Notre Dame Press, Notre Dame, Ind.

Seamon, David

1993 *Dwelling, Seeing, and Designing: Toward a Phenomenological Ecology.* State University of New York Press, Albany.

Seamon, David, and Robert Mugerauer

1985 *Dwelling, Place, and Environment: Towards a Phenomenology of Person and World.* Kluwer Academic, Dordrecht and Boston.

Tibbetts, Paul

1969 *Perception; Selected Readings in Science and Phenomenology.* Quadrangle Books, Chicago.

Torres Rubio, Diego de

1754 *Arte, y vocabulario de la lengua quicua general de los indios de el Perú.* Imprenta de la Plazuela de San Chriftoval, Lima.

Tuan, Yi-fu

1990 *Topophilia: A Study of Environmental Perception, Attitudes, and Values.* Columbia University Press, New York.

Urton, Gary

1981 *At Crossroads of the Earth and the Sky.* University of Texas Press, Austin.

Virilio, Paul

1986 *Speed and Politics: An Essay on Dromology.* Columbia University, New York.

1994 *Bunker Archaeology.* Translated by George Collins. Princeton Architectural Press, New York.

2002 *Desert Screen: War at the Speed of Light.* Athlone Contemporary European Thinkers. Continuum, London and New York.

Wright, Kenneth R., Jonathan M. Kelly, and Alfredo Valencia Zegarra

1997 Machu Picchu: Ancient Hydraulic Engineering. *Journal of Hydraulic Engineering* 123(10):838–843.

Wright, Kenneth R., Gordon Francis McEwan, and Ruth M. Wright

2006 *Tipon: Water Engineering Masterpiece of the Inca Empire.* American Society of Civil Engineers (ASCE) Press, Reston, Va.

Wright, Kenneth R., and Alfredo Valencia Zegarra

2000 *Machu Picchu: A Civil Engineering Marvel.* American Society of Civil Engineers (ASCE) Press, Reston, Va.

Wright, Kenneth R., Gary D. Witt, and Alfredo Valencia Zegarra

1997 Hydrogeology and Paleohydrology of Ancient Machu Picchu. *Ground Water* 35(4):660–666.

Wright, Kenneth R., Ruth M. Wright, M. E. Jensen, and Alfredo Valencia Zegarra

1997 Machu Picchu: Ancient Agricultural Potential. *Applied Engineering in Agriculture* 13(1):39–47.

Zuidema, R. Tom

1964 *The Ceque System of Cuzco: The Social Organization of the Empire of the Inca.* E. J. Brill, Leiden.

8

Time and the Other

The Early Colonial Mythohistorical Landscapes of the Huarochirí Manuscript

JALH DULANTO

THIS ESSAY IS A FIRST APPROACH TO THE study of the early colonial temporalities and historicities[1] of the Huarochirí people—that is, to the study of the experiences and understandings, constructions and representations, of time and history by the people who inhabited the Huarochirí region[2] of the western slopes of the Peruvian central highlands immediately after the Spanish conquest of the central Andes (Figure 8.1). The particular case I analyze here is the famous Huarochirí Manuscript (Figure 8.2),[3] or, to be more precise, the narrative of the first few chapters of the document, which record the successive actions of several ancestral heroes, beginning in a more "mythical" pre-Hispanic past and ending in a more "historical" colonial present.

In this essay, I show that a syntagmatic-paradigmatic analysis of the seemingly disparate events of the ancestral heroes mentioned in the Huarochirí Manuscript allows us to discover some basic spatial, temporal, and social principles by which the huarochiranos of the sixteenth and seventeenth centuries constructed and represented myth and history and, thus, experienced and understood the "situation of contact" in which they were inscribed. In particular, I show that the actions of these ancestral heroes, and especially their movement on the landscape, account, through a complex system of successive dual divisions, alternations, and reversals, for the ordering of places (space), events (time), and relations (society) of the indigenous huarochiranos and their definition vis-à-vis the places, events, and relations of the nonindigenous Spanish.

The Huarochirí Manuscript: An Early Colonial Testament of Andean Religion

As Frank Salomon (1991:1) has correctly pointed out, the Huarochirí Manuscript, written in Lengua General Quechua[4] around 1608 CE,[5] is the single most important early colonial indigenous document on Andean religion that is written in a language of

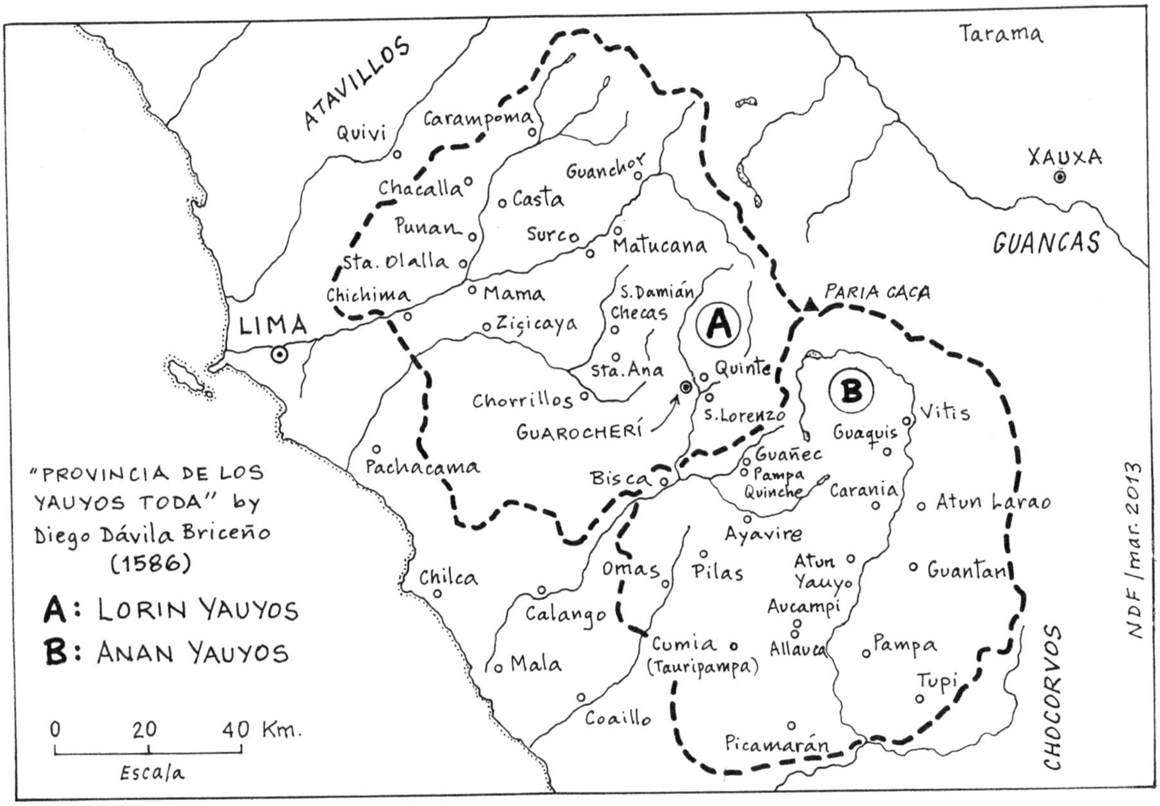

figure 8.1
"Provincia de los Yauyos Toda" according to Dávila Brizeño (1965 [1586]). (Drawing by Nicanor Dominguez.)

the Andes. Often compared with the Popol Vuh of the Maya (e.g., Millones 2008:147; Millones and Tomoeda 2007:XIII; Úzquiza Gonzalez 2011), the Huarochirí Manuscript records the myths and rituals being narrated and celebrated by the Huarochirí people during the last years of the sixteenth century and the first years of the seventeenth century. Although some authors consider the manuscript to be a rather disorganized collection of stories and descriptions (e.g., Millones 2008:147; Millones y Tomoeda 2007:XIII), the document (unfinished and in the process of being edited as is) clearly has a totalizing ambition (Salomon 1991:2). Here, I argue that in some ways, and to some extent, the Huarochirí Manuscript can be treated as a unified mythohistorical account of the situation of contact in which the Huarochirí people were inscribed—that is, as a mythical and historical record of a pre-Hispanic past (remote and recent), a colonial present, and, in a few instances, even a future, as these were being experienced and understood, constructed and represented, by the Huarochirí people at the beginning of the seventeenth century. Treating the Huarochirí Manuscript as a unified account is an essential step toward exploring the interconnections between narrativity, temporality, and historicity, and exploring these interconnections is an essential step toward understanding the Huarochirí Manuscript as an account of the "colonial situation" in which the Huarochirí people were in practice defining themselves vis-à-vis other Andean peoples and the Spanish.

The process of composition of the Huarochirí Manuscript is probably the best expression of the colonial situation in which the document was written. The manuscript brings together at least three distinct types of actors: first, several mostly anonymous local indigenous informants from different settlements in the Huarochirí region; second, at least one anonymous, local indigenous editor,

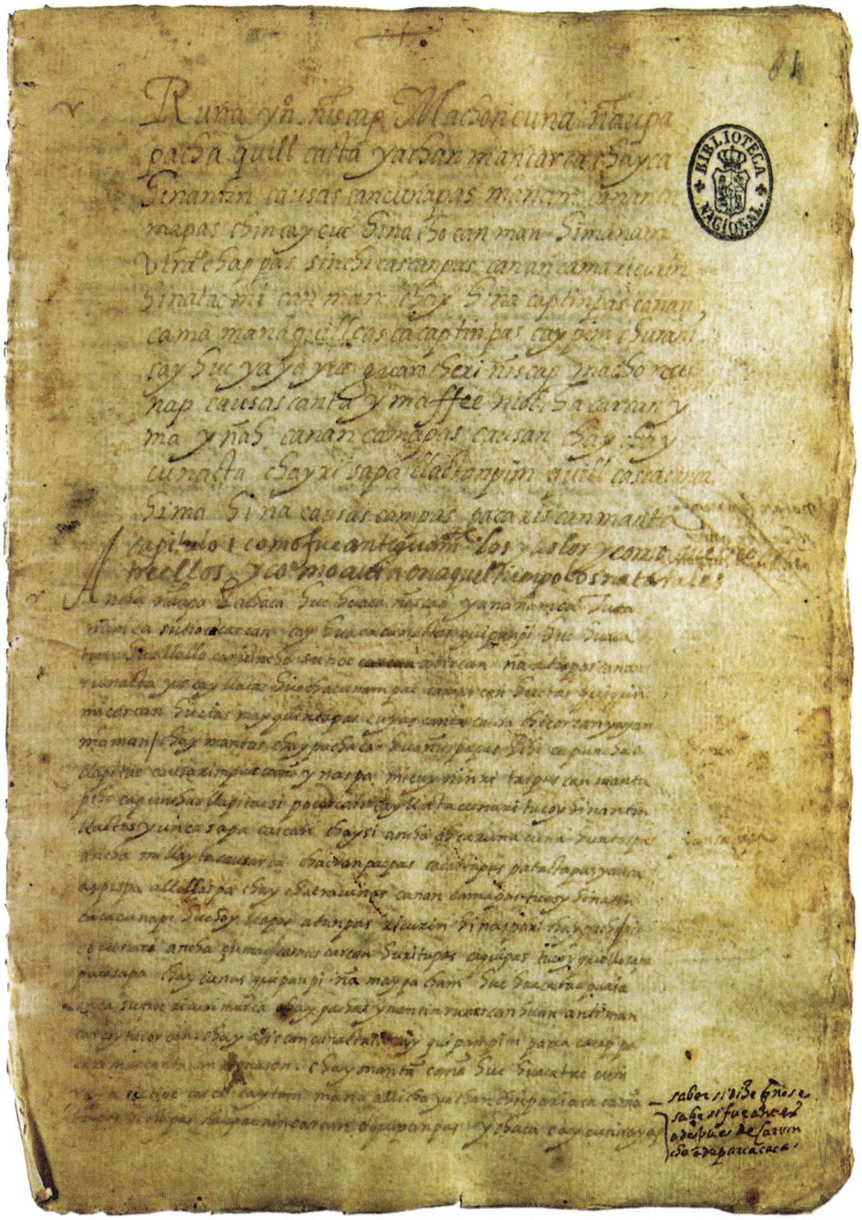

figure 8.2
First page of the Huarochirí Manuscript (Manuscrito 3169) in the Biblioteca Nacional, Madrid (Arguedas and Uzquiza González 2011).

very likely a converted Indian nobleman from one of these settlements (San Damián de Checa); and third, a well-known nonlocal and nonindigenous supervisor—the priest of San Damián, Father Francisco de Ávila (see Figures 8.1 and 8.3).

Indeed, we know that the document was commissioned by Ávila and the work was given to at least one of his assistants (the converted Indian nobleman from San Damián de Checa), who seems to have been in charge of collecting, compiling, and organizing the information provided by several informants (who were being questioned, often forcefully, by Ávila and others about their non-Christian beliefs and practices).

Thus, it is not surprising that the Huarochirí Manuscript shows strong similarities with the Bible. These similarities are of form as well as content.[6] As Salomon (1991:2–3) has pointed out, however, it would be incorrect to explain these resemblances simply as the result of imposition or acculturation. At least three not mutually exclusive factors should be taken into account to explain these similarities:

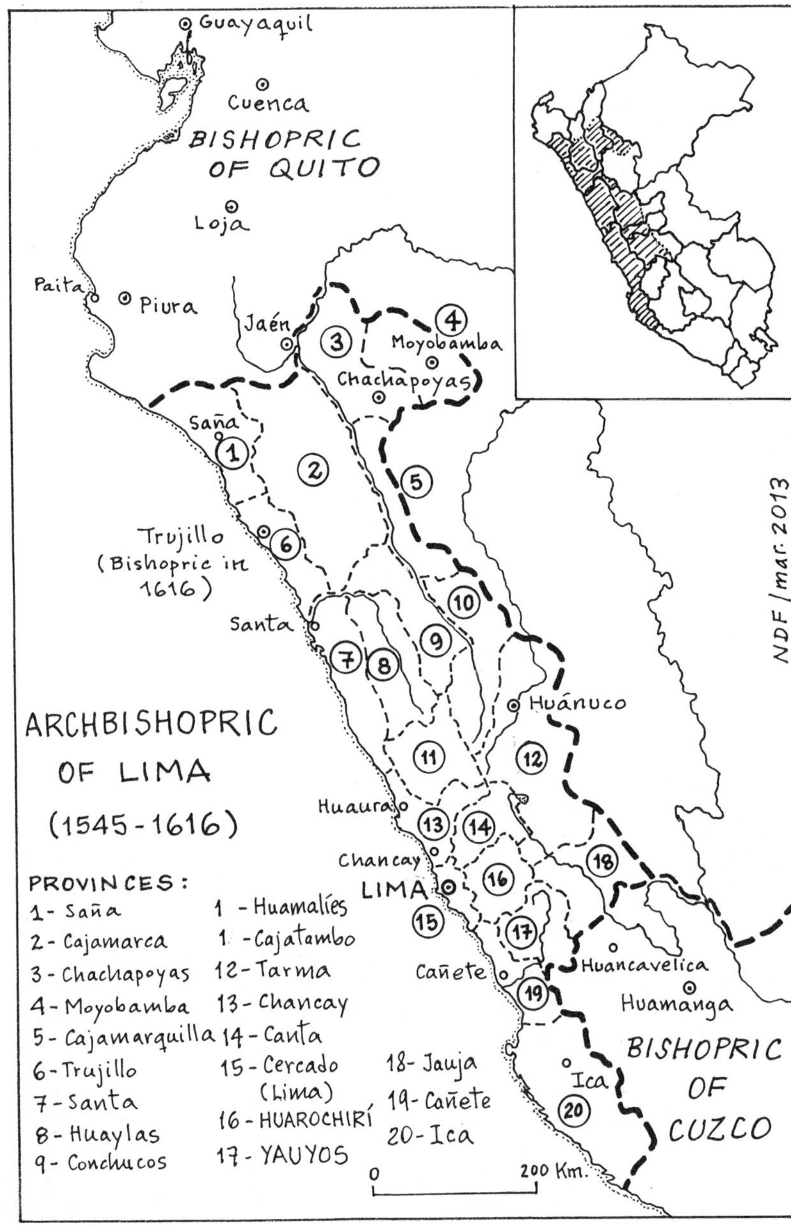

figure 8.3
Archbishopric of Lima (1545–1616). (Drawing by Nicanor Dominguez.)

(1) the imposition of a foreign European mode of historical and temporal consciousness by Ávila, the supervisor of the whole project; (2) the reconceptualization of local Andean mythical and ritual traditions by the informants, who had already been exposed for several decades to Christianity; and (3) the existence of certain commonalities between what Terence Turner (1988:249) calls Andean and Hebraic modes of historical and temporal consciousness resulting from very general and basic similarities in the particular histories of societies that have been repeatedly exposed to conquest and domination by larger foreign expansive states. Indeed, we could argue that this last factor is probably behind the eagerness with which the peoples of the central Andes have been appropriating and reconceptualizing, again and again for almost five hundred years, apocalyptic, millenarian, and messianic visions from Old World religious traditions.

In any case, whatever the causes of these similarities, the Huarochirí Manuscript also shows important differences from the accounts of Old World religious traditions (Salomon 1991:4). These differences are particularly evident not only in the syntagmatic structure of the individual stories compiled in the manuscript—that is, in the way events are articulated within these stories—but also, as I argue here, in the syntagmatic structure of the compilation itself, in the way these seemingly disparate stories are articulated within the manuscript as a whole.

The Syntagmatic-Paradigmatic Approach to the Analysis of Narratives

Traditional approaches to the study of narratives in the central Andes, especially those by archaeologists, usually ignore their colonial status. If not in theory, at least in practice, they treat these narratives as relics from a pre-Hispanic past surviving, though corrupted, in a colonial present. In part, because of this mistaken assumption, they ignore the syntagmatic axis of the narrative and focus only on the paradigmatic one. Typical examples of these traditional approaches are a number of studies of "mythical cycles" that work under the assumption that the actual narratives they work with are the partial, and largely contaminated, pieces of a complete, pure, archetypical, essential, and arguably official myth.[7]

The problem with these studies is that in order to reconstruct these "mythical cycles," they usually extract "mythical events" from different narratives, sometimes without even paying attention to the historical contexts in which these narratives were produced. Insofar as the "mythical hero," often identified mainly by name, is the same, the "mythical events" in which he/she participates are compared and the archetypical "mythical cycle" is reconstructed. What these traditional approaches overlook is, of course, not only the complexity of power and identity that is at the heart of these narratives but also the social realities in which they were told. What they accomplish is to kill the storyteller.

The syntagmatic-paradigmatic approach that I follow in this paper is different. Inspired by Turner's (1977, 1985) critique and reformulation of the structuralist analysis of myth,[8] I focus, on the one hand, on the integration of seemingly disparate events—events that are regarded as parts of the "mythical cycles" of different "ancestral heroes"—and, on the other hand, on those events that appear together within the same narrative context. Moreover, I compare these events in the order in which they are introduced into the narrative. Instead of assuming the existence of an ideal context, such as a "mythical cycle," I focus on the actual narrative context in which the events I analyze happen to be found together—in this case, the Huarochirí Manuscript. I show that by following this approach, we are able to understand not only why seemingly disparate events appear together within the same narrative context, but also why they show the particular characteristics they show when they are found together.

The Analysis

The following analysis is based on what I consider to be the two best translations and editions of the Huarochirí Manuscript available today: Frank Salomon and George Urioste's (1991) English translation and Gerard Taylor's (1999) Spanish translation (for other translations and editions see Adelaar 1988; Arguedas and Duviols 2012 [1966]; Galante 1942; Szeminski 1985; Taylor 1980, 1987b, 2008; Trimborn 1939; Urioste 1983; and Úzquiza Gonzalez 2011, among others). In this presentation, I mainly paraphrase and summarize the English translation, trying to keep, as far as possible for the sake of accuracy, the exact words used by the translator. I use the Spanish translation only to clarify a few discrepancies. All references to the original text are made using the section numbers of the English translation.

In order to facilitate the analysis, I first present a summary of the progressive segmentation and differentiation of space, time, and society that results from the successive actions of several ancestral heroes—Cuni Raya, Huatya Curi, and Paria Caca—and after that I show how it is that the specific actions of these ancestral heroes account for such segmentation and differentiation. Finally, I deal with the actions of two more heroes—Maca

table 8.1
Spatial and temporal relations defined in the mythical cycles of Cuni Raya Vira Cocha, Huatya Curi, and Paria Caca.

MYTHICAL CYCLE	SPACE	TIME
Cuni Raya Vira Cocha	Vertical segmentation of LAND and WATER	Separation of DAYS and NIGHTS
Huatya Curi	Vertical segmentation of HIGHLANDS and LOWLANDS	Separation of DRY SEASON and RAINY SEASON
Paria Caca	Horizontal differentiation of space along CARDINAL DIRECTIONS (NORTH–SOUTH and EAST–WEST)	Segmentation along AGRICULTURAL CYCLES and ANCESTOR CYCLES

table 8.2
Social relations defined in the mythical cycles of Cuni Raya Vira Cocha, Huatya Curi, and Paria Caca.

MYTHICAL CYCLE	SOCIAL RELATIONS
Cuni Raya Vira Cocha	Unsuccessful, unsettled relations between groups located on the LAND (center) and those coming from the WATER (periphery)
Huatya Curi	Successful, settled relations between groups moving from the LOWLANDS (periphery) and groups remaining in the HIGHLANDS (center)
Paria Caca	Alliance and cooperation between highland groups against lowland ones

Huisa and Cuni Raya again—who are involved in events linked to the Inca and Spanish conquests of the Huarochirí region.

In the first episode, the episode of Cuni Raya, space is segmented into land and water; concurrently, time is segmented into day and night. In the second episode, the episode of Huatya Curi, space is segmented by a basic Andean division of the landscape into highlands and lowlands. Simultaneously, time is segmented into dry and rainy seasons. Finally, in the third episode, the episode of Paria Caca, space is even further extended and segmented along cardinal directions, while time is further segmented through the introduction of a series of calendrical rituals that are associated with agricultural cycles (Table 8.1).

Each of these spatiotemporal frameworks defines a specific type of intergroup relationship. In the episode of Cuni Raya, intergroup relations are defined in terms of an unsuccessful, unsettled relationship between those groups located on the land, the center, and those who come from the water, the periphery. In the episode of Huatya Curi, intergroup relations are defined in terms of an initial confrontation between those located in the highlands, which is now the center, and those who come from the lowlands, which is now the periphery. This confrontation, however, eventually turns into a successful, exogamic relation. Finally, in the episode of Paria Caca, intergroup relations are defined in terms of political alliances and cooperation between highland groups and confrontation between these highland groups and the lowland ones (Table 8.2).

The spatiotemporal frameworks and the types of intergroup relationships they define are revealed in the actions of our first three mythical ancestral heroes.

Initial Situation: Primordial Darkness and Undifferentiated Unity

The beginning of the story is clearly set up in a space-time of primordial darkness that the Huarochirí Manuscript describes as the space-time of Yana Ñamca and Tuta Ñamca (HM 3). *Yana* means "black," and *Tuta* means "night" in Quechua (Salomon and Urioste 1991:43n17; Taylor 1995), which makes mentioning these two classes of deities a sort of general reference to an undifferentiated space-time of primordial unity. It is in this undifferentiated space-time of primordial unity that Cuni Raya Vira Cocha appears for the first time (HM 7–8).[9]

First Episode: Cuni Raya Vira Cocha

In the first episode, Cuni Raya is presented as a male trickster who goes around posing as a poor and friendless man while performing all kinds of wonders, particularly making villages and agricultural terraces just by speaking, forming irrigation canals just by tossing down the flower of a reed, and putting some of the local huacas[10] to shame (HM 9).

Briefly, the story tells us that a local female named Caui Llaca, who is a virgin (HM 10), is weaving under a *lúcuma* tree (*Pouteria lucuma*). Cuni Raya turns himself into a bird and climbs into the tree. He puts his semen into a fruit and drops it next to the woman (HM 11), who swallows it down delightedly. In this way, Cuni Raya manages to secretly impregnate her. After nine months, she gives birth to a son and nurses him for a year, wondering, "Whose child could this be?" (HM 12). At the end of that year, when her son is crawling around on all fours, she summons all the huacas in order to identify the child's father. When the day of the contest comes, all the huacas of the region attend in their best clothes (HM 13). The gathering takes place in Anchi Cocha. The huacas sit down along a bench. Then, Caui Llaca asks all the males who made her pregnant, but nobody answers (HM 14). Cuni Raya, who dresses like a beggar, is seated at the end of the bench. Caui Llaca tells her son to identify his father himself (HM 15). The child crawls along the bench on all fours, from one end to the other, without climbing up on anyone until he reaches the other end, where Cuni Raya is seated. When Caui Llaca's son identifies his father by climbing onto his father's knees, she is indignant at having given birth to the child of a beggar and flees with her child from Anchi Cocha in the highlands to the ocean, in the west (HM 16). At that moment, Cuni Raya rises dressed in gold garments and starts to chase her (HM 17).[11] Caui Llaca heads straight out into the ocean, where she and her son turn themselves into two stones right in front of Pacha Camac (HM 18; see Duviols 1976). In his chase, Cuni Raya meets several animals—a condor (HM 19), a skunk (HM 20), a puma (HM 21), a fox (HM 22),[12] a falcon (HM 23), and a group of parakeets[13]—conferring a good fortune on those who give him good news and cursing those who give him bad news (HM 24). When Cuni Raya reaches the seashore, he enters the ocean and goes to Spain. In the old days, people said he went to another land (HM 25—crossed out).[14]

After reaching the seashore, Cuni Raya, angry, turns back toward Pacha Camac, where Pacha Camac's two daughters live guarded by a snake. While Urpuy Huachac, the mother of the two girls, is away visiting Caui Llaca in the ocean, Cuni Raya rapes the oldest daughter. The youngest daughter escapes by turning herself into a dove and flying away (HM 25). He then destroys a pond where Urpuy Huachac breeds fish, releasing in this way the fish into the ocean. That is the origin of fish in the sea (HM 26). When Urpuy Huachac returns, she gets angry and chases Cuni Raya along the coast (HM 27). When she reaches him with the excuse of picking his lice, she causes a huge abyss to open up next to him, with the intention of knocking Cuni Raya down into it. But Cuni Raya escapes with the excuse of going to relieve himself (HM 28).[15]

These two stories about Cuni Raya—the first one about Cuni Raya and Caui Llaca and the second one about Cuni Raya and Urpuy Huachac—mirror each other in several ways. In the first story, a foreign male (Cuni Raya) descends from the outside (the sky, up/above) to the inside (the land). He seduces a local female (Caui Llaca). The local female escapes, moving from the inside (the land) to the outside (the ocean, down/below). The foreign male (Cuni Raya) chases her, moving vertically from the center (Anchi Cocha) to the periphery (Pacha Camac). The local female goes outside (the ocean, down/below) and turns into stone. She is fixed in the border between the inside (the land) and the outside (the ocean, down/below). In the second story, a foreign male (Cuni Raya) ascends from the outside (the ocean, down/below) to the inside (the land). He rapes a local female (Pacha Camac's oldest daughter). Another local female (Pacha Camac's youngest daughter) turns into a dove and escapes from the inside (the land) to the outside (the sky, up/above). The foreign male (Cuni Raya) does not chase but is chased by a local female (Urpuy Huachac, the mother of the two sisters seduced by Cuni Raya), moving horizontally along the periphery (along the coast, from Pacha Camac to the north or to the south). The foreign male (Cuni Raya) remains in the inside (the land) and does not turn into stone. He is not fixed and moves freely between the inside (the land) and the outside (the ocean, down/below). A first differentiation of the world is thus accomplished on the basis of a dual division between the inside (the land) and the outside (the sky, up/above, and the ocean, down/below). Local females, seduced or raped by Cuni Raya, move from the inside to the outside, from "this world" to the worlds of "up/above" and "down/below" and are, in a sense, fixed to mark the new limits. At the same time, birds and fish are moved from the land to the sky and to the ocean, respectively. The foreign male Cuni Raya, in contrast, remains unfixed and moves freely between the inside (the land) and the outside (the sky, up/above, and the ocean, down/below; Table 8.3 and Figure 8.4).

Moreover, Cuni Raya's vertical movement—first descending from the sky (outside, up/above) to the land (inside, here), impregnating Caui Llaca, and later moving from the land (inside, here) to the ocean (outside, down/below) while chasing Caui Llaca, who finally turns into stone—defines a vertical axis that goes from up/above to here to down/below, connecting in a sort of primitive hydraulic system the waters of the sky, land, and ocean.[16] It also defines a horizontal extension of the land (inside, here) from Anchi Cocha (in the center) to Pacha Camac (in the periphery). Likewise, Cuni Raya's ascension in the highlands and his subsequent movement from the center in the east to the periphery in the west marks at the same time the first appearance of the sun, illuminating the world and separating in this way days and nights. The blessing and cursing of animals by Cuni Raya during his movement from the center to the west—in a sort of double alternation from black/white animals (condor, skunk) to red (?) animals (puma, fox) and from diurnal animals (condor, puma) to nocturnal animals (skunk, fox)—further emphasizes this temporal separation.

The spatiotemporal framework resulting from these actions, however, is still quite unstable. Cuni Raya is a *celeritas* foreign male from the outside who never succeeds in establishing a relationship with any of the *gravitas* local females he encounters on the inside. In the first story, Cuni Raya comes from the "world of up/above" and secretly impregnates a local female in Anchi Cocha, in the center of "this world." The female rejects him and escapes. He chases her. The female turns into stone. In the second story, Cuni Raya comes from the "world of down/below" and rapes another local woman in Pacha Camac, in the periphery of "this world." Another female rejects him and escapes. He is chased by another woman who wants to expel him and turn him into stone. He escapes. In other words, he rapes a woman in the center, he rapes a woman in the periphery, he marries none. The basic relations introduced so far in the narrative are clearly of opposition without complementarity between: a celeritas foreign male side identified with the "worlds of up/above and down/below" and a gravitas local female side identified with "this world."

In the two chapters (chapters three and four) that immediately follow the chapter of Cuni Raya (chapter two), the world is described as unstable. It suffers sudden floods (HM 29–34) and eclipses (HM 35) that result in the obliteration and inversion of the limits established to this point and threaten to return the world to the initial situation of primordial darkness in which the beginning of the story is set up. The border established between the ocean and the land is almost obliterated by a flood. The border between days and nights is almost obliterated by an eclipse. Cuni Raya, however, still travels freely between the land and the sea, between day and night, in other words, between "this world" (inside) and the "other world" (outside).

table 8.3
Main events in the cycle of Cuni Raya Vira Cocha.

	EPISODE 1, STORY 1: CUNI RAYA AND CAUI LLACA	EPISODE 1, STORY 2: CUNI RAYA AND URPUY HUACHAC
1	A foreign male (CRVC) descends from the outside (the sky, up/above) to the inside (the land)	A foreign male (CRVC) ascends from the outside (the ocean, down/below) to the inside (the land)
2	The foreign male (CRVC) seduces a local female (CLL)	The foreign male (CRVC) seduces a local female (PCOD)
3	The local female escapes from the inside (the land), to the outside (the ocean, down/below)	Another local female (PCYD) escapes from the inside (the land), to the outside (the sky, up/above)
4	The foreign male (CRVC) chases the local female (CLL) moving vertically from the center (Anchi Cocha) to the periphery (Pacha Camac)	The foreign male (CRVC) is chased by a local female (UH) moving horizontally along the periphery (along the coast, from Pacha Camac to the north or to the south)
5	The local female (CLL) goes outside (the ocean, down/below) and turns into stone. She is fixed in the border between the inside (the land) and the outside (the ocean, down/below)	The foreign male (CRVC) remains in the inside (the land) and does not turn into stone. He is not fixed and moves freely between the inside (the land) and the outside (the ocean, down/below)

CRVC = Cuni Raya Vira Cocha; CLL = Caui Llaca; PCOD = Pacha Camac's Oldest Daughter; PCYD = Pacha Camac's Youngest Daughter

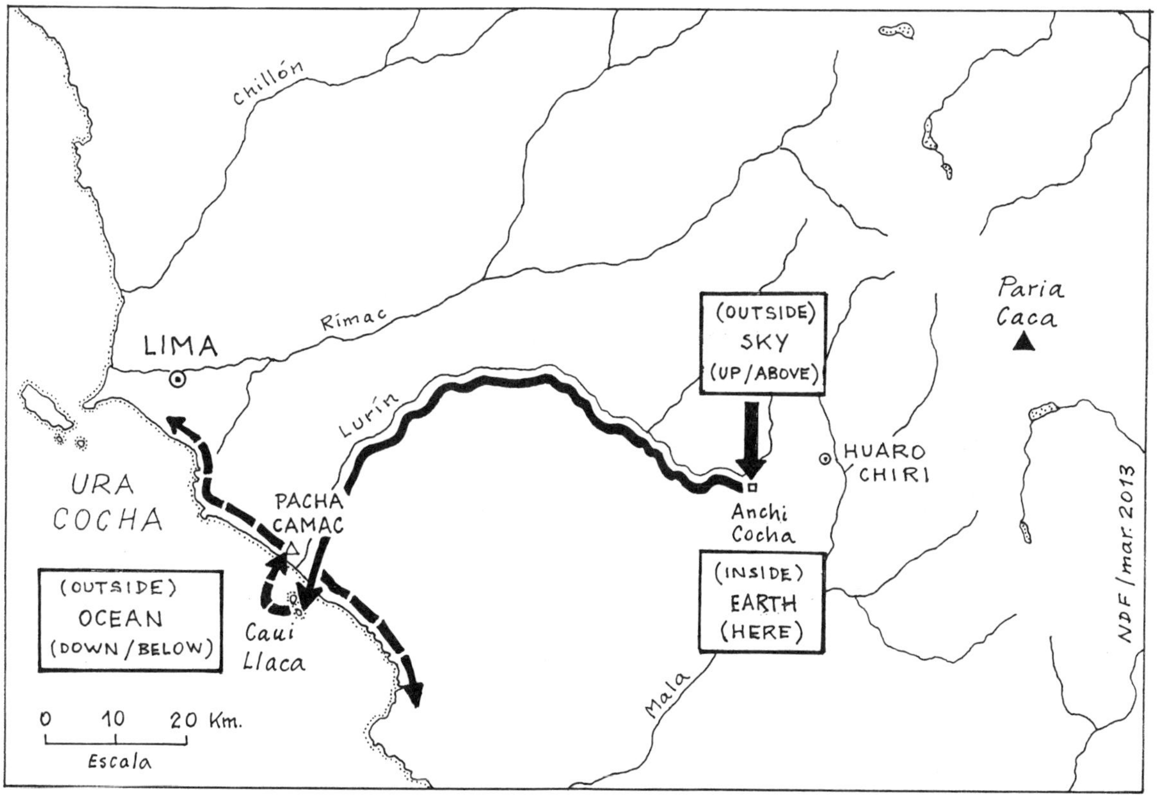

figure 8.4
The mythical cycle of Cuni Raya. (Drawing by Nicanor Dominguez.)

Second Episode: Huatya Curi

In the second episode, Huatya Curi is also described as a male trickster who dresses like a poor and friendless man and goes around performing all kinds of wonders (HM 38–39). Briefly, the story tells us that in contrast to Cuni Raya, Huatya Curi moves from Pacha Camac, in the lowlands, to Anchi Cocha, in the highlands. Midway, he stops to sleep in a mountain called Latab Zaco, and while he sleeps, two foxes, one that is coming from down below and one that is coming from up above, meet face to face (HM 42). The one from up above reports that a lord from Anchi Cocha is ill and is dying because of a woman, the lord's wife, who committed adultery (HM 43–44). The one from down below reports that a lady from Pacha Camac, a lord's daughter, has almost died because of a man who raped her (HM 45).[17] Once in the highlands, Huatya Curi meets the local lord named Tamta Ñamca. Huatya Curi offers to cure him if, in exchange, the lord allows him to marry his youngest daughter (HM 49). Tamta Ñamca accepts, and Huatya Curi cures him by revealing that Tamta Ñamca's wife has committed adultery and that as a result two snakes dwell on his house roof (made of beautiful feathers) and a two-headed toad dwells under his grinding stone (HM 50).[18] He dismantles Tamta Ñamca's beautiful house, kills the two snakes, and expels the two-headed toad (HM 53–54). Once cured, Tamta Ñamca gives Huatya Curi his youngest daughter (HM 56). When Huatya Curi cures Tamta Ñamca, however, he makes evident that Tamta Ñamca's power is illegitimate. Huatya Curi's brother-in-law (the one married to Tamta Ñamca's oldest daughter) challenges Huatya Curi to a series of contests. Huatya Curi goes to the mountain Condor Coto to ask Paria Caca for help. Paria Caca, who has not been born yet, dwells in that mountain in the form of five eggs (HM 57). Aided by Paria Caca, Huatya Curi wins all the contests, and in each of them he is awarded sacred objects that (according to later episodes of the manuscript) are clearly pieces of paraphernalia used in calendrical rituals that mark the transitions between dry and rainy seasons. They compete in drinking and dancing (HM 58–62), wearing splendid costumes (HM 63), putting on puma skins (HM 64), building (HM 65) and roofing a house (HM 66), and dancing with blue tunics and white breechcloths (HM 67).[19] During the last contest, Huatya Curi charges down on his brother-in-law from the outside, screaming. The man panics, turns himself into a deer, and flees. Tamta Ñamca's oldest daughter follows him (HM 68), and Huatya Curi chases both. Midway to the lowlands, he catches the woman, stands her upside down on her head, with her vagina exposed, and right then and there she turns herself into stone (HM 69). At that moment, in Condor Coto, Paria Caca is born. He flies from the five eggs in the shape of five falcons. These five falcons turn into five humans and begin to roam around (HM 72).

Like Cuni Raya of the first episode, Huatya Curi is also a foreign male who moves along a vertical dimension. In contrast to Cuni Raya, however, who moves from up/above to down/below, from "this world" to the "world of down/below," Huatya Curi moves from down/below to up/above, from Pacha Camac to Anchi Cocha. As in the case of Cuni Raya's actions, Huatya Curi's actions also lead to a local female being turned into a stone to mark a spatial separation (Table 8.4 and Figure 8.5). In contrast to Cuni Raya, however, whose actions lead to Caui Llaca turning into stone to mark the separation between this world and the underworld, Huatya Curi's actions lead to Tamta Ñamca's older daughter turning into stone, marking the separation between highlands and lowlands. Moreover, in contrast to Cuni Raya, whose actions introduce the division of time into years (the introduction of the solar cycle), Huatya Curi's actions result in the division of time into dry and rainy seasons. In each of the contests between Huatya Curi and Tamta Ñamca's son-in-law, Huatya Curi uses a different piece of ritual paraphernalia that later will be associated with one of the three main festivities that mark the beginning, the climax, and the end of the dry season—that is, the festivities in honor of Paria Caca, Chaupi Ñamca, and Tutay Quiri (see following). Furthermore, and also in contrast to Cuni Raya's actions, Huatya Curi's actions result in a successful relation with a local female. Huatya Curi

table 8.4

Main events in the cycle of Huatya Curi.

	EPISODE 2, STORY 1: HUATYA CURI AND TAMTA ÑAMCA
1	A foreign male (HC) ascends from the periphery (Pacha Camac) to the center (Anchi Cocha)
2	The foreign male (HC) marries a local female (TÑYD)
3	Another local female (TÑOD) and her husband escape from the center (Anchi Cocha), to the periphery (the lowlands, down/below)
4	The foreign male (HC) chases the local female (TÑOD) moving vertically from the center (Anchi Cocha) to the periphery (Pachacamac)
5	The local female (TÑOD) goes outside (the lowlands, down/below) and midway turns into stone. She is fixed in the border between the inside (the highlands) and the outside (the lowlands, down/below)

HC = Huatya Curi; TÑYD = Tamta Ñamca's Youngest Daughter; TÑOD = Tamta Ñamca's Oldest Daughter

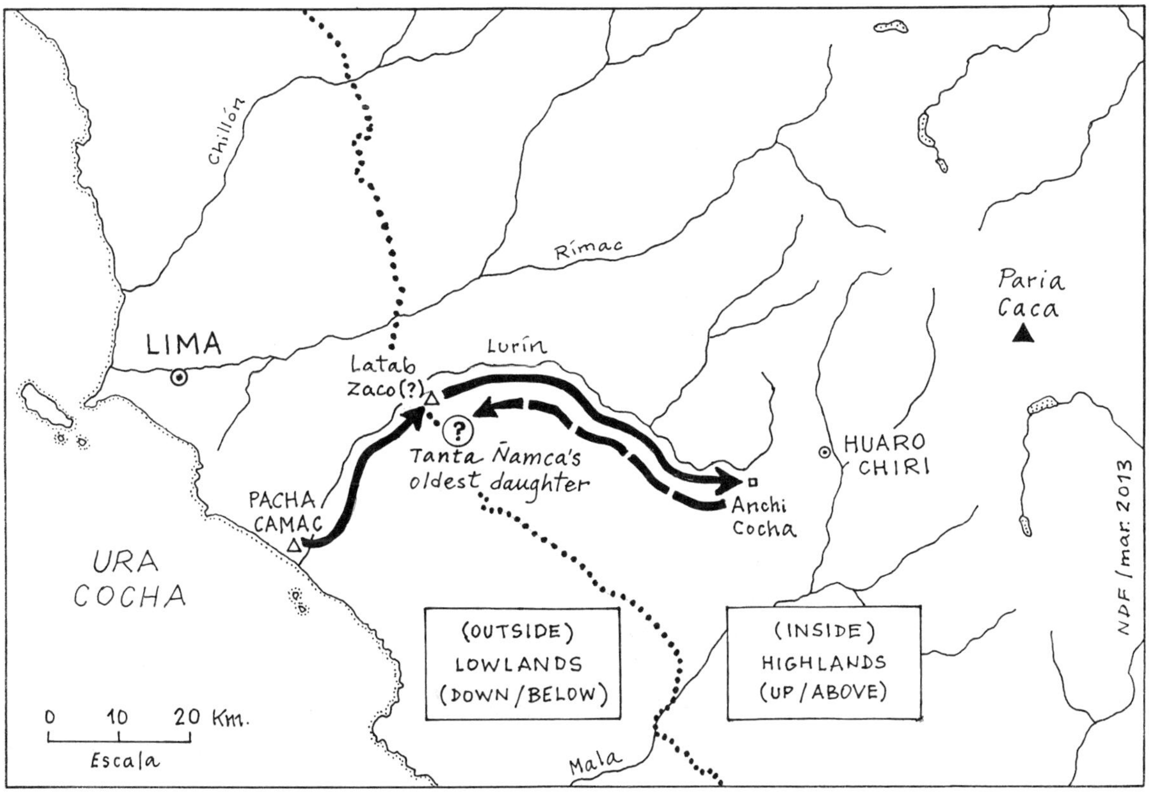

figure 8.5

The mythical cycle of Huatya Curi. (Drawing by Nicanor Dominguez.)

accomplishes what Cuni Raya could not: he marries Tamta Ñamca's youngest daughter.

The basic relationships introduced so far in the narrative are clearly of opposition and complementarity between a *celeritas*, foreign, male younger brother who comes from the outside (lowlands) and occupies the inside (highlands) and a *gravitas*, local, female older sister who goes from the inside (highlands) and is expelled to the outside (lowlands). The inversion of highlands/lowlands, inside/outside, is clear.

Third Episode: Paria Caca

In the third episode, Paria Caca is also described as a trickster who moves around, fooling everyone else. But in contrast to Cuni Raya and Huatya Curi, who move mainly along a vertical dimension, Paria Caca moves mainly around a horizontal dimension. He begins his journey in the center of the highlands.

After being born from five eggs on top of the mountain Condor Coto in the form of five falcons, Paria Caca turns into five humans and begins to roam around. Rising up as rain, he descends on Anchi Cocha and flushes Tamta Ñamca and his people away to the ocean (HM 72). Midway between Anchi Cocha and the ocean (over the Lurin River) was a gigantic tree, known as Pullao, that bridged two mountains, Llantapa and Vichoca, and was inhabited by lowland animals. Paria Caca flushes all these animals away to the ocean as well (HM 73).

Next in the story, Paria Caca appears in Huauqui Usa, a village of lowland people to the south of Anchi Cocha, where an important festival is being celebrated. Paria Caca enters the village and sits down at the end of the banquet (HM 76). After staying there for a whole day and not being invited to drink by anybody, he is finally invited to drink by a woman who offers him a drink in a large white gourd (HM 77). Thanking her, Paria Caca reveals himself and tells the woman that he is going to destroy the village in five days (HM 78). Five days later, he climbs the mountain Matao Coto that overlooks the village (HM 79). Once in the mountain, Paria Caca starts a torrential rainstorm and rising up as yellow and red hail, he washes all the people from Huauqui Usa to the ocean. The mudslide created in this way shapes the high slopes and the valleys of Huarochirí. After that, Paria Caca crosses over (the Mala River) and moves east to the village of the Cupara (HM 80).

In the village of the Cupara, which suffers from a lack of water (HM 81), Paria Caca helps a woman named Chuqui Suso to water her maize plants by widening an irrigation canal that goes from Coca Challa to Cupara (HM 82–87). After widening the canal, Paria Caca and Chuqui Suso go to the mouth of the canal. Midway, they sleep together in a place called Yana Caca (HM 88). When they arrive at Coca Challa, Chuqui Suso turns herself into a stone (HM 89–90). At this point, the calendrical rituals associated with the cleaning of irrigation canals are introduced for the first time in the narrative (HM 91–95).

Next, Paria Caca continues to climb (HM 89) north and east to Mullo Cocha, where a male fire deity named Huallallo Carhuincho lives (HM 96–97). He travels, whirling his hunting bolas (HM 98). Midway, in Ocsa Pata, he runs into a man who is carrying one of his children, a thorny oyster shell, coca, and other goods to make a drink offering to Huallallo Carhuincho. Paria Caca convinces the man to leave his offerings with him and return with his son to his village (HM 99), and he tells the man to return in five days to witness his combat against Huallallo Carhuincho (HM 100). Paria Caca eats the thorny oyster shell and the other offerings (HM 101), and from Ocsa Pata he goes to Mullo Cocha. Turning into red and yellow rain and flashing like lightning, he descends on Mullo Cocha from five directions, trying to drown Huallallo Carhuincho (HM 102–103). But at the very last moment, Huallallo Carhuincho flees to the lowlands to the east (HM 104). One of the five Paria Cacas (Paria Carco) chases him and midway to the lowlands turns himself into stone to prevent Huallallo Carhuincho from returning (HM 105).

From Mullo Cocha, Paria Caca moves north and west to somewhere below Mama, where a female fire deity known as Mana Ñamca has appeared. Midway, in Tumna, he attacks this female deity and expels her into the ocean (HM 106). In this place,

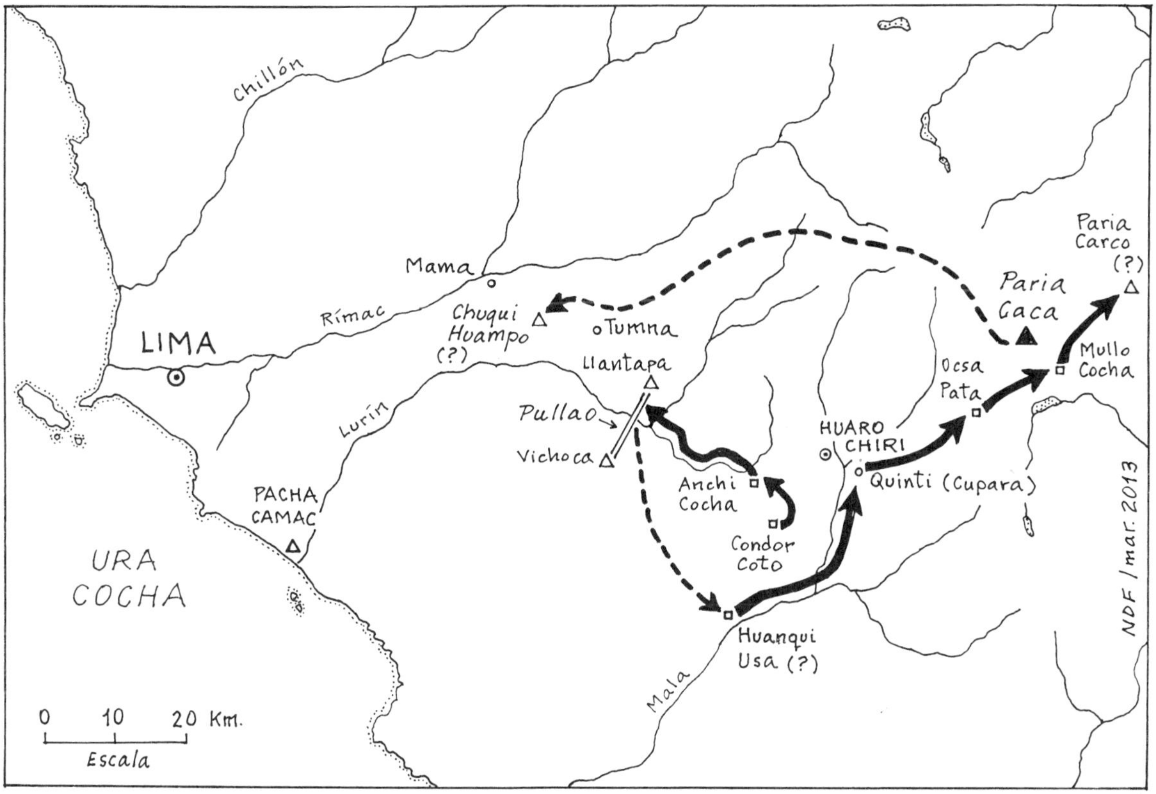

figure 8.6
The mythical cycle of Paria Caca. (Drawing by Nicanor Dominguez.)

another of the Paria Cacas (Chuqui Huampo), whose foot was injured, turns himself into stone to prevent Mana Ñamca from returning (HM 107). At this point, the calendrical rituals associated with the harvest of coca fields (?) are introduced for the first time in the narrative (HM 108–109).

Finally, Paria Caca establishes his dwelling in the heights (in the center of the Huarochirí region; HM 116). From there, Paria Caca begins to order his cult, describing all the rituals highland people should observe (HM 117). Indeed, through the next three whole chapters (nine through eleven), the author of the manuscript describes in detail these rituals: the Auquisna festivals celebrated in April in honor of Paria Caca, the Chaycasna festivals celebrated at the end of May and the beginning of June in honor of Chaupi Ñamca, and the Chanco festivals, celebrated in honor of another hero, Tutay Quiri, at the end of November.[20] The author explicitly states that the dates of the rituals are established by ritual specialists who observe the course of the sun (HM 118).

In this manner, through Paria Caca's actions, the territory is extended horizontally and segmented according to cardinal directions. Beginning in the west and continuing counterclockwise to the south, to the east, and to the north, lowland peoples are expelled from the highlands. Clear limits between lowland and highland regions are marked in the east and the west, but no significant limits are marked in other highland regions to the north (?) and to the south. Relations of confrontation and competition are established with the lowlands, while relations of cooperation are established with the highlands. Meanwhile, a further segmentation of time is created by the introduction of calendrical rituals associated with agricultural cycles and cycles of ancestor veneration. In this manner, the regional integration of the local highland groups is finally established (Figure 8.6).

Fourth and Fifth Episodes: The Inca and the Spanish Conquests

The events of the ancestral heroes Cuni Raya, Huatya Curi, and Paria Caca and the relationships they embody between space, time, and society serve as frames of reference in subsequent descriptions of the events of the Inca and Spanish conquests of the Huarochirí region. It is only when the accounts of these events are taken into consideration that the reasons for the integration of the events discussed previously become clear.

The Huarochirí Manuscript accounts for the Inca and Spanish conquests of the Huarochirí region in distinct spatial and temporal terms. The Inca conquest is described as a period of further horizontal expansion and segmentation of highland territories, as well as further segmentation of time resulting from the introduction of state calendrical rituals in honor of the local deity Paria Caca. The Spanish conquest, in contrast, is described as a period of obliteration of the spatial and temporal limits introduced in the previous cycles, including the separation of "this world" and the "world of down/below" (Table 8.5).

The Huarochirí Manuscript also accounts for the Inca and Spanish conquest of the Huarochirí region in terms of distinct types of intersocietal relations. The Inca conquest is described as a period of treaties and alliances—in short, cooperation—between the Huarochirí groups and the Inca against the groups from the lowlands. The Spanish conquest, in contrast, is described as a period of confrontation between those located in the highlands, those in the center, and those who come from the ocean, the "world of down/below" (Table 8.6).

The Inca occupation of the Huarochirí region is described in one specific account that tells us about an Inca king, Tupay Inga Yupanqui, who could not defeat a group of rebels that had appeared in the periphery of his empire in the Alancu Marca, Calanco Marca, and Chaque Marca provinces (HM 285). After sending thousands of men and being defeated again and again for a period of about twelve years (HM 286), the Inca king decides to summon all the huacas he serves in his empire (HM 287). Paria Caca responds to the call by sending Maca Huisa, one of his children (HM 288). Maca Huisa moves from Huarochirí to Aucay Pata, the plaza in the very heart of Cuzco, the capital of the Inca empire. Upon his arrival, the Inca begins to speak and asks the huacas to come to his aid, but the huacas do not speak up (HM 289). As a result, the Inca threatens to burn them immediately if they do not help him (HM 290–291). Finally, Pacha Camac speaks up, excusing himself and

table 8.5
Spatial and temporal relations defined in the mythical cycles of Maca Huisa and Cuni Raya Vira Cocha.

Maca Huisa	Further horizontal expansion and segmentation of highland territories	Further segmentation of time (introduction of new rituals)
Cuni Raya Vira Cocha	Collapsed of existing segments and limits. Return to a basic segmentation of land and water	Cancellation of native calendar rituals

table 8.6
Social relations defined in the mythical cycles of Maca Huisa and Cuni Raya Vira Cocha.

Maca Huisa	Alliance and cooperation between highland groups (Huarochirí and Inka) against lowland groups
Cuni Raya Vira Cocha	Return to a time of unsettled relations between groups located on the land (center) and those coming from the water (periphery)

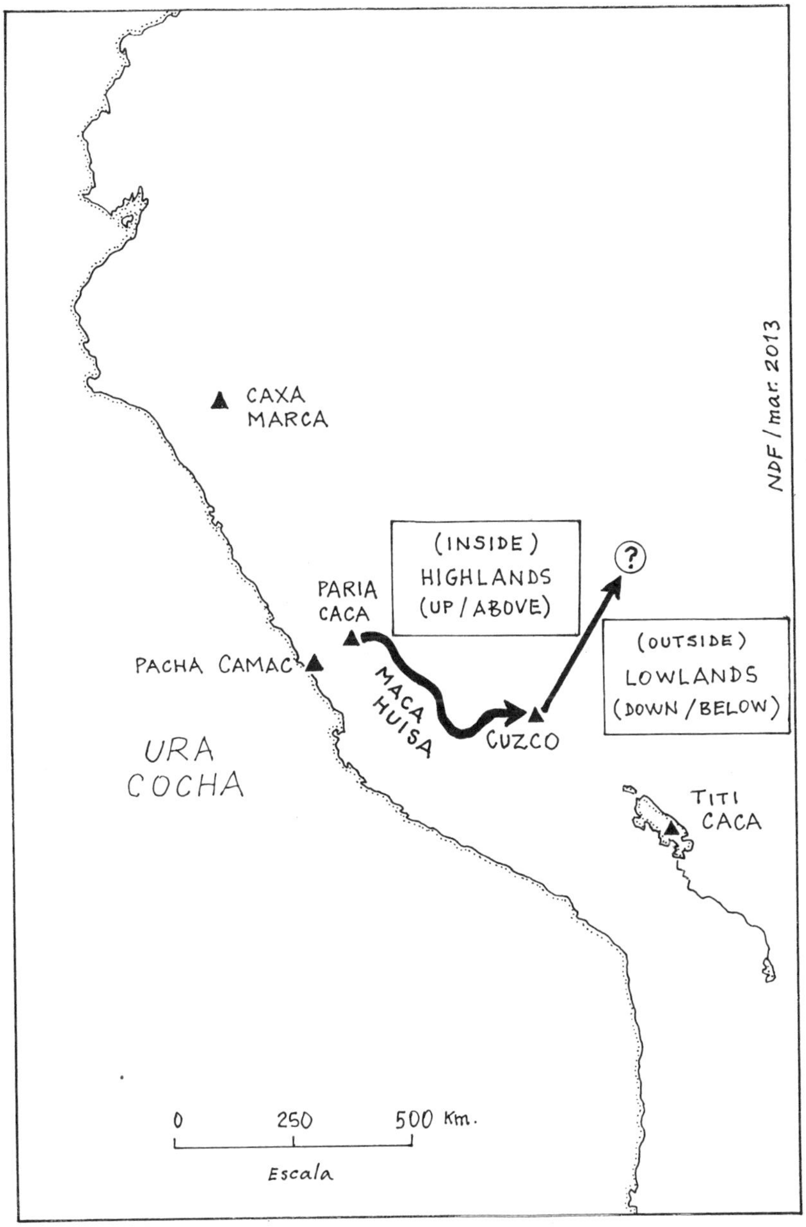

figure 8.7
The mythical cycle of Maca Huisa and the Inca. (Drawing by Nicanor Dominguez.)

explaining that if he moves he will destroy not only the Inca's enemies but the whole world (HM 292). As the other huacas remain silent, Maca Huisa speaks up and offers to subdue the Inca's enemies (HM 293). Traveling in the Inca's litter, Maca Huisa moves from Cuzco to the provinces, while the Inca remains in the city (HM 294). As soon as he arrives uphill, Maca Huisa turns into rain and flushes the rebels away. After that, he returns to Cuzco with some prisoners (HM 295–296). To compensate, the Inca offers Maca Huisa whatever he wants, and Maca Huisa asks the Inca to serve as a *huacsa*[21] priest to Paria Caca (HM 297). Afraid (HM 298), the Inca has some food served to Maca Huisa, but he only accepts thorny oyster shell to eat (HM 299). The Inca offers him some of the Inca ladies of his nobility, but Maca Huisa refuses them. Since then, the Inca has acted as a huacsa priest of Paria Caca in Xauxa (HM 300) and in Cuzco (HM 301). From earlier references in the narrative, we know that the Inca had thirty people—fifteen from upper Yauyos and fifteen from lower Yauyos—assigned

figure 8.8
The mythical cycle of Cuni Raya Vira Cocha and the Spanish Vira Cochas. (Drawing by Nicanor Dominguez.)

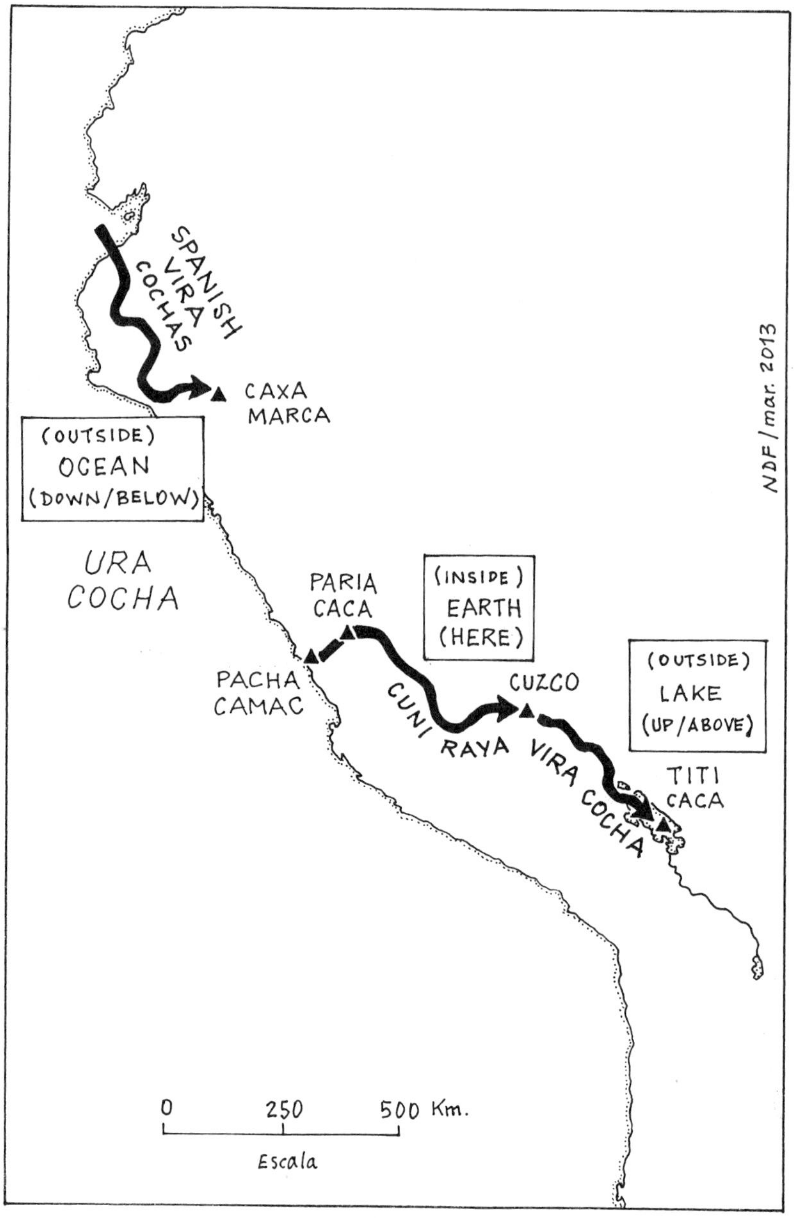

to celebrate these rituals in turns of fifteen days according to the phases of the moon (HM 220, 229).

In sum, space is further expanded and segmented so as to include Cuzco. Time is further expanded and segmented through the introduction of rituals marked by the cycle of the moon. Society is further expanded and segmented so as to include the Inca. The relations between the Huarochirí and the Inca are exactly the same as those defined in the cycle of Paria Caca: cooperation between highland groups against lowland ones. By refusing to take a bride from the Inca, Maca Huisa preserves his autonomy (Salomon 1991:10), and Huarochirí maintains, at least in this local version of the story, a more egalitarian relationship with the Inca (Figure 8.7).

The Spanish invasion of Huarochirí is described in another account. In this account, Cuni Raya Vira Cocha goes to Cuzco to see the Inca king, Huayna Capac, right before the appearance of the Spanish. Once in Cuzco, Cuni Raya convinces the Inca to move with him from Cuzco to Titi Caca (HM 190). The author of the Huarochirí

Manuscript is quite explicit in describing Titi Caca and Pacha Camac as sacred places located, respectively, on the upper and lower foundations of the world (Figure 8.8). In his own words: "The world ends somewhere in the waters of Ura Cocha that are below Titi Caca, and somewhere past the place they call Pacha Camac. It looks like there is no village beyond these points, possibly nothing at all" (HM 278). What the author is describing is, of course, a cosmological model in which Titi Caca (the lake, the body of water from up/above) and Ura Cocha (the ocean, the body of water from down/below) are connected, with Titi Caca lifting the water from Ura Cocha through some sort of underground canals (Salomon and Urioste 1991:111–112n553; see also Earls and Silverblatt 1978).

Once in Titi Caca, Cuni Raya tells the Inca to send his magicians and all sorts of shamans (condor, falcon, and swift shamans) to Ura Ticsi, the world's lower foundations (HM 191), to bring him a bride (HM 192). After a failed attempt (HM 193–194), the swift shaman finally returns with the bride (HM 195) exactly on the fifth day, and Cuni Raya and the Inca receive her overjoyed. Then, in what is probably one of the most dramatic moments of the narrative, Cuni Raya tells the Inca: "Inca! Let's draw a line across this world. I'll go into this space and you go into this other space with my sister. You and I mustn't see each other anymore" (HM 196).[22]

Cuni Raya divides the world (HM 196), and the Inca and his lady disappear forever. And Cuni Raya does the same. At that exact moment, the Spanish Vira Cochas appear in Caxa Marca (HM 197). Later, the Spanish Vira Cochas arrive at Huarochirí, where they set fire to one of the huacsa priests in charge of Paria Caca's cult—an action that clearly evokes the time when the highlands were inhabited by lowland people and were ruled by fire deities (HM 225).

The Spanish arrival, thus, involves a return to circumstances defined in the first mythical cycle, the Cuni Raya cycle, a presocial time of unsuccessful relations between those who occupied the land and those who came from the sea. By accepting a bride from Cuni Raya Vira Cocha, the celeritas male trickster who moves between "this world" (inside) and the "world of down/below" (outside), the Inca king knowingly or unknowingly but in any case unavoidably causes a reversal of the world. The world is now upside down. Time has ended. The world is out of time.

In sum, the Inca conquest is seen as a continuation and expansion of the spatial, temporal, and social limits and relations defined by Paria Caca. The Spanish conquest, in contrast, is seen at best as a break from, and at worst as a definite obliteration of, such limits and relations and a return to the space and time of Cuni Raya at the beginning of time.

Conclusions

The foregoing analysis shows us how the seemingly disparate episodes of the first few chapters of the Huarochirí Manuscript are, in reality, part of a unified narrative. It is only when these episodes are analyzed with reference to the narrative context in which they occur that the reasons for their specific details of form and structure become clear. In particular, the analysis reveals that the integration of the first few chapters of the manuscript is regarded in terms of a sequence of actions carried out by different heroes on the landscape. It is in the segmentation of space by means of these actions that distinct times are created, and it is in the segmentation of time that distinct spaces are created.

Moreover, this sequence of actions is organized incrementally. As we move from one episode to the next, more and more segmentations and expansions of space, time, and intergroup relations are introduced. Furthermore, these segmentations and expansions are usually introduced following a sequence that goes from a relation of opposition to a relation of complementarity and in which the terms of opposition and complementarity are alternated or switched from one episode to the next.[23]

Even more important, these segmentations and expansions of space, time, and society provide a basic framework for measuring the extent and degree of the changes introduced by the two expansive states that conquered the Huarochirí region: first the Inca and later the Spanish. While the Inca conquest is regarded as time, the Spanish conquest is regarded as out of time.

As a unified narrative concerned with the definition of space, time, and identity within the colonial situation in which it was written, the Huarochirí Manuscript is therefore a historical account that, like any account that considers itself to be historical, is concerned with the explanation of the present as an outcome of the past and a projection into the future. In clear cosmogonic style, it represents the colonial present as a crisis. Because of this representation, the Huarochirí Manuscript itself—not only the myths and rituals that it records—is also a rather active medium in which the experience and understanding, the construction and representation, of history and time—historicity and temporality themselves—are being reformulated.

Acknowledgments

This essay is based on several different papers presented at different meetings and conferences (Dulanto 1999, 2000, 2009, 2012). I would like to thank Anthony Aveni, Susan Gillespie, David Haskell, Nestor Quiroa, Frank Salomon, Carl Wendt, R. Tom Zuidema, and two anonymous reviewers for their valuable commentaries and suggestions regarding this or previous work. Nicanor Dominguez provided all the maps included in this essay.

NOTES

1 The terms *temporality* and *historicity* are used here not in the traditional sense of "the linear progression of past, present, and future" and "the actuality of events and their ordering in time," respectively, but in the sense of "the culturally determined manners of experiencing and understanding, constructing and representing time and history" (Dulanto and Haskell 2009:1; see Hirsch and Steward 2005:262).

2 The Huarochirí region corresponds roughly to what today is the Huarochirí Province on the western slopes of the Andes, east of Lima. During late pre-Hispanic and early colonial times, this region was part of the Yauyos Province. According to Dávila Brizeño (1965 [1586]), the Yauyos were organized in two Inca-style moieties: the Anan, or upper, Yauyos to the south and the Lorin, or lower, Yauyos to the north. Anan Yauyos covered the upper sections of the Asia and Cañete Valleys, while Lorin Yauyos covered the upper sections of the Mala, Lurin, Rimac, and Chillón Valleys. The Huarochirí region corresponds to the northern moiety. At least in Inca times (the late fifteenth and early sixteenth centuries), each of these moieties seems to have been further subdivided into several *huarangas*, or tributary units of a thousand households each (see Spalding 1984). It is important to note, however, that the stories of the Huarochirí Manuscript cover a larger area, including the lowland areas of the *yuncas* to the west and to the east, as well as other highland areas to the south and to the north, and even Spain, beyond the sea.

3 It is the fourth of six documents bound together in a volume, Manuscrito 3169, in the Biblioteca Nacional in Madrid. Among the other documents included in the volume are some of the most important early colonial sources on Andean religion available today, such as Cristóbal de Molina's (1959 [1575?]) *Relación de las fábulas y ritos de los incas* and Juan de Santa Cruz Pachacuti Yamqui Salcamaygua's (1879 [1613]) *Relación de antigüedades de este reyno del Perú*. All these manuscripts clearly belonged to Father Francisco de Ávila (1573?–1647), who annotated several of them. Ávila was a secular cleric from Cuzco, who between 1596 and 1598 was appointed priest of San Damián de Checa, and vicar and ecclesiastical judge of the Huarochirí Province and later became one of the first *extirpadores de idolatrías* of the Archbishopric of Lima. Although most researchers are reluctant to identify the author of the Huarochirí Manuscript, they agree that the author was not Ávila but probably one of his assistants, a converted and educated Indian, very likely a nobleman from San Damián de Checa, who wrote

the document upon Ávila's request and under his supervision (Duviols 1966:234–235; Salomon 1991:2; Taylor 1987b:16). Following Rowe (Taylor 1987b:17), Durston (2007) identifies Cristóbal Choquecasa as the author of the manuscript.

4 That is, the Huarochirí Manuscript is written in a variant of Quechua A or Quechua II used by the Inca and later by the Spanish as a lingua franca. The language of the manuscript is most certainly not the language spoken regularly in the Huarochirí region at the time of the Spanish conquest and even at the time when the manuscript was written. Among the candidates are several dialects of the Quechua and Aymara families (Salomon 1991:30–31).

5 The date of the Huarochirí Manuscript is still being debated. Different authors argue either in favor of an early date, shortly after Ávila's arrival at the *doctrina* of San Damián de Checa in 1597 (e.g., Durston 2007:230; Duviols 1966:235–236), or a later one, shortly before Ávila's (1966 [1608]) composition of his *Tratado de los errores y falsos dioses . . .* , which is based on the Huarochirí Manuscript (e.g., Acosta 1987:596; Taylor 1987b:16). Regardless of the date around which it was written, however, the stories included within the Huarochirí Manuscript clearly were compiled and edited around seventy years or roughly three generations after the arrival of the Spanish in the area. Therefore, all of those involved in its production, whether Spanish or indigenous, clearly grew up within the colonial situation initiated by the Spanish invasion.

6 As Salomon (1991:2) has pointed out, as in the case of the Bible, the Huarochirí Manuscript "begins with myths that contrast the human condition with an imagined alternative"; contains "a flood myth [that] signals the end of this era"; "pictures antiquity as the story of hero-ancestors who share a common descent and a covenant like relation to an ethnic deity"; has as "its collective subject . . . a set of groups, each of which considered itself the progeny of a focalized ancestor"; describes how "these groups relate to each other, at least in ideology, approximately as a phratry"; describes how "their disparate traditions of origin and separate cults have been welded *ex post facto* onto the unifying argument of kinship and imperfectly articulated with apical priestly cults"; relates how "their story . . . is intensely concerned with control over specific resources in a sacralized landscape"; contains myths that "encode political struggles with surrounding peoples and even internecine struggles as mythic combats with superhuman intervention"; "is greatly concerned with the relation between the local sacra and the leaders and priests of immense invading empires—first the Inca, later the Spanish"; contains "texts about priesthood, sacrifice, ritual law, and prophecy [that] jostle with vernacular myth, claims concerning land and water, and mythicized remembrance of historic events"; demonstrates that "bits of oracular response, religious formulas, and perhaps songs have become embedded in the text, too"; and gives one a sense of "the pressure of contemporary political defeat on religious testimony."

7 Probably one of the best examples of this traditional approach to the analysis of Andean narratives is the study of Urbano (1981). Common to these studies is the assumption that deities identified by the same name in different narratives—notably, Vira Cocha—are one and the same, independently of the particular narrative in which they are inscribed.

8 "Turner's method is thoroughly explained and applied mainly in two articles: 'Narrative Structure and Mythopoesis: A Critique and Reformulation of Structuralist Concepts of Myth, Narrative, and Poetics' (1977) and 'Animal Symbolism, Totemism, and the Structure of Myth' (1985). The first article provides an extensive critique of Lévi-Straussian structuralism and a detailed reformulation of structuralist theory and method, as well as a brief analysis of the Oedipus myth that was analyzed by Lévi-Strauss (1967) himself. Turner's basic critique of Lévi-Strauss is that he misunderstood the insights of the Prague School and erroneously discarded the syntagmatic or combinatorial axis of narrative. The second article provides a complete and detailed application of the method, as reformulated by Turner, to the analysis of the Kayapó myth of the origin of cooking fire. Briefly, in Turner's method the basic unit of analysis is the 'episode'—that is, a unit of narrative in which at least one relationship between the characters who participate in the story changes. These relationships change because either the characters themselves or their characteristics (e.g., gender, age, status, etc.) change. Furthermore, these changes from one episode to the next are likely to follow patterns, by which changes in one set of relations

are switched with changes in other sets of relations. More important, the characters themselves are constructed as the story is narrated, to the point that they can hardly be understood out of the specific narrative context in which they are built. Turner's method thus allows for the identification of how characters, relations, and categories in general are constructed and explained. In doing so, it opens a window to explaining how people use narratives to construct and represent events and their ordering and reordering in time" (Dulanto and Haskell 2009:5).

9 In sections HM 3–8, the author also summarizes the chronological order in which different heroes appear in the story—Cuni Raya Vira Cocha, Huallallo Carhuincho, Paria Caca—and describes in some detail the world of Huallallo Carhuincho. But it is clear that the initial situation is not the world of Huallallo Carhuincho, but the world of Yana Ñamca and Tuta Ñamca, who, according to the author, are defeated by Huallallo Carhuincho. The author makes an explicit decision to locate Cuni Raya before Huallallo Carhuincho and Huallallo Carhuincho before Paria Caca (HM 8).

10 *Huaca* is a complex and difficult term that broadly refers to sacred persons, places, and objects. See Salomon 1998 for a detailed analysis of the metaphysical and ontological standing of huacas. See also Salomon 1991:17–19.

11 In Taylor's (1999:21) translation, "al verlo todos los huacas locales se asustaron mucho. 'Hermana Cahuillaca' la llamó, '!mira aquí!, ahora soy muy hermoso' y se enderezo iluminando la tierra."

12 Note the double alternation in the sequence of encounters: condor (black/white, diurnal), skunk (black/white, nocturnal), puma (red [?], diurnal), fox (red [?], nocturnal). This first dual division of time is presented as an alternation from black/white to red (?), and a second division of time is presented as an alternation from day to night. Later in the manuscript (HM 64), the author refers to the puma skin used by Huatya Curi in a contest as red in color.

13 It is particularly interesting that Cuni Raya's encounters with these two last animals—the falcon and the parakeets—are introduced to the story in a marginal addition in Quechua. It is possible that an original story narrated by an informant was later corrected by the editor of the manuscript. This possibility reinforces the idea that the movement of Cuni Raya from Anchi Cocha to Pacha Camac follows the double alternation mentioned in the previous note.

14 The disappearance of Cuni Raya into the sea is crossed out in the original in order to match two different stories—the first one is the story of Cuni Raya and Caui Llaca, the second one is the story of Cuni Raya and Urpuy Huachac.

15 In Taylor's (1999:31) translation, Urpuy Huachac "hizo crecer una gran peña para que le cayera encima." Therefore, Urpuy Huachac's intention could be read as being either to expel Cuni Raya from the inside (the land, here) to the outside (the ocean, down/below) or to turn him into stone, fixing him like Caui Llaca on the very limit between these two realms.

16 This primitive hydraulic system, and the interconnection of the waters in the sky (outside, up/above), the land (inside, the springs), and the ocean (outside, down/below), is thoroughly explained by the author of the manuscript in a later chapter (chapter twenty-nine). According to him, a llama called the Yacana, explicitly identified with the dark constellation of the same name that can be seen in the Milky Way (HM 372), drinks at night the water from the springs (HM 373) and from the ocean, keeping in this way the ocean from drowning the whole world (HM 374). For a detailed analysis of the Yacana in Andean cosmology, see Urton 1981 and Urton and Zuidema 1976.

17 Here, the author of the manuscript states explicitly that, as in the previous episode, this episode is also divided into two stories: one story about a lord in the highlands, Tamta Ñamca, who suffers because of his female partner's actions, and the second about a lady in the lowlands, Pacha Camac's daughter (?), who suffers because of her male partner's actions. The author, however, states that he is not going to elaborate on the second story. In other words, the opposition between highlands and lowlands is expressed in gender terms: a local male in the highlands (up/above) and a local female in the lowlands (down/below), both suffering because of their partners of the opposite gender. Note the inversion of the opposition as represented in the previous episode: A local female in the land (inside, here) and a foreign male from the sky and the ocean (outside up/above and down/below), the local female suffering because of the foreign male actions.

18 That is, lowland animals are hidden in a highland house.

19 Some patterns are clear here. Cuni Raya encounters six different animals; Huatya Curi participates in six different contests. Both the animals Cuni Raya encounters and the animals and objects involved in the contests are, in some way, related to the three main calendrical rituals described later in the Huarochirí Manuscript (see next note).

20 The rituals in honor of Paria Caca, Chaupi Ñamca, and Tutay Quiri were celebrated, respectively, sometime between April 1 and April 15, sometime between May 25 and June 9, and on November 30. Thus, these rituals seem to cover the period between the beginning of harvest and the end of the planting and probably even to the beginning of rainy season (see Taylor 1987a).

21 According to Salomon (1991:18), huacsas were mainly impersonators of huacas who reenacted huacas' myths in festivals. The huacsa office seems to have been nonhereditary and rotated among those who recognized each other as kin—at least in relation to a particular ritual duty.

22 Following the logic used in several stories of the Huarochirí Manuscript, we see that if at the beginning of the story Cuni Raya disappears into Ura Cocha (the lower foundations of the world), at the end of the story he should disappear into Titi Caca (the upper foundations of the world).

23 Salomon (2002) provides a very interesting analysis of a similar pattern of alternating dualism and its connection with historicity in rituals of political regeneration in the contemporary community of Pacota (Huarochirí). As Salomon (2002:17–20) points out, this alternation between opposed and complementary halves plays a crucial role in the articulation of past and future events. It is in this alternation that a present—"a time out of time" and "a space out of space"—is defined.

REFERENCES CITED

Acosta, Antonio

1987 Francisco de Ávila 1573(?)–Lima 1647. In *Ritos y tradiciones de Huarochirí: Manuscrito quechua de comienzo del siglo XVII*, edited by Gerald Taylor, pp. 551–616. Instituto de Estudios Peruanos, Instituto Francés de Estudios Andinos, Lima.

Adelaar, Willem

1988 *Het Boek van Huarochirí: Myhten en Riten het Oude Peru zoals opgetekend in de zestiende eeuw voor Francisco de Avila, bestrijder van afgoderij.* Meulenhoff, Amsterdam.

Arguedas, José María, and Pierre Duviols

2007 [1966] *Dioses y hombres de Huarochirí: Narración quechua recogida por Francisco de Ávila (¿1598?).* Universidad Antonio Ruiz de Montoya, Lima.

2012 [1966] *Dioses y hombres de Huarochirí: Narración quechua recogida por Francisco de Ávila (¿1598?).* Instituto de Estudios Peruanos, Lima.

Arguedas, José María, and José Ignacio Uzquiza González

2011 *El manuscrito de Huarochirí: Libro sagrado de los Andes peruanos.* Biblioteca Nueva, Universidad de Extremadura, Madrid.

Ávila, Francisco de

1966 [1608] Tratado de los errores y falsos dioses y otras supersticiones y ritos diabólicos en que vivían antiguamente los indios de las provincias de Huaracheri Mama y Chaclla y hoy también viven engañados con gran perdición de sus almas. In *Dioses y hombres de Huarochirí*, edited by José María Arguedas, pp. 198–217. Museo Nacional de Historia, Instituto de Estudios Peruanos, Lima.

Dávila Brizeño, Diego

1965 [1586] Descripción y relación de la provincia de Yauyos. In *Relaciones geográficas de Indias*, edited by Marcos Jiménez de la Escapada, vol. 1, pp. 155–165. Ediciones Atlas, Madrid.

Dulanto, Jalh

1999 Between Cosmogonies and Chronologies: The Huarochirí Manuscript and the Cosmogonic Mapping of History onto the Landscape. Paper presented at the 98th Annual Meeting of the American Anthropological Association, Chicago.

2000 Cosmogonies as Cosmologies: A Syntagmatic Analysis of the Mythical Cycles of the Huarochirí Manuscript. Paper presented at the 77th Annual Meeting of the Central States Anthropological Society, Bloomington, Ind.

2009 The Children of Paria Caca and the Materialization of History on the Mythical and Ritual Landscapes of Huarochirí. Paper presented at the Annual Meeting of the American Society for Ethnohistory, New Orleans, La.

2012 Inca! Let's Draw a Line across This World: Central Andean Notions of the End of Time. Paper presented at the University of Pennsylvania Center for Ancient Studies Symposium: The End of Time, Philadelphia.

Dulanto, Jalh, and David L. Haskell

2009 Syntagmatic/Paradigmatic Approaches to Ethno-Ethnohistory: An Introduction. Paper presented at the Annual Meeting of the American Society for Ethnohistory, New Orleans, La.

Durston, Alan

2007 Notes on the Authorship of the Huarochirí Manuscript. *Colonial Latin American Review* 16(2):227–241.

Duviols, Pierre

1966 Estudio bibliográfico. In *Dioses y hombres de Huarochirí: Narración quechua recogida por Francisco de Ávila*, edited by José María Arguedas y Pierre Duviols, pp. 218–240. Museo Nacional de Historia, Instituto de Estudios Peruanos, Lima.

1978 Un symbolisme andin du double: La lithomorphose de l'ancetre. *Actes du XLII Congrès International des Américanistes* 4:359–364. Société des Américanistes, Paris.

Earls, John, and Irene Silverblatt

1978 La realidad física y social en la cosmología andina. *Actes du XLII Congrès International des Américanistes* 4:299–325. Société des Américanistes, Paris.

Galante, Hipólito

1942 *Francisco de Avila de priscorum Huaruchiriensium origine et institutis . . .* Instituto Gonzales de Oviedo, Madrid.

Hirsch, Eric, and Charles Steward

2005 Introduction: Ethnographies of Historicity. *History and Anthropology* 16(3):261–274.

Lévi-Strauss, Claude

1967 *Structural Anthropology.* Anchor Books, New York.

Millones, Luis

2008 Dioses y demonios de los Andes. In *Dioses del norte y dioses del sur: Religiones y cosmovisión en Mesoamérica y los Andes*, edited by Alfredo Lopez Austin and Luis Millones, pp. 145–272. Instituto de Estudios Peruanos, Lima.

Millones, Luis, and Hiroyasu Tomoeda

2007 Estudio preliminar. In *Dioses y hombres de Huarochirí: Narración quechua recogida por Francisco de Ávila*, edited by José María Arguedas y Pierre Duviols, pp. XIII–XXXII. 2nd ed. Universidad Antonio Ruiz de Montoya, Lima.

Molina, Cristóbal de

1959 [1575?] *Relación de las fábulas y ritos de los incas.* Edited by Ernesto Morales. Editorial Futuro, Buenos Aires.

Salomon, Frank

1991 Introductory Essay: The Huarochirí Manuscript. In *The Huarochirí Manuscript: A Testament of Ancient Colonial Andean Religion*, edited by Frank Salomon and George Urioste, pp. 1–38. University of Texas Press, Austin.

1998 How the Huacas Were: The Language of Substance and Transformation in the Huarochirí Quechua Manuscript. *RES: Anthropology and Aesthetics* 33:7–16.

2002 "Huayra Huayra Pichcamanta!": Augurio, risa y regeneración en la política tradicional (Pacota, Huarochirí). *Bulletin de l'Institut Français d'Études Andines* 31(1):1–22.

Salomon, Frank, and George Urioste (editors)

1991 *The Huarochirí Manuscript: A Testament of Ancient Colonial Andean Religion.* University of Texas Press, Austin.

Santa Cruz Pachacuti Yamqui Salcamaygua, Juan de

1879 [1613] Relación de las antigüedades deste reyno del Piru. In *Tres relaciones de antigüedades peruanas*, edited by Marcos Jiménez de la Espada, pp. 231–328. Ministerio de Fomento, Madrid.

Spalding, Karen

1984 *Huarochirí: An Andean Society under Inca and Spanish Rule.* Stanford University Press, Stanford, Calif.

Szeminski, Jan (editor)

1985 *Bogowie i ludzie z Huarochirí.* Wydawnictwo Literackie, Kraków.

Taylor, Gerald

1980 *Rites et traditions de Huarochirí: Manuscrit quechua du 17e siécle.* L'Harmattan, Paris.

1987a Cultos y fiestas de la comunidad de San Damián (Huarochirí) según la Carta Annua de 1609. *Bulletin de l'Institut Français d'Études Andines* 16:85–96.

1987b *Ritos y tradiciones de Huarochirí: Manuscrito quechua de comienzos del siglo XVII.* Instituto Francés de Estudios Andinos, Instituto de Estudios Peruanos, Lima.

1999 *Ritos y tradiciones de Huarochirí.* 2nd ed. Instituto Francés de Estudios Andinos, Banco Central de Reserva del Perú, Universidad Ricardo Palma, Lima.

2008 *Ritos y tradiciones de Huarochirí.* Instituto de Estudios Peruanos, Instituto Francés de Estudios Andinos, Universidad Nacional Mayor de San Marcos, Lima.

Trimborn, Herman

1939 *Francisco de Avila: Dämonen und Zauber im Inkareich.* Quellen und Forshungen zur Geschichte der Geographie und Völkerjunde 4. K. F. Koehler, Leipzig.

Turner, Terence

1977 Narrative Structure and Mythopoesis: A Critique and Reformulation of Structuralist Concepts of Myth, Narrative, and Poetics. *Arethusa* 10(1):103–163.

1985 Animal Symbolism, Totemism, and the Structure of Myth. In *Animal Myths and Metaphors in South America*, edited by Gary Urton, pp. 49–106. University of Utah Press, Salt Lake City.

1988 Ethno-Ethnohistory: Myth and History in Native South American Representation of Contact with Western Society. In *Rethinking History and Myth: Indigenous South American Perspectives on the Past*, edited by Jonathan D. Hill, pp. 235–281. University of Illinois Press, Urbana-Champaign.

Urbano, Henrique

1981 *Wiracocha y Ayar: Héroes y funciones en las sociedades andinas.* Centro de Estudios Rurales Andinos "Bartolomé de las Casas," Cuzco.

Urioste, George

1983 *Hijos de Pariya Qaqa: La tradición oral de Waru Chiri.* Maxwell School of Citizenship and Public Affairs, Syracuse, N.Y.

Urton, Gary

1981 *At the Crossroads of the Earth and the Sky: An Andean Cosmology.* University of Texas Press, Austin.

Urton, Gary, and Tom Zuidema

1976 La constelación de la llama en los Andes peruanos. *Allpanchis phuturinqa* 9:59–120.

Úzquiza Gonzalez, José Ignacio (editor)

2011 *El manuscrito de Huarochirí: Libro sagrado de los Andes peruanos.* Biblioteca Nueva y Universidad de Extremadura, Madrid.

9

Linearity and Cyclicity in Pre-Columbian Maya Time Reckoning

VICTORIA R. BRICKER AND HARVEY M. BRICKER

AMONG THE SIGNAL ACHIEVEMENTS FOR which the civilization of the Pre-Columbian Maya is known is its pervasive concern with calendrical time. Several different calendrical systems—lunar, solar, and nonastronomical—were in use simultaneously, and a major function of the writing system of the Classic period was recording, with to-the-day precision, when various political and natural events occurred. Because Classic Maya calendrical notations can be translated, as it were, into our modern Western calendar, and because the substantial decipherment of the hieroglyphic writing system present on abundant archaeological monuments has been accomplished during the last four decades or so, we now have very rich and well-dated dynastic histories of many of the Classic city-states. Calendrical practices and emphases changed during the Postclassic and early Colonial periods, but the concern with tracking the passage of time and situating important events within it remained a constant. The survival of a few of the Postclassic books or codices, which deal with matters other than dynastic history, including astronomy, shows how the to-the-day precision of the Classic calendar, exemplified by the so-called Long Count, had been preserved even after the demise of the great city-states of the Petén. The codices emphasize also the uses to which the 260-day sacred calendar or *tzolkin* was put for ritual and ceremony. This emphasis on the cyclicity of time is what is most apparent in early Colonial manifestations of Maya calendrical practice in Yucatan and in highland Guatemala of the present day. Maya time reckoning is and has been for centuries a combination of linear and cyclic elements. That such combination was important in the Classic period is emphasized by Markus Eberl in his contribution to this volume. Linear time is best exemplified in the Long Count, the best-known Classic Maya calendrical component.

The Maya Long Count is a count of the days in the current era that have been completed. As such, it refers to a span of time nested within a much longer temporal concept that is only rarely mentioned on Classic monuments, but it is of central importance in the Postclassic Dresden Codex. During the year 2012, most discussion of Maya time reckoning understandably focused on the recently completed Maya era, which began in 3114 BC and ended in December 2012 on the winter solstice. The dictionary defines an *era* as "a period of time measured from some occurrence or date," but it is instructive to place the Maya era in the broader context of an *eon*, defined generally and nongeologically as "an extremely long, unlimited period of time." If time is really unlimited, only part of it is likely to be known to and used by any given group or culture, and that is true of the Pre-Columbian Maya.

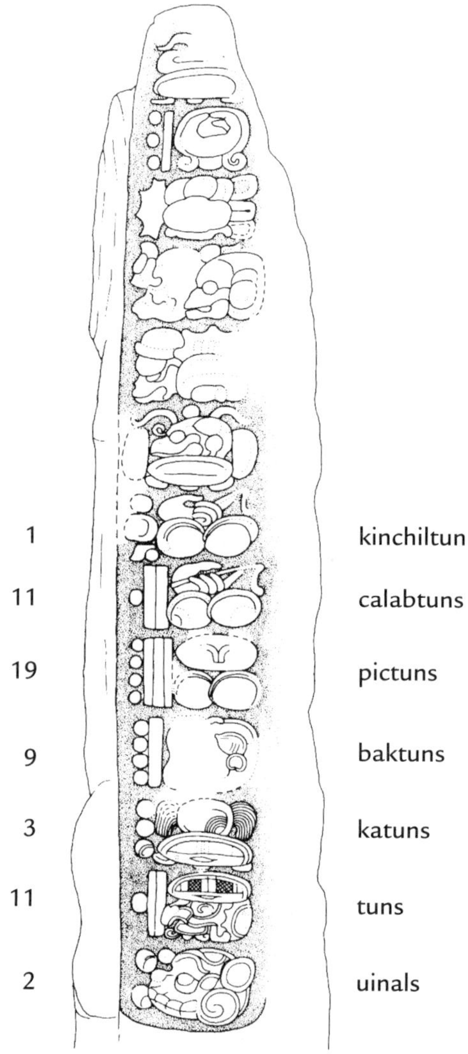

figure 9.1
Inscription on the right edge of Stela 10 at Tikal, showing three calendrical positions (1 *kinchiltun*, 11 *calabtuns*, 19 *pictuns*) preceding the specification of *baktuns* and the other elements of an Initial Series date. (Modified after Jones and Satterthwaite 1982:fig. 15a; courtesy of The University of Pennsylvania Museum of Archaeology and Anthropology.)

Looking Back in Time

The outer limits of the Maya eon cannot be specified precisely. If some numerical notations on Classic monuments are treated as chronological statements, Maya time reckoning can be claimed to have extended back beyond the Big Bang, but this idea is more amusing than informative. The earliest temporal limit that can be considered operational in a calendrical sense was recognized in the early twentieth century (Morley 1915:115, fig. 60) on Stela 10 at Tikal, a Classic period site in Guatemala (Figure 9.1). The date recorded on Stela 10 can be written 1.11.19.9.3.11.2.[6?]. Each place position represents a different kind of calendrical unit, the shortest (days) at the right extremity, increasing in length, usually vigesimally, as one moves leftward. The rightmost value is missing and has been reconstructed, but the maximum possible error here is only nineteen days.

The topmost value on the stela (the leftmost value as conventionally written) is a 1 (Figure 9.1), and this unit has been called by scholars the *kinchiltun*, a period of 1,152,000,000 days. One such period had been completed at the start of the chronological statement. The next value is 11 so-called *calabtuns*. A *calabtun* is a period of 57,600,000 days, and 11 of them represent an additional 633,600,000 days that have been completed. The next value is 19 *pictuns*. One *pictun* contains 2,880,000 days, and 19 *pictuns* add another 54,720,000 completed days to the total.

The next is that of the *baktun*, and it is in this position that the day count of the Maya era begins. It is important to realize that before the beginning of the recently completed era in 3114 BC, when the Initial Series day count began, Maya chronology reckoned that 1.11.19 days—or, in our decimal system, 1,840,320,000 days—of the eon had already been completed. This is a period of over five million years, and if we wanted to relate it to modern geological time reckoning, it would fall in the early Pliocene Epoch, earlier than the famous australopithecine fossil "Lucy."

Tikal Stela 10 provides an appropriate example of the Maya time periods before the recently completed era because it is unambiguous and because it has been important since 1915 in the history of research on this topic (Morley 1915:116).[1] It is not unique, however; there are a few other Classic monuments on which pre-eral time periods are recorded. Several other examples are to be found in the hieroglyphic book known as the Dresden Codex (SLUB 2010). Although the physical object dates to the Late Postclassic (indeed, the time of the Conquest), much of its content, including all the astronomical tables, refers explicitly to the centuries of the Classic period. The references that reach farthest back in time are the so-called serpent numbers.

Serpent numbers (Figure 9.2) are six-place bar-and-dot numerical values written in the coils of snakes that occur in the introductions to four of the long, complex, heavily astronomical tables in the codex, the seasonal tables (Bricker and Bricker 2011:489–550), and the water tables (Bricker and Bricker 2011:398–469). One of these serpent numbers from the seasonal tables, the left one in the figure, with the solid black bar-and-dot numbers, can be written 4.6.9.15.12.19. It is associated with two so-called Calendar Round dates, permutations of the 260-day sacred calendar or *tzolkin* and the 365-day vague year or *haab*. The two dates are 9 Kan 12 Kayab, near the top of the figure, and 13 Akbal 1 Kankin, near the bottom. Such calendar dates recur every 18,980 days or approximately fifty-two years. It has been known for more than a century that 9 Kan 12 Kayab is the so-called serpent base and that a serpent number is a count of the days completed since that base date (Förstemann 1891:18, 1904:422). This kind of date is very different from an Initial Series date, and the serpent base lies earlier in time than the beginning of the recently completed era. However, because a Calendar Round date recurs every fifty-two years, the exact chronological placement of the serpent base was not known

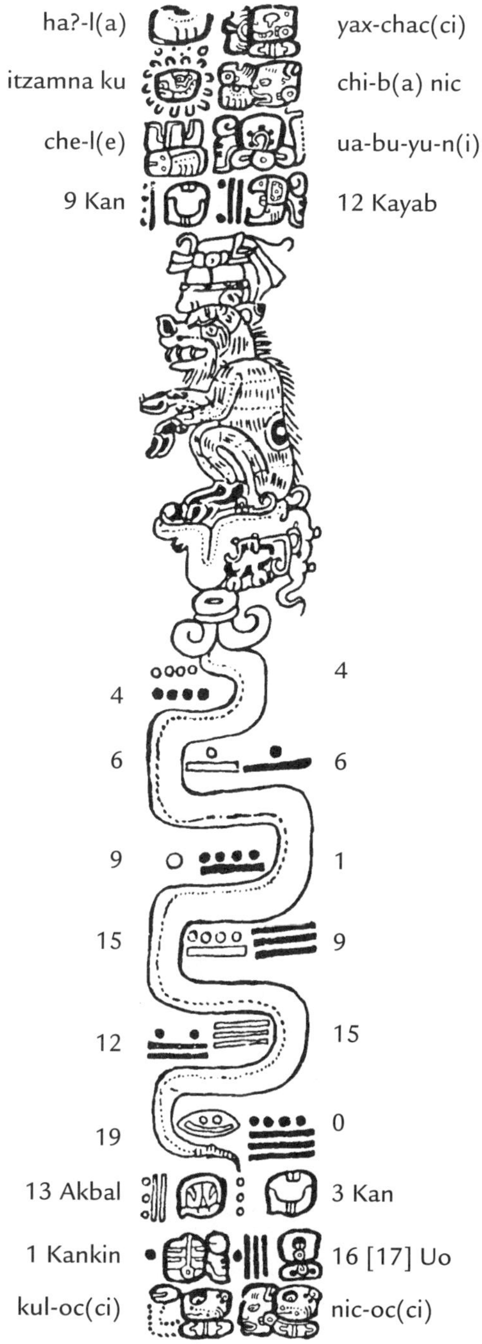

figure 9.2
Serpent-number dates and accompanying text in the introduction to the seasonal tables on page D.62 of the Dresden Codex. (After Bricker and Bricker 2011:497, fig. 11-9; drawing modified after Villacorta C. and Villacorta 1976:134.)

for a long time. A brilliant piece of research, published posthumously by Hermann Beyer (1943), formerly of Tulane's Middle American Research Institute, showed that the date of the serpent base is 1.11.15.3.6.14.11.4 9 Kan 12 Kayab. In a back-reckoned Gregorian calendar, this date would correspond to 8 July 33,142 BC, the time of Cro-Magnon people in southwestern Europe and the spectacular art of Chauvet Cave. The addition of the serpent number to this base produces a Maya date in the Terminal Classic period of the recently completed era. The Terminal Classic date results from the following calculation:

	1. 11. 15. 3. 6. 14. 11. 4	9 Kan 12 Kayab
+	4. 6. 9. 15. 12. 19	
	1. 11. 19. 10. 6.10. 6. 3	13 Akbal 1 Kankin.

The result is 1 September AD 958 (Gregorian). Beyer's solution was made possible by the fact that the same 13 Akbal 1 Kankin date appears elsewhere in the Dresden Codex, written in the better understood ring-number format (to be discussed). This date pair presented a Rosetta Stone–like translational tool. Although serpent numbers had been known previously only from the pages of the Dresden Codex, a very fragmentary stuccoed ceramic vessel recently recovered from the Classic site of Ek Balam in Yucatan bears two serpent numbers (Alfonso Lacadena, personal communication 14 November 2008). Neither is complete enough to be read with certainty, but one date is probably in the eleventh-century range of one of the Dresden serpent numbers.

The recently completed Maya era began with the completion of a *baktun* on a Calendar Round date of 4 Ahau 8 Cumku. In the full calendrical format we have been using, this date, the eral base, is 1.11.19.0.0.0.0.0 4 Ahau 8 Cumku. The five final positions, represented here by zeros, are, from left to right, those of the *baktun* (144,000 days), *katun* (7,200 days), *tun* (360 days), *uinal* (20 days), and *kin* (1 day). For reasons not fully understood, the era was considered to have only thirteen *baktuns*, not twenty.[2] Therefore, the first day of the era, when none of its days had yet been completed, could be expressed (in short form) as 13.0.0.0.0 4 Ahau 8 Cumku, which is exactly the same as the computationally more transparent 0.0.0.0.0. In either form, the eral base of 4 Ahau 8 Cumku dates to 11 August 3114 BC (Gregorian), equivalent to Julian Day Number 584,283 in the system used by modern Western chronologists.

In Pre-Columbian Maya mythological and astrological traditions, various events were considered to have occurred in the last years and months of the previous era, shortly before 4 Ahau 8 Cumku, and two different notational formats were used to specify the dates of such events. One is the use of a so-called ring number (Figure 9.3), subtracted from 13.0.0.0.0 4 Ahau 8 Cumku in order to arrive at a date in the last *baktun* of the previous era, usually designated "12." The "9 Ix" located just above the ring number in the figure is the *tzolkin* portion of

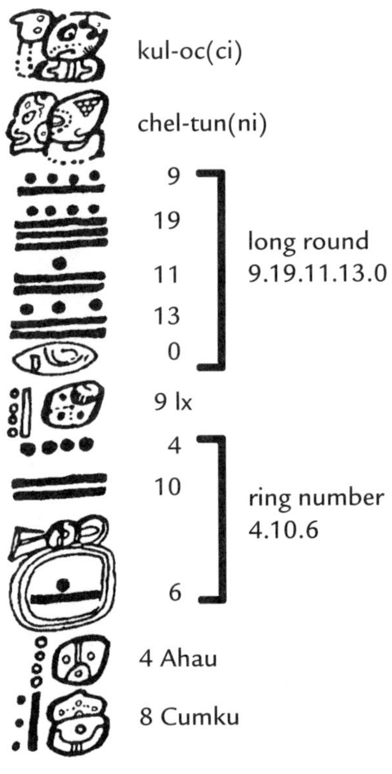

figure 9.3
Ring-number date and accompanying text in the introduction to the water tables on page D.70 of the Dresden Codex. (Bricker and Bricker 2011:409, fig. 10-25; drawing modified after Villacorta C. and Villacorta 1976:150.)

the date arrived at by the subtraction. This date, 9 Ix 2 Ch'en, is 4.10.6 days—that is, 1,646 days or about 4 1/2 years—before the start of the recently completed era. To complete the calculation, a quantity known as the long round is added to the date before the eral base to obtain a date in the recently completed era. The complete calculation is:

	13. 0. 0. 0. 0	4 Ahau 8 Cumku	(eral base)
-	4. 10. 6		(ring number)
	12. 19. 15. 7. 14	9 Ix 2 Ch'en	(before eral base)
+	9. 19. 11. 13. 0		(long round)
	9. 19. 7. 2. 14	9 Ix 17 Ch'en	(a base date of the upper water table).

This ring-number format is used frequently in the Dresden Codex.

In some carved inscriptions of the Classic period, a briefer format—a simple reference to a *baktun* 12 date—is used. For example, a date found in Temple XIX at Palenque is 12.10.1.13.2 9 Ik 5 Mol (Figure 9.4; Stuart 2005:60). That the *baktun* "12" in this expression refers to the previous era rather than the recently completed one is indicated not only by the context of the inscription, but also the fact that the same long-count date in the latter (in AD 1817) would have a Calendar Round date of 9 Ik 0 Zodz rather than 9 Ik 5 Mol.

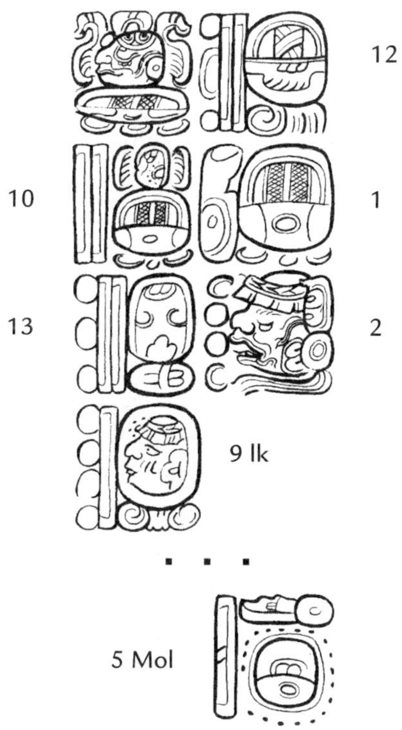

figure 9.4
Inscribed date, 12.10.1.13.2 9 Ik 5 Mol, from Temple XIX at Palenque, referring to the *baktun* 12 preceding the eral base at 4 Ahau 8 Cumku. (After Stuart 2005:61, fig. 34.)

Looking Forward in Time

The other extremity of the recently completed era, the completion of thirteen *baktuns* in December 2012, was almost never recorded in Classic or Postclassic records. One definite, unambiguous mention (Figure 9.5) occurs at the Classic site of Tortuguero in Mexico (Gronemeyer and MacLeod 2010). A Calendar Round date of 4 Ahau 3 Kankin refers to the day of completion when the Long Count expression was once again 13.0.0.0.0, the day 21 December AD 2012 in the Gregorian calendar. Another example of this date has recently been reported from the Classic site of La Corona in Guatemala (Canuto n.d.; Stuart n.d.).

Looking ahead to the end of the era was, however, not the farthest view into the future found in the Pre-Columbian record. Despite the convention that limited the era to thirteen *baktuns*, the fact that a *pictun* contains twenty *baktuns* was not ignored. Recorded at the Classic site of Palenque are elements of a calculation that added one *pictun* to the eral base of 4 Ahau 8 Cumku (Lounsbury 1991:817–818). The result is recorded only as the Calendar Round 10 Ahau 13 Yaxkin (Figure 9.6). The calculation in its full form is

	1. 11. 19. 0. 0. 0. 0. 0	4 Ahau 8 Cumku
+	1. 0. 0. 0. 0. 0	
	1. 12. 0. 0. 0. 0. 0. 0	10 Ahau 13 Yaxkin.

The date reached, if rendered in the Gregorian calendar, would be 13 October AD 4772. The Dresden Codex contains several instances in which a *pictun* is used in a calendrical calculation, and one, in the

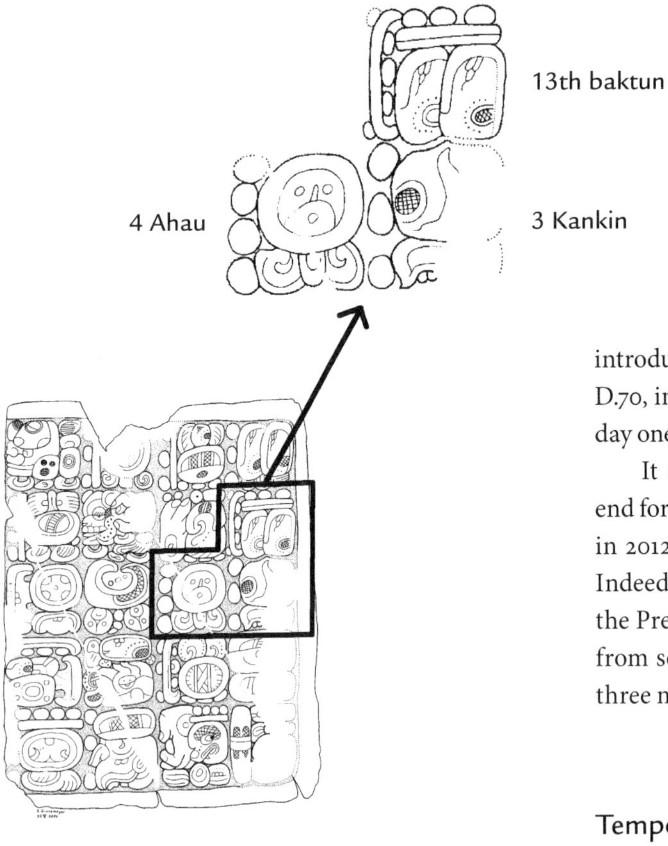

figure 9.5
Inscriptional reference to the completion of the thirteenth *baktun* on a day 4 Ahau 3 Kankin, part of the text on the right panel of Monument 6 at Tortuguero. The reference is to the end of the Maya era on 21 December AD 2012. (After Gronemeyer and MacLeod 2010:2, fig. 1a.)

figure 9.6
Portion of the inscription on the west panel of the Temple of the Inscriptions at Palenque referring to a day 10 Ahau 13 Yaxkin (13 October AD 4772, Gregorian), which is one *pictun* after the eral base at 4 Ahau 8 Cumku. (After Maudslay 1889–1902:4:pl. 62.)

introduction to the water tables on pages D.69 and D.70, includes a reference to 10 Ahau 13 Yaxkin, the day one *pictun* after the eral base.

It should then be obvious that time did not end for the Pre-Columbian Maya with the era's end in 2012, nor did it start with that era's beginning. Indeed, the part of the unlimited eon within which the Pre-Columbian Maya reckoned time extended from some five million years in the past to nearly three millennia into the future.

Temporal Notation: Maya Regional Variants

In the southern lowlands and the adjacent highlands of the Maya area, the Initial Series was the principal method for specifying a date within the era. This notation, recording how many days, expressed as *baktuns*, *katuns*, *tuns*, *uinals*, and *kins*, had been completed since 13.0.0.0.0 4 Ahau 8 Cumku, was usually accompanied by a Calendar Round permutation and placed at the beginning of a historical inscription. It served as the temporal anchor of the inscription, to which other dates recorded as Calendar Round permutations were connected by distance numbers.

The Initial Series notation was less common in the northern lowlands, where a completely different notation emerged during the Late Classic period and became the dominant one in that region. This so-called *tun*-Ahau notation consisted of three sets of elements (Figure 9.7): (1) a Calendar Round permutation (e.g., 9 Lamat 11 Yax), (2) a reference to a specific *tun* in a *katun* (e.g., 13 *tuns*), and (3) the name of the day on which the *katun* in question ended (e.g., 1 Ahau). This information, recorded on one of the

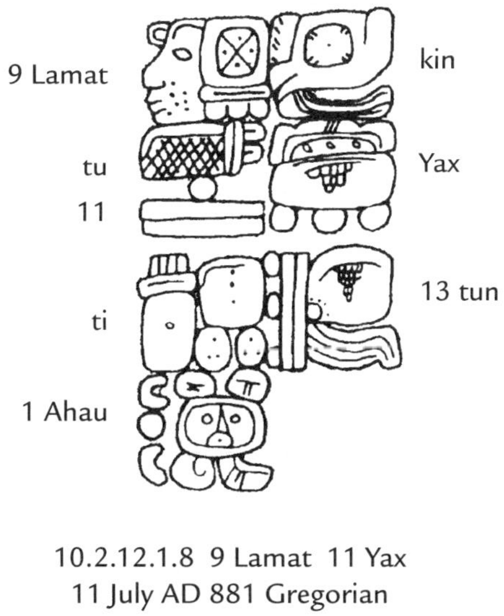

figure 9.7
A *tun*-Ahau date—a day 9 Lamat 11 Yax in the thirteenth *tun* of a *katun* ending on 1 Ahau—on Lintel 4 of the Temple of the Four Lintels at Chichen Itza. (After Krochock 1989:13, fig. 7.)

figure 9.8
A *tun*-Ahau date—a day (?) Akbal 16 Zac in the sixth *tun* of a *katun* ending on 11 Ahau—on page P.4 of the Paris Codex. (Full-page sketch after Gates 1909; enlarged detail adapted from Bricker and Bricker 2011:359, fig. 9-55b.)

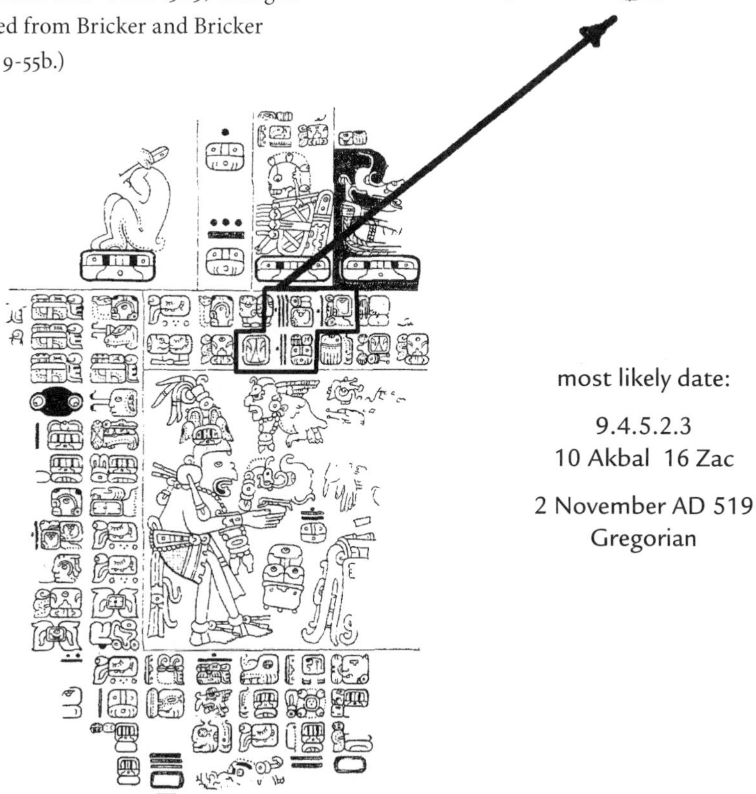

lintels at Chichen Itza, is sufficient for calculating the Initial Series position of the date in question:

> 10.2.12.1.8 9 Lamat 11 Yax (= 11 July AD 881, Gregorian).

The largest component of the *tun*-Ahau notation is composed of thirteen *katuns* or about 256 years, resulting from the fact that the coefficients of the last day of each *katun*, always Ahau, are numbered from 1 to 13 and occur in a fixed sequence: 13, 11, 9, 7, 5, 3, 1, 12, 10, 8, 6, 4, 2. For this reason, this notation has sometimes been called the "short count." However, J. Eric S. Thompson (1937), the scholar who explained how this notation worked, pointed out that it specifies a date uniquely in an eighteen-*baktun* cycle, which is five *baktuns* larger than the era! An example of this notation (Figure 9.8) appears on one of the so-called *katun* pages of the Paris Codex (Codex Peresianus 1968), which refer to a series of *katuns* extending from the fifth to the middle of the eighth centuries AD (Bricker and Bricker 2011:357–363). The presence of a *tun*-Ahau date, a characteristically "northern" notation, in the Paris Codex suggests that the codex may have been created somewhere in the northern lowlands of the Yucatan Peninsula, a suggestion in agreement with other lines of evidence.

Some elements of the "northern" notation are of historical significance because they continued to be used after the end of the Classic period, when Initial Series dates ceased to be recorded. They were still in use when the Spaniards arrived in northern Yucatan during the sixteenth century, making it possible to correlate the Maya calendar with the European one and therefore to determine the Western equivalents of the Initial Series dates recorded on Classic period monuments.

Temporal Notation: Adoption of Central Mexican Conventions

After the collapse of Classic Maya civilization in the southern lowlands between AD 900 and 1000, the *tun* and the Long Count based on it declined in importance,[3] and the 365-day *haab* and the fifty-two-year Calendar Round based on it became more significant units of time in later Maya contexts. This change may have been one aspect of the adoption in the Maya area during the Postclassic of some cultural elements from Central Mexico, where there had never been a Long Count based on the 360-day *tun*. This difference is most evident in the Madrid Codex (Codex Tro-Cortesianus 1967), which contains neither serpent dates, dates derived from ring numbers and long rounds, nor any *tun*-Ahau dates. Instead, there are only almanacs based on the 260-day *tzolkin* and its multiples and *haab* dates embedded in the captions and pictures of some of the almanacs (a feature almost completely limited to the Madrid Codex). The codex also contains several full or partial *in extenso* almanacs, which are a common feature of Central Mexican codices (Just 2004). The listed days in such almanacs are not separated by distance numbers; instead, every day in the *tzolkin* is mentioned explicitly (Figure 9.9). The one from the Madrid Codex is only a partial *in extenso* almanac, combining the structure of such an almanac with the traditional structure of a Maya almanac, but its basic layout is strikingly similar to that of the *in extenso* almanac on pages B.1 to B.8 of the Central Mexican Codex Borgia (Códice Borgia 1993) and its cognates in Codices Vaticanus B (Codex Vaticanus 3773 1972) and Cospi (Codex Cospi 1968).

In one of the captions above the Madrid's *in extenso* layout is a reference to Cumku (spelled syllabically as **cu-ku**), the eighteenth month of the *haab* (Figure 9.9). In the previous caption appears a collocation that can be deciphered as *ik-kat*, which is a phonetic spelling of the name of the second month of the *haab*, whose name in Colonial Yucatec was Uo (Vail and Bricker 2004:208–209). Other month names can be found in the text and pictures of the New Year's pages (M.34–M.37) of the Madrid Codex (Figure 9.10): Pop, Yaxkin, Yax, and Ceh, in several cases with bar-and-dot numbers as coefficients (Graña-Behrens 2002:107; Hernández and Bricker 2004:306–309; Vail and Bricker 2004:199–201). The presence of *haab* dates in these and other almanacs in the Madrid Codex contrasts with their

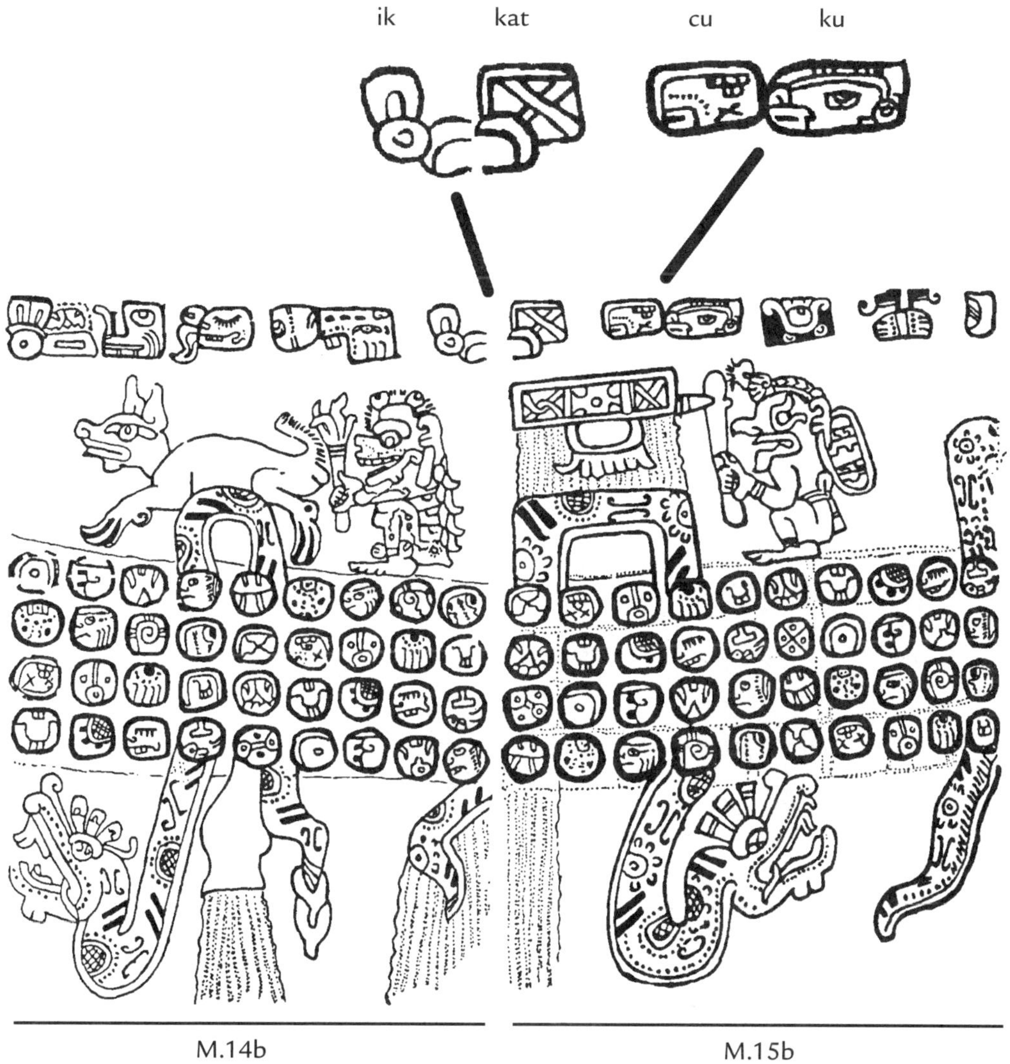

figure 9.9
Two pages of the *in extenso* almanac in the Madrid Codex containing glyphic references to months of the *haab*. (Bricker and Bricker 2011:574, fig. 11-47; drawing modified after Villacorta C. and Villacorta 1976:252, 254.)

near absence in almanacs in the Dresden and Paris Codices and attests to the greater interest in the *haab* cycle by the compilers of the Madrid Codex.[4]

An even stronger link between the Madrid Codex and Central Mexico can be found in the almanacs on pages M.24 to M.31 of the codex (Figure 9.11). The two almanacs on page M.31 replicate the layout and calendrical structure of the right side of the planting almanac on page B.27 of the Codex Borgia (Hernández and Bricker 2004:284–287). In the lower registers, the starting date in the Aztec calendar, 1 Crocodile, is the day after 13 Ahau in the Maya calendar, and in the upper registers, 1 Death in the Aztec calendar is one day after 13 Chicchan in the Maya calendar. Other almanacs in the same section of the Madrid Codex have been shown to represent a reworking of the layout and calendrical and iconographic information on Borgia pages B.27 and B.28 in terms of Maya conventions for almanacs (Bricker and Milbrath 2011). This discovery was facilitated by Gabrielle Vail's (2002) crucial observation that the distance numbers in Maya codical

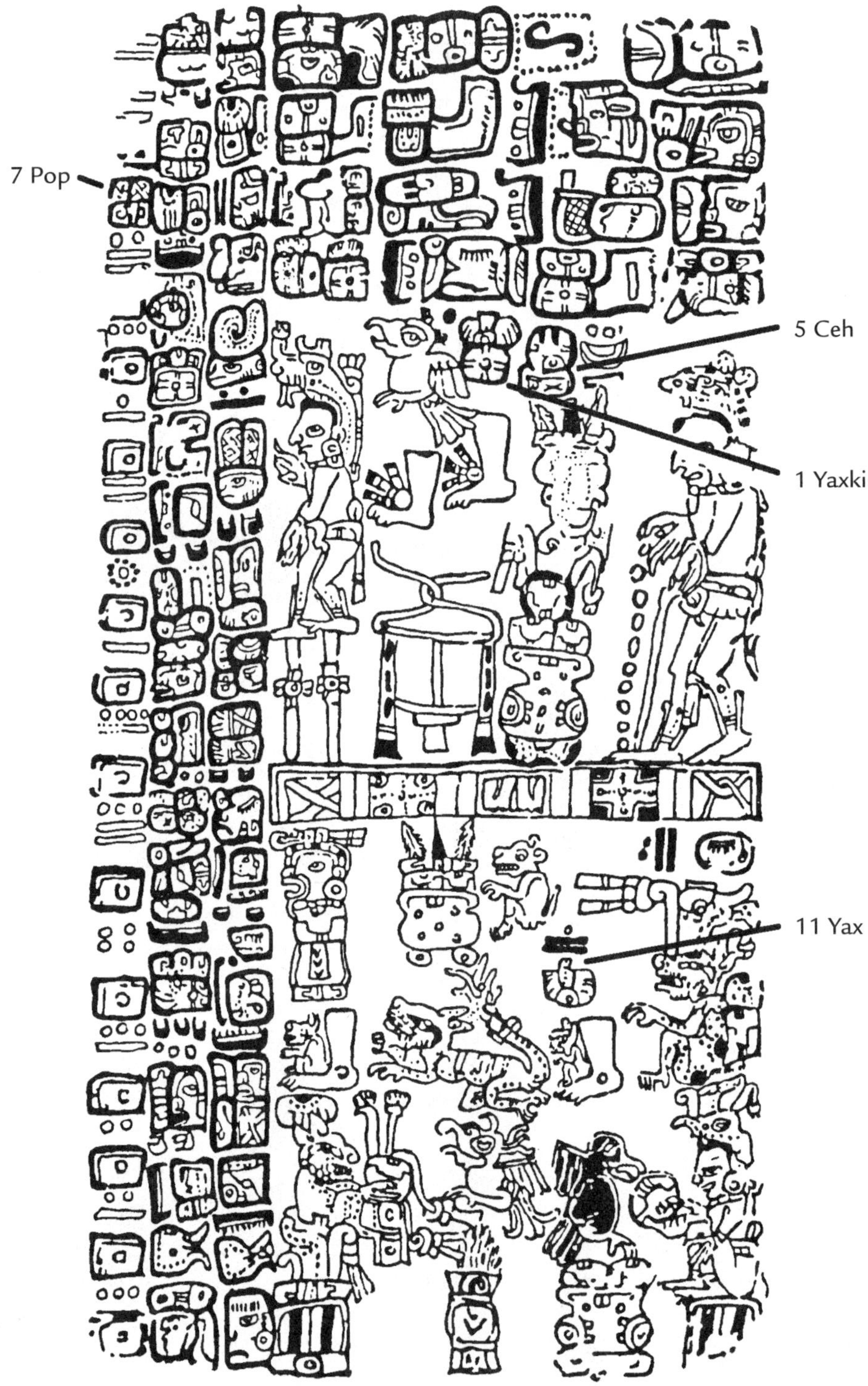

figure 9.10

One of the New Year's pages (M.36) of the Madrid Codex containing several *haab* dates. (Bricker and Bricker 2011:141, fig. 6-21; drawing modified after Villacorta C. and Villacorta 1976:296.)

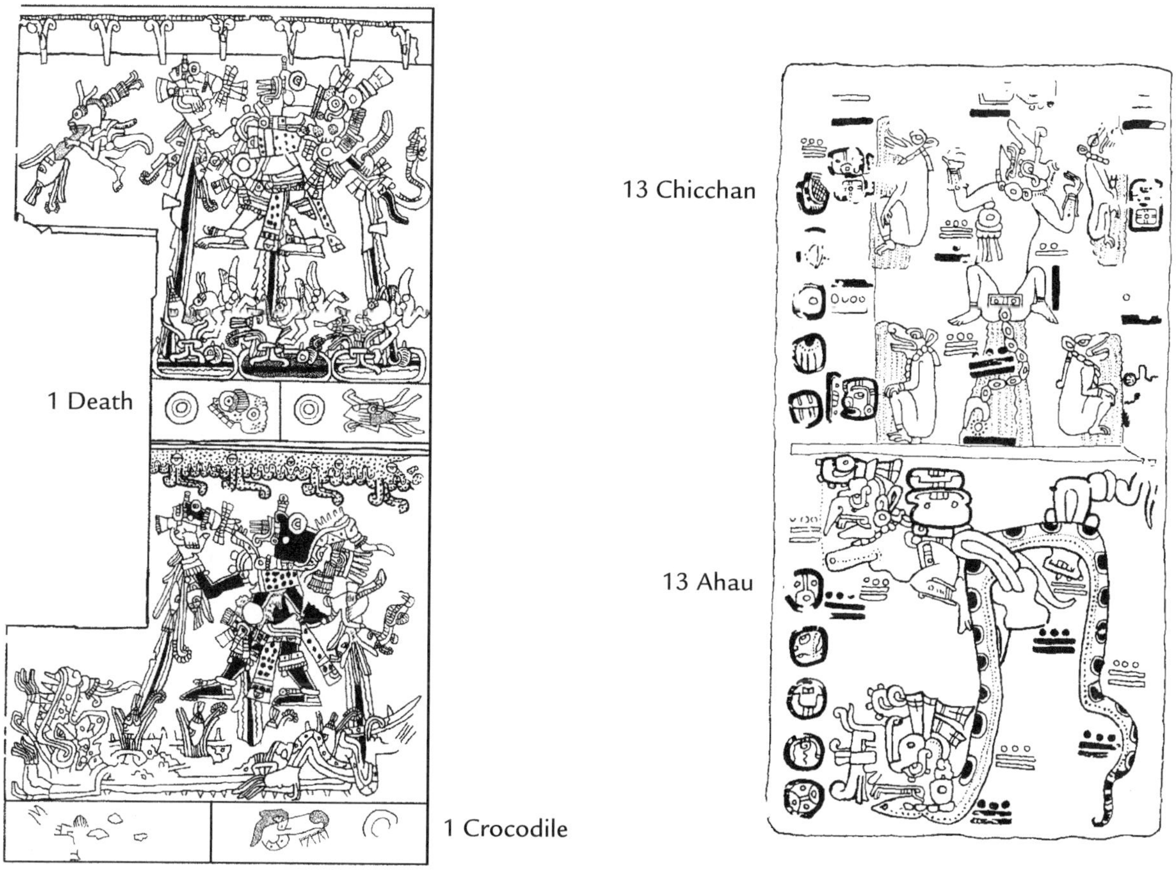

figure 9.11
Comparison of structurally similar almanacs on page B.27 of the Codex Borgia (drawing by Christine Hernández) and page M.31 of the Madrid Codex (After Villacorta C. and Villacorta 1976:286.)

almanacs are inherently ambiguous because they do not include period glyphs, and some could have represented *haabs* rather than days.

Another Central Mexican link in the Madrid Codex is what we have called the "Calendar Round almanac," which is found on pages M.65a–M.73b (Figure 9.12). This almanac ends with a full Calendar Round, 13 Ahau 13 Cumku, an unusual feature for a Maya codical almanac, but one that is consistent with the increasing importance of the *haab*. It is also an *in extenso* almanac, in which every day in the 260-day *tzolkin* is mentioned. The almanac contains thirty-three pictures, thirty-two of which are associated with eight *tzolkin* days; the remaining four days, including the Calendar Round, are in the last picture and caption. This is an uncommon format for a Maya 260-day almanac, which is usually evenly divided into either five rows of fifty-two days, four row of sixty-five days, or ten rows of twenty-six days, without any leftover days.

Because each and every day in the 260-day period is mentioned in the Madrid's Calendar Round almanac, there is no need for distance numbers to express the interval between them. Nevertheless, some distance numbers *are* included, but they are in the captions over the pictures, which is a unique placement in a Maya codical almanac. The almanac contains twenty-nine distance numbers. One of them refers to lunations, six to *uinals*, one to *tuns* (spelled phonetically as [tu]**tun**), and probably one to days. The remaining twenty distance numbers are represented by the collocation that represented the 260-day *tun* during the Classic period but here represents the 365-day *haab* (Bricker and Bricker 2011:811–812). The *haab* is clearly more significant than the *tun* in

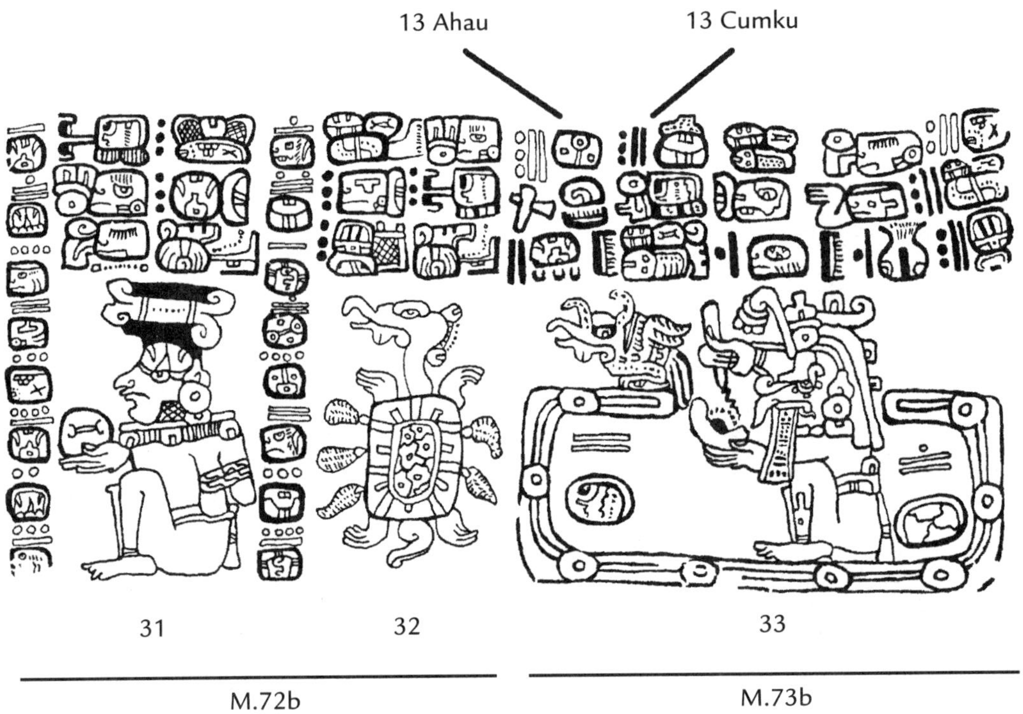

figure 9.12
The last three pictures, captions, and calendrical notations of the Calendar Round almanac in the Madrid Codex. (Bricker and Bricker 2011:771, fig. 12-46; drawing modified after Villacorta C. and Villacorta 1976:368, 370.)

this Postclassic almanac that has as its format the Central Mexican *in extenso* principle.

Cosmological Cyclicity?

Having now documented the influence of Central Mexican time reckoning in the Late Postclassic Madrid Codex, we move on to another Central Mexican temporal concept that found its way into Maya worldview sometime during the Postclassic period. We are referring here to the notion of multiple creations, which is best exemplified in the famous K'iche' manuscript, the Popol Vuh, of highland Guatemala (Christenson 2007; Edmonson 1971; Tedlock 1996), but is manifested also in the oral tradition of Chamula in highland Chiapas (Gossen 2002). We are aware of no clear evidence that the Classic period Maya believed in multiple creations (a point made decades ago by Mary Miller [1980:60]), and indeed, the centrality of the eon in their time reckoning argues otherwise. Furthermore, there is no evidence of multiple creations in any of the Maya codices and no comparable account of multiple creations in the northern lowlands today.

According to the Anales de Cuauhtitlán (1945:5), the Aztecs of Central Mexico believed that time was divided into five Suns or ages. The first age was called the Sun of Water because it was associated with the day 4 Water, and it ended in a flood. The second age was called the Sun of the Jaguar because it was associated with the day 4 Jaguar, and it was brought to an end by people being devoured by animals. The third age was called the Sun of Rain because it was associated with the day 4 Rain. The "rain" in question was the rain of fire from a volcanic eruption, which brought the Sun of Rain to an end. The fourth age was called the Sun of Wind because it was associated with the day 4 Wind. At the end of the age, people were turned into monkeys and were dispersed throughout the mountains by a hurricane. The fifth age was called the Sun of Movement because it was associated with the day 5 Movement. The Aztecs were living in the fifth Sun or age when the Spaniards arrived in the sixteenth century, and they believed that it would end with earthquakes.

The Popol Vuh is also concerned with multiple creations, but it has reduced the number of creations from five to four, as has the creation myth in Chamula (Gossen 2002). It begins with an account of the first three creations by Quetzalcoatl, a deity of Central Mexican origin. In the first creation, he made animals. But because the animals could not speak or pray, they were destroyed. In the second creation, Quetzalcoatl made people of mud, but their bodies were too soggy and kept coming apart, so he destroyed them. In the third creation, the people he created were made of wood. They had no hearts or minds, and their faces were dry, so they were killed in a flood, a rain of glue or resin (an allusion to a volcanic eruption?). In addition, the animals attacked them, as did their tools and pots, and they became the monkeys in the forest today.

The account of the four creations is then interrupted by two stories belonging to the fourth creation. It continues in the next section with the creation of humans, representing the fourth and present creation.

The K'iche' account of the four creations shares several traits with the Aztec myth of multiple creations. In both accounts, there is an age of animals and the world being brought to an end by a volcanic eruption. Both accounts describe the origin of monkeys. The Aztec account is structured in terms of five days in the 260-day ritual calendar; the K'iche' account is not. The Aztec account forms a continuous narrative. The K'iche' account does not, suggesting that it was pieced together at a later date, combining themes from the K'iche' heritage with those received from elsewhere. The differences between these two bodies of mythic material, particularly as they relate to multiple creations, were examined in detail by Miller (1980).

The rest of the Popol Vuh exhibits the same piecing together of elements from Central Mexico with those of highland Guatemala. It contains a migration history that mentions "Seven Caves" as the place of origin of the K'iche' people (Edmonson 1971:161). But "Seven Caves," or Chicomostoc, is also the homeland of the Aztecs. It refers to the Yaqui of northern Mexico (Edmonson 1971:160). It also says that the K'iche' traveled from Seven Caves to a place called Tula-Zuyua. Tula is the name of a Late Classic city in Central Mexico, lying between Chicomostoc, the Aztec homeland, and the shores of Lake Texcoco, where the Aztecs eventually settled. It says that the K'iche' were dressed in hides, in this respect resembling the Chichimec ancestors of the Aztecs on their journey to the Valley of Mexico. Finally, it mentions that Tohil, the deity who accompanied the K'iche' on their migration from Seven Caves to their present location in highland Guatemala, was carried on the back of a migrant, as was Huitzilopochtli, the patron deity of the Aztecs, who accompanied them on their migration from Chicomostoc to the Valley of Mexico. These similarities suggest that the K'iche' account of their migration was heavily influenced by Central Mexican views of creation and migration in the Late Postclassic and/or the early Colonial period, as Robert Hill (2012) has recently demonstrated for the Kaqchikel counterpart of the Popol Vuh. For this reason, we do not regard the Popol Vuh as a reliable source on lowland Classic Maya concepts of time as it relates to creation or multiple creations. Indeed, the obvious mixture of Maya and Mexican elements in the Popol Vuh means that it cannot be accepted uncritically as an outline of Maya mythology and culture during the Classic period anywhere in the Maya area or during the Postclassic in the Maya lowlands.[5]

A final consideration about linearity and cyclicity in Pre-Columbian time reckoning is a closer look at the meaning of the eral base date 4 Ahau 8 Cumku. It has become common practice to refer to this mythical day in 3114 BC as the day of Maya "creation," but this is certainly a misnomer and the application of the wrong concept. To understand how 4 Ahau 8 Cumku was considered, based on Classic inscriptions themselves, we need to pull away from the highly imaginative poetic interpretations of Linda Schele as expressed in *Maya Cosmos* (Freidel et al. 1993:59–107). One of the crucial texts about this day appears on Stela C from Quiriguá, Guatemala. A discussion of this text by Stuart (2011:216–222), reflecting modern advances in epigraphy, clarifies its content. The stela was erected by the ruler of Quiriguá in commemoration of the contemporary period ending (9.17.5.0.0 6 Ahau 13 Kayab or 27 December AD 775, Gregorian), and the account of what was said to have happened in the deep past, on 4 Ahau 8 Cumku, describes elements of period-ending ritual. This ritual included the binding together of three stones, probably stelae, described by Stuart (2011:219) as "an important ceremony of stone dedication, common in history during a period ending." According to the Stela C inscription, the ceremony was carried out or presided over by two deities (dubbed "paddler gods" in some of the literature) who were "godly patrons of period endings in historical time." (Stuart 2011:219). Another part of the Quiriguá inscription says that on 4 Ahau 8 Cumku "the face image changed," and the same statement occurs in a text at Palenque describing the events of 4 Ahau 8 Cumku. Although the exact meaning of this phrase is not clear, Stuart (2011:220–221) suggests that "at the very least it refers to some type of replacement or switch from a previous state." The message here is that on 4 Ahau 8 Cumku something ended and was replaced by something else. There was no *ex nihilo* "creation." The world was not created, a "present world" was not created, the gods were not created, and—most certainly—time itself was not created. What ended was one thirteen-*baktun* era, and it was replaced by another one, the one that ended and was replaced by a new one on the winter solstice in AD 2012. What was said to have happened on 4 Ahau 8 Cumku is an excellent example of temporal cyclicity nested within the linearity of a limitless eon. The eral base was not the day of Maya creation; it was a day of *inauguration* of a new era.[6]

Summary

How, then, can we summarize the essential linear and cyclical characteristics of Pre-Columbian Maya time reckoning? First, the fullest calendrical system was one of repeating cycles, including the era, that followed each other within a nonrepeating eon without beginning or end. Although mythology dealt with the births of gods or dynasties at specific moments in the distant past, time itself and its terrestrial and celestial contents existed before the recently completed era and continues to exist now that a new era has started.

Second, there was a regional difference during the Classic between the southern and northern parts of the Maya area in how time was tracked within the recently completed era. In the south, this tracking was accomplished most precisely with Initial Series dates, whereas in the north the same degree of precision was accomplished with *tun*-Ahau dates, which were used in addition to Initial Series dates.

Third, new kinds of cycles of time achieved importance for some Maya groups during the Postclassic as a result of the adoption of ideas and practices from Central Mexico. One result was the increasing importance of the *haab* as the principal calendrical unit at the expense of the *tun*. Another, limited to the Maya highlands, was the importation of the mythology of multiple creations.

Nevertheless, in spite of the differences and changes that can be seen throughout the centuries, enough of the fundamental elements of Pre-Columbian Maya time reckoning remained in place after the Conquest to allow the determination of the correlation between the European calendar and that of the Classic Maya.

Acknowledgments

We are grateful to Alfonso Lacadena for making us aware of the serpent-number vessel from Ek Balam. Anthony Aveni, Susan Milbrath, and John Justeson offered very helpful comments on the version of this study presented at the Dumbarton Oaks symposium, and we thank them for their much appreciated collegiality.

NOTES

1 Thompson (1950:314–315) suggested that the numerical series beginning with 1.11.19 was not a date at all but rather a distance number connecting an 8 Manik *tzolkin* date at the top of the inscription with a date far in the future, which is, however, not found on the stela. In a detailed discussion of the Stela 10 Inscription, Jones and Satterthwaite (1982:25–29) firmly rejected Thompson's suggestion about a distance number, accepting instead Morley's original interpretation of the series as an extended Initial Series date. They noted that the series was immediately preceded by an Initial Series Introductory Glyph and that it moved from larger values to smaller ones, features expected of an Initial Series date rather than of a distance number. They observed further that in order to make his distance-number interpretation work, Thompson was obliged to regard a very clear "11" coefficient for the *calabtun* value as a scribal error for a putative "12." In addition, the Initial Series date as traditionally read falls in the Maya month of Muan, consistent with the variable element in the Initial Series Introductory Glyph. Finally, the Western-calendar equivalent of this date is in the year AD 506, unambiguously within the life-span of the nineteenth ruler of Tikal, who is portrayed on the front of Stela 10 (Martin and Grube 2000:38–39).

2 Perhaps, as suggested to us by John Justeson, what is recognized here is a kind of quantitative consonance between the 260 days of the *tzolkin* and the 260 *katuns* of the thirteen-*baktun* era.

3 The latest currently known Initial Series Long Count date on a carved stone monument is from Monument 101 at the site of Tonina, Chiapas, Mexico; the date is 10.4.0.0.0 or 18 January AD 909, Gregorian (Sharer and Traxler 2006:471). However, Initial Series dates in Long Count format did not just disappear. The Late Postclassic hieroglyphic book known as the Dresden Codex contains several such dates, the latest of which, in the introduction to the eclipse table, is 10.19.6.1.8 or 25 September AD 1210, Gregorian.

4 *Haab* dates are, of course, present in all the *tables* (as opposed to the almanacs) in the Dresden Codex. One of these tables, the Venus table on pages D.24, D.46–D.50, contains a very explicit link with Postclassic Central Mexico in the form of Nahuatl deity names written phonetically in Maya hieroglyphs (as discussed in Bricker and Bricker 2011:188–204). And, indeed, a concern with the Venus cycle in relation to the solar year has earlier roots in the Central Mexican Classic at Teotihuacan (Milbrath n.d.).

5 In this matter we agree with the position of David Stuart, who expressed strong reservations about the use to which the Popol Vuh has been put in the interpretation of Classic Maya culture. After detailing the very limited extent to which the content of the Popol Vuh overlaps with the content of Classic Maya art or mythology, Stuart criticizes the current tendency to see the Popol Vuh as a survival of Classic Maya religion. He concludes: "Some of my colleagues may not like me saying this, but I'm of the strong opinion that the Popol Vuh has been overused as *the* document on Maya religious thought" (Stuart 2011:215).

6 In our own calendrical system, Y2K was not the "creation" of a new millennium; it was the *inauguration* of that new period.

REFERENCES CITED

Anales de Cuauhtitlán

1945 *Códice Chimalpopoca: Anales de Cuauhtitlan y Leyenda de los Soles.* Translated by Primo Feliciano Velázquez. Universidad Nacional Autonoma de México, Instituto de Historia, Mexico City.

Beyer, Hermann

1943 The Long Count Position of the Serpent Number Dates. Proceedings of the 27th International Congress of Americanists, Mexico City 1939, 1:401–405.

Bricker, Harvey M., and Victoria R. Bricker

2011 *Astronomy in the Maya Codices.* American Philosophical Society Memoirs 265. American Philosophical Society, Philadelphia.

Bricker, Victoria R., and Susan Milbrath

2011 Thematic and Chronological Ties between the Borgia and Madrid Codices Based on Records of Agricultural Pests in the Planting Almanacs. *Journal of Anthropological Research* 67:497–531.

Canuto, Marcello

n.d. The "2012" Block. Electronic document, http://mari.tulane.edu/PRALC/news.html, accessed July 22, 2012.

Christenson, Allen J.

2007 *Popol Vuh: The Sacred Book of the Maya.* University of Oklahoma Press, Norman.

Codex Cospi

1968 *Codex Cospianus (Codex Cospi, Calendario messicano 4093); Biblioteca Universitaria Bologna (mit Einleitung und Summary von Karl A. Nowotny).* Codices Selecti XVIII. Akademische Druck- und Verlagsanstalt, Graz.

Codex Peresianus

1968 *Codex Peresianus (Codex Paris); Bibliothèque Nationale Paris (mit Einleitung und Summary von Ferdinand Anders).* Codices Selecti IX. Akademische Druck- und Verlagsanstalt, Graz.

Codex Tro-Cortesianus

1967 *Codex Tro-Cortesianus (Codex Madrid); Museo de America, Madrid (mit Einleitung und Summary von Ferdinand Anders).* Codices Selecti VIII. Akademische Druck- und Verlagsanstalt, Graz.

Codex Vaticanus 3773

1972 *Codex vaticanus 3773 (Codex vaticanus B); Biblioteca Apostolica Vaticana (mit Einleitung, Summary, und Resumen von Ferdinand Anders).* Codices Selecti XXXVI. Akademische Druck- und Verlagsanstalt, Graz.

Códice Borgia

1993 *Códice Borgia.* Códices Mexicanos V. Akademische Druck- und Verlagsanstalt, Graz.

Edmonson, Munro S.

1971 *The Book of Counsel: The Popol Vuh of the Quiche Maya of Guatemala.* Middle American Research Institute Publication 35. Middle American Research Institute, Tulane University, New Orleans.

Förstemann, Ernst W.

1891 *Zur Entzifferung der Mayahandschriften. II.* Privately printed pamphlet. Druck von C. Heinrich, Dresden.

1904 Pages 61 to 64 and 69 to 73, Dresden Codex. In *Mexican and Central American Antiquities, Calendar Systems, and History,* edited by Charles Bowditch, pp. 409–422. English translation of *Zur Entzifferung der Mayahandschriften II.* Bureau of American Ethnology Bulletin 28. Smithsonian Institution, Washington.

Freidel, David, Linda Schele, and Joy Parker

1993 *Maya Cosmos: Three Thousand Years on the Shaman's Path.* Perennial/HarperCollins, New York.

Gates, William E.

1909 *Codex Perez, Maya-Tzental.* Privately printed, Point Loma. Facsimile ed., 2001, by Ecological Linguistics as part of a collection entitled "The Paris Codex, By Learned Mayans."

Gossen, Gary H.

2002 *Four Creations: An Epic Story of the Chiapas Mayas.* University of Oklahoma Press, Norman.

Graña-Behrens, Daniel

2002 *Die Maya-Inschriften aus Nordwestyukatan, Mexiko.* Inaugural-Dissertation zur Erlangung der Doktorwürde der Philosophischen Fakultät, Rheinische Friedrich-Wilhems-Universität, Bonn.

Gronemeyer, Sven, and Barbara MacLeod

2010 What Could Happen in 2012: A Re-analysis of the 13-*Bak'tun* Prophecy on Tortuguero Monument 6. *Wayeb Notes* 34:1–68.

Hernández, Christine L., and Victoria R. Bricker

2004 The Inauguration of Planting in the Borgia and Madrid Codices. In *The Madrid Codex: New Approaches to Understanding an Ancient Maya Manuscript*, edited by Gabrielle Vail and Anthony F. Aveni, pp. 277–320. University Press of Colorado, Boulder.

Hill, Robert M., II

2012 *Pictograph and Alphabet—and Back: Reconstructing the Pictographic Origins of the Xajil Chronicles.* American Philosophical Society Transactions, Vol. 102, No. 5. American Philosophical Society, Philadelphia.

Jones, Christopher, and Linton Satterthwaite

1982 *The Monuments and Inscriptions of Tikal: The Carved Monuments.* University Museum Monographs 44, Tikal Reports 33, Part A. University Museum, University of Pennsylvania, Philadelphia.

Just, Bryan R.

2004 *In Extenso* Almanacs in the Madrid Codex. In *The Madrid Codex: New Approaches to Understanding an Ancient Maya Manuscript*, edited by Gabrielle Vail and Anthony F. Aveni, pp. 255–276. University Press of Colorado, Boulder.

Krochock, Ruth

1989 *Hieroglyphic Inscriptions at Chichén Itzá, Yucatán, México: The Temples of the Initial Series, the One Lintel, the Three Lintels, and the Four Lintels.* Research Reports on Ancient Maya Writing 23, pp. 7–14. Center for Maya Research, Washington.

Lounsbury, Floyd G.

1991 Distinguished Lecture: Recent Work in the Decipherment of Palenque's Hieroglyphic Inscriptions. *American Anthropologist* 93:809–825.

Martin, Simon, and Nikolai Grube

2000 *Chronicle of the Maya Kings and Queens.* Thames and Hudson, London.

Maudslay, Alfred P.

1889–1902 *Biologia Centrali-Americana: Archaeology.* 5 vols. Porter and Dulau, London.

Milbrath, Susan

n.d. The Venus Almanac in Mesoamerica. Paper presented in the symposium "New Perspectives on Ancient Maya Astronomy," 72nd Annual Meeting of the Society for American Archaeology, Austin, April 2007.

Miller, Mary Ellen

1980 Central Mexican Myth in the Popol Vuh. *Mexicon* 2:60–63.

Morley, Sylvanus G.

1915 *An Introduction to the Study of the Maya Hieroglyphs.* Bureau of American Ethnology Bulletin 57. Smithsonian Institution, Washington.

Sharer, Robert J., with Loa P. Traxler

2006 The Ancient Maya. 6th ed. Stanford University Press, Stanford.

SLUB

2010 Codex Dresdensis–Mscr.Dres.R.310. High-resolution facsimile of the Dresden Codex, electronic document, http://digital.slub-dresden.de/sammlungen/titeldaten/280742827/, accessed March 9, 2010.

Stuart, David

2005 *The Inscriptions from Temple XIX at Palenque: A Commentary.* Pre-Columbian Art Research Institute, San Francisco.

2011 *The Order of Days: The Maya World and the Truth about 2012.* Harmony Books, New York.

n.d. Notes on a new text from La Corona. Electronic document, http://decipherment.wordpress.com/2012/06/30/notes-on-a-new-text-from-la-corona/, accessed November 20, 2012.

Tedlock, Dennis

1996 *Popol Vuh: The Mayan Book of the Dawn of Life*. Rev. ed. Touchstone/Simon & Schuster, New York.

Thompson, J. Eric S.

1937 *A New Method of Deciphering Yucatecan Dates with Special Reference to Chichen Itza*. Carnegie Institution of Washington Publications 483; Contributions to American Archaeology 22. Carnegie Institution of Washington, Washington.

1950 *Maya Hieroglyphic Writing: An Introduction*. Carnegie Institution of Washington, Publications 589. Carnegie Institution of Washington, Washington.

Vail, Gabrielle

2002 Haab' *Rituals in the Maya Codices and the Structure of Maya Almanacs*. Research Reports on Ancient Maya Writing 53, pp. 69–116. Center for Maya Research, Washington.

Vail, Gabrielle, and Victoria R. Bricker

2004 *Haab* Dates in the Madrid Codex. In *The Madrid Codex: New Approaches to Understanding an Ancient Maya Manuscript*, edited by Gabrielle Vail and Anthony F. Aveni, pp. 171–214. University Press of Colorado, Boulder.

Villacorta C., J. Antonio, and Carlos A. Villacorta

1976 *Codices mayas*. 2nd ed. Tipografía Nacional, Guatemala City.

10

Divine Reckoning

The Calendrical Ground of Mexican Dynastic Imagery

WILLIAM L. BARNES

The year is 8 Reed (1487), and the day is 7 Reed.[1] The imperial center of Tenochtitlan is alight with torches and other fires, illuminating the way for one of the most spectacular processions the city has ever seen. Despite the grandeur of this event, the city and its inhabitants have, in recent years, faced a number of challenges that have shaken the great *altepetl* to its core. Only recently, the *huei tlatoani*, or great speaker, Tizocic Tlalchiuhtonatiuh (Durán 1994:304) died after only a few cycles in office, barely enough time to complete most of the accession rites required of him. The ruling nobles and priests know that what the city and the greater *huei altepetl* needs is continuity and a renewed sense of divine guidance. Where, then, can they turn for such a thing? I argue that in this case, and in most others, their initial impulse was to turn to the Mesoamerican calendar. Within the calendar's incredibly complex yet predictable measure, all the polity's inhabitants could witness the irrefutable essence of the divine. In our modern world, most great acts of divine providence are seen as things of the distant past, preserved only in sacred texts and relics, while the manner in which the Tenochca-Mexica, or Aztecs, employed their calendars allowed them to argue for a conflation of the sacred and the mundane, so that even those with only a moderate understanding of the calendar's workings could recognize a guiding supernal hand at work in almost every recorded imperial event. This divine reckoning, as I call it here, was then broadcast by carefully encoded references to specific festivals and periods of time on public monuments.

This essay will look into two such public monuments. One, the so-called Tizoc Stone, was likely made for the great celebration just mentioned. Yet it is quite unusual, for as a monument that commemorates a great calendrical fiesta, it paradoxically bears no glyphic date. The other monument to be addressed is one of the most complex calendrical Tenochca-Mexica monuments yet recovered and also one that carries the most calendrical glyphs of any known Mexica monument, the Aztec Calendar Stone (also known as Piedra del Sol,

"Stone of the Sun"). I will argue that this monument, too, was commissioned with a particular festival in mind, but one that had as much historical relevance as it did contemporary significance.

Yet despite this contemporary significance, both of these monuments were retrospective in that they both maintained links to the past. The Tizoc Stone contains references to past battles won by previous rulers as well as stylistic links to the past in the work's resemblance to the earlier so-called Ex-Archbishop's Stone.[2] While the Calendar Stone is another type of sun stone commissioned by Mexica rulers, it is stylistically different from these earlier works. In order to reference the past, its artists relied upon the Mesoamerican calendar.

In order to understand these two monuments and the calendrical tropes encoded within them, one must be aware of the intricacies inherent in the dual Mesoamerican calendars employed by the Tenochca-Mexica.[3] These two independent calendars operated simultaneously while having variant 260-day and 365-day counts. The shorter of the two, and the only one that regularly appears in the central Mexican visual record, was the 260-day divinatory calendar, or *tonalpohualli*, made of twenty thirteen-day periods (usually called *trecena*). The longer was the Aztec ritual count that approximated a vague solar year, or *xihuitl*. It divided time into eighteen twenty-day periods, commonly called *veintena*, and included a small five-day period at the end. The count of this year may have been called the *cemilhuitlapohualli* or *xiuhtlapohualli* (see Umberger 2002:89n5). In this longer count, each veintena concluded with a significant festival that celebrated significant deities or supernaturals related to particular phenomenon (i.e., "water deities," "maize deities"). Operating in tandem, these two calendars provided unique identification for every day in any fifty-two-year period (the so-called Calendar Round).

In the course of each year were also variable festivals that sixteenth-century chroniclers likened to Catholic moveable feasts (Sahagún 1950–1982:2:35). These periodic feasts were, principally, the name days of various supernaturals reckoned by use of the 260-day ritual calendar (Caso 1959). These moveable feasts would fall once or twice in different veintena each year, and their placement varied in each year of the fifty-two-year cycle. Sixteenth-century chronicler Bernardino de Sahagún (1950–1982:2:41) noted that "these moveable feasts in some years displace the feasts of the calendar." This passage is interesting, for if one looks at a tabular view of the calendar, at specific points in the fifty-two-year Calendar Round these moveable feasts would indeed fall on the twentieth and final day of the veintena, which usually served as its principal feast day (Caso 1971).

During some of the more minor veintenas, it may well have been the case that these moveable feasts outstripped their more regular veintena-ending counterparts in terms of religious significance. Two noteworthy points should be mentioned here, though. Unlike in the European calendar, these moveable feasts only appeared to float throughout the year. In reality, they were fixed in their occurrence within the airy confines of the fifty-two-year (18,980-day) Calendar Round. If the moveable feast for the goddess Tlazolteotl, 9 Reed, fell as twentieth day of the veintena of Tlaxochimaco [X][4] in the year 13 Reed, it would do so every time that year came around (every fifty-two years). In the years before and after 13 Reed, this feast would fall at different (and often less propitious) times. The second thing to keep in mind is that veintena feasts could only fall on one of the four year-bearer days (a Rabbit, Reed, Flint, or House date); a great number of other moveable feasts with non-year-bearing day names could never serve as a veintena-ending feast day. As I (Barnes 2009) have argued elsewhere, it is unlikely that, in most cases, the veintena feast day, or the entirety of the veintena itself, would be superseded by the moveable feast. It is much more likely that the celebration of the "moveable" feast would be superimposed upon the more regular veintena festival, thereby modifying or, in some cases, magnifying its celebration. Indeed, I find it quite likely that the propitious conjunction of various veintena and moveable feasts were looked upon as significant, perhaps even historic, events.

Take, for example, the principal moveable feast of the Tenochca-Mexica tutelary deity Huitzilopochtli: 1 Flint. In certain years, the day 1 Flint would fall in the veintena of Panquetzaliztli [XVI], the principal veintena dedicated to this deity. Only once in the count of fifty-two years would it fall as the twentieth day, the principal feast day, of this deity: year 2 Flint, the fifteenth of the fifty-two-year count. Aside from this rarity, the day 1 Flint would fall somewhere in Panquetzaliztli [XVI] at three other times: in 4 Reed (30/52) it falls on the fifth day of the veintena, in 12 Rabbit (25/52) it falls on the tenth, and in 7 House (20/52) it falls on the fifteenth.[5] The cult of Huitzilopochtli and the veintena of Panquetzaliztli [XVI] would have increased considerably in those years due to the coincidence of these festival periods.[6] Historically, the last round of such festivals before the arrival of the Spaniards fell in the years 1468, 1473, 1478, and 1483. The first would have been during the reign of the first Moteuczoma (ca. 1441–1469); the second and third both fell within the reign of Axayacatl (ca. 1470–1481); and the fourth fell during the relatively short reign of Tizoc (ca. 1482–1486).

Most sixteenth-century accounts of veintena festivals are not fine-grained enough to allow us to differentiate between the annual variations in such past festivals. Yet at times one can read between the lines of the ethnohistorical accounts. For example, as noted previously, the last 2 Flint year before the arrival of the Spaniards fell in 1468 at the close of the reign of the first Moteuczoma. I would argue that the preparations for this propitious Panquetzaliztli [XVI], where Huitzilopochtli's name day fell as its veintena feast, may have fueled Moteuczoma's interest in locating Huitzilopochtli's birthplace at Teoculhuacan and his desire to send lavish gifts to Huitzilopochtli's mother as famously recorded by Diego Durán (1994:212–222).[7] The role that the calendar played in this tale also had relevance to Durán's sources in the last quarter of the sixteenth century. The passage wherein Huitzilopochtli's mother, Coatlicue, recalls the words her son spoke to her at his departure from the mount in the mythic past were written in Durán's day with a full knowledge of their calendrical irony. Coatlicue notes that Huitzilopochtli told her,

> During this time [away from you] I shall wage war against provinces and cities, towns and villages. All of these will become my subjects. But in the same way that I conquered them they will be torn from me. Strangers will take them from me, and I shall be expelled from that land. Then I shall return, then I shall return here because those whom I subjected with my sword and shield will rise against me. They will pick me up by the feet and cast me down head first. My weapons and I will roll upon the floor. It will be then, O mother, when my time has come, that I shall return for you to shelter me. (Durán 1994:219)

If these words were spoken in 1 Flint 1468, as I suggest, the author of Durán's source chronicle clearly would have known that a single fifty-two-year cycle later, in 1 Flint 1520, the city of Tenochtitlan would be under siege by the Spanish and their allies, and the divine figures of the Templo Mayor would have already been cast down by Hernán Cortés (2001:106).

In the relationship between veintena festivals and moveable feasts, some coincidental occurrences were surely seen as more propitious than others. For example, the placement of the day 1 Flint in the year 1 Flint was not as significant as one might expect, considering that the name day of Huitzilopochtli served as the year-bearer day (its final day and the day for which the year was named; Table 10.1). In this year, however, the day 1 Flint did not fall as the feast day of Panquetzaliztli [XVI] but rather as the veintena feast days of the maize-related Hueitozoztli [V] and the final feast of the year, Titit [XVIII], neither being festivals of pointed import to the cult of Huitzilopochtli. But at least one calendrical coincidence did mark the year 1 Flint as significant: 1 Death, the principal name day of the major central Mexican deity Tezcatlipoca, falls in Huitzilopochtli's veintena of Panquetzaliztli [XVI].

The day 1 Death is also the first of its trecena, and as such its patrons also presided over the entire thirteen-day period. In this year, then, Titlacahuan Tezcatlipoca, the young warrior, would serve as the principal patron (along with Tecciztecatl [Anders et al. 1991:134–135]) for the concluding thirteen days of the festival (as the veintena and this trecena

table 10.1
The year 1 Flint in tabular format

1 FLINT (27/52)	*VEINTENA DAY*	IZCALLI	ATLCAHUALO	TLACAXIPEHUALIZTLI	TOZOZTONTLI	HUEITOZOZTLI	TOXCATL	ETZALCUALIZTLI
		I	II	III	IV	V	VI	VII
Rain / Quiahuitl (19)	*1*	6	13	7	1	8	2	9
Flower / Xochitl (20)	*2*	7	1	8	2	9	3	10
Alligator / Cipactli (1)	*3*	8	2	9	3	10	4	11
Wind / Ehecatl (2)	*4*	9	3	10	4	11	5	12
House / Calli (3)	*5*	10	4	11	5	12	6	13
Lizard / Cuetzpallin (4)	*6*	11	5	12	6	13	7	1
Serpent / Coatl (5)	*7*	12	6	13	7	1	8	2
Death / Miquiztli (6)	*8*	13	7	1	8	2	9	3
Deer / Mazatl (7)	*9*	1	8	2	9	3	10	4
Rabbit / Tochtli (8)	*10*	2	9	3	10	4	11	5
Water / Atl (9)	*11*	3	10	4	11	5	12	6
Dog / Itzcuintli (10)	*12*	4	11	5	12	6	13	7
Monkey / Ozomatli (11)	*13*	5	12	6	13	7	1	8
Grass / Malinalli (12)	*14*	6	13	7	1	8	2	9
Reed / Acatl (13)	*15*	7	1	8	2	9	3	10
Jaguar / Ocelotl (14)	*16*	8	2	9	3	10	4	11
Eagle / Cuauhtli (15)	*17*	9	3	10	4	11	5	12
Buzzard / Cozcacuauhtli (16)	*18*	10	4	11	5	12	6	13
Movement / Ollin (17)	*19*	11	5	12	6	13	7	1
Flint / Tecpatl (18)	*20*	12	6	13	7	1	8	2

are coterminal). I believe that such calendrical coincidences played a part in the close relationships among, and, at times, the conflation of, various deities. In this case, the relationship would be between the powerful deities Tezcatlipoca and Huitzilopochtli, a relationship borne out by their close iconographic and theological associations (see Nicholson 1971:412, 426).[8]

While I have been addressing the occurrences of various ritual periods, festivals, and days, we have, as Richard Townsend (1979) reminds us, yet another factor to consider when addressing the use of these dates on Mexica monuments: the work's historical context. While some events would certainly come to be regardless of the date on which they fell, when an event happened to correspond with a propitious date or fell in a significant period—say, advantageous rainstorms occurring during the rain god–related veintena of Atemoztli (The Water Falls) [XVII]—it would surely have had

TECUILHUITONTLI	HUEITECUILHUITL	TLAXOCHIMACO	XOCOTLHUETZI	OCHPANIZTLI	TEOTLECO	TEPEILHUITL	QUECHOLLI	PANQUETZALIZTLI	ATEMOZTLI	TITITL	NEMONTEMI
VIII	IX	X	XI	XII	XIII	XIV	XV	XVI	XVII	XVIII	Ø
3	10	4	11	5	12	6	13	7	1	8	2
4	11	5	12	6	13	7	1	8	2	9	3
5	12	6	13	7	1	8	2	9	3	10	4
6	13	7	1	8	2	9	3	10	4	11	5
7	1	8	2	9	3	10	4	11	5	12	6
8	2	9	3	10	4	11	5	12	6	13	
9	3	10	4	11	5	12	6	13	7	1	
10	4	11	5	12	6	13	7	1	8	2	
11	5	12	6	13	7	1	8	2	9	3	
12	6	13	7	1	8	2	9	3	10	4	
13	7	1	8	2	9	3	10	4	11	5	
1	8	2	9	3	10	4	11	5	12	6	
2	9	3	10	4	11	5	12	6	13	7	
3	10	4	11	5	12	6	13	7	1	8	
4	11	5	12	6	13	7	1	8	2	9	
5	12	6	13	7	1	8	2	9	3	10	
6	13	7	1	8	2	9	3	10	4	11	
7	1	8	2	9	3	10	4	11	5	12	
8	2	9	3	10	4	11	5	12	6	13	
9	3	10	4	11	5	12	6	13	7	1	

greater resonance among contemporary witnesses and historians. And much as we rank and mark the occurrence of regular celebrations, say, July 4, 2076, versus the more mundane July 4, 2017, the celebration of key rituals themselves could become historic events. The co-occurrences of dates already marked as momentous with significant contemporary events might even have been seen as a principal catalyst for imperial commissions, as I would argue the two monuments under study here attest.

The festival evoked at the start of this study, a festival set in the year 8 Reed (34/52), or 1487, on the day 7 Reed, was the great veintena feast of Panquetzaliztli [XVI] (Table 10.2). In 1487, this festival also likely marked the culmination of the great celebrations surrounding the completion of the Templo Mayor renovations, begun four years earlier by the deceased Tizoc.[9] This veintena was significant not only due to its patron Huitzilopochtli but also because, in 8 Reed years, it happened to

table 10.2

The year 8 Reed in tabular format

8 REED (34/52)	VEINTENA DAY	IZCALLI	ATLCAHUALO	TLACAXIPEHUALIZTLI	TOZOZTONTLI	HUEITOZOZTLI	TOXCATL	ETZALCUALIZTLI
		I	II	III	IV	V	VI	VII
Jaguar / Ocelotl (14)	*1*	13	7	1	8	2	9	3
Eagle / Cuauhtli (15)	*2*	1	8	2	9	3	10	4
Buzzard / Cozcacuauhtli (16)	*3*	2	9	3	10	4	11	5
Movement / Ollin (17)	*4*	3	10	4	11	5	12	6
Flint / Tecpatl (18)	*5*	4	11	5	12	6	13	7
Rain / Quiahuitl (19)	*6*	5	12	6	13	7	1	8
Flower / Xochitl (20)	*7*	6	13	7	1	8	2	9
Alligator / Cipactli (1)	*8*	7	1	8	2	9	3	10
Wind / Ehecatl (2)	*9*	8	2	9	3	10	4	11
House / Calli (3)	*10*	9	3	10	4	11	5	12
Lizard / Cuetzpallin (4)	*11*	10	4	11	5	12	6	13
Serpent / Coatl (5)	*12*	11	5	12	6	13	7	1
Death / Miquiztli (6)	*13*	12	6	13	7	1	8	2
Deer / Mazatl (7)	*14*	13	7	1	8	2	9	3
Rabbit / Tochtli (8)	*15*	1	8	2	9	3	10	4
Water / Atl (9)	*16*	2	9	3	10	4	11	5
Dog / Itzcuintli (10)	*17*	3	10	4	11	5	12	6
Monkey / Ozomatli (11)	*18*	4	11	5	12	6	13	7
Grass / Malinalli (12)	*19*	5	12	6	13	7	1	8
Reed / Acatl (13)	*20*	6	13	7	1	8	2	9

host a number of significant moveable feasts, the greatest of which was the feast of the sun, Nahui Ollin (4 Movement). Also in this veintena were the moveable feasts of the maize goddess Chicomexochitl (7 Flower) or Xochiquetzal (Sahagún 1950–1982:4:7), the Cihuateteo (on 1 Deer), the pulque god Ometochtli (2 Rabbit), and the Quetzalcoatl-related day 7 Reed (Nicholson 1955:4; Nicholson and Quiñones-Keber 1983:52–55; Umberger 1981:101, 129), which served as the Panquetzaliztli [XVI] feast day. Such a conjunction happens in each Reed year, albeit in different veintenas.[10]

The Tizoc Stone

Of interest to us here is the great sacrificial stone that likely played a central role in these 8 Reed

TECUILHUITONTLI	HUEITECUILHUITL	TLAXOCHIMACO	XOCOTLHUETZI	OCHPANIZTLI	TEOTLECO	TEPEILHUITL	QUECHOLLI	PANQUETZALIZTLI	ATEMOZTLI	TITITL	NEMONTEMI
VIII	IX	X	XI	XII	XIII	XIV	XV	XVI	XVII	XVIII	Ø
10	4	11	5	12	6	13	7	1	8	2	9
11	5	12	6	13	7	1	8	2	9	3	10
12	6	13	7	1	8	2	9	3	10	4	11
13	7	1	8	2	9	3	10	4	11	5	12
1	8	2	9	3	10	4	11	5	12	6	13
2	9	3	10	4	11	5	12	6	13	7	
3	10	4	11	5	12	6	13	7	1	8	
4	11	5	12	6	13	7	1	8	2	9	
5	12	6	13	7	1	8	2	9	3	10	
6	13	7	1	8	2	9	3	10	4	11	
7	1	8	2	9	3	10	4	11	5	12	
8	2	9	3	10	4	11	5	12	6	13	
9	3	10	4	11	5	12	6	13	7	1	
10	4	11	5	12	6	13	7	1	8	2	
11	5	12	6	13	7	1	8	2	9	3	
12	6	13	7	1	8	2	9	3	10	4	
13	7	1	8	2	9	3	10	4	11	5	
1	8	2	9	3	10	4	11	5	12	6	
2	9	3	10	4	11	5	12	6	13	7	
3	10	4	11	5	12	6	13	7	1	8	

festivals: the so-called Tizoc Stone (Umberger 1998, 2002:100). This famous work did not exist in isolation; each Mexica sculpture carried with it considerable historical baggage, and while many commemorated current events, they most often did so by referencing significant periods in the past. Tizoc had acceded to the throne after the death of his brother Axayacatl around 1481.[11] Axayacatl had been a relatively successful ruler by most accounts and presided over a number of significant victories. But his coronation had set off a bitter civil war wherein Tlatelolco, with the help of its allies, attempted to supersede its sister city Tenochtitlan as the principal polity in central Mexico. The decisive defeat and punishment of Tlatelolco in 1473 (Barlow 1987:107–117) was one of Axayacatl's great victories.

After vanquishing Tlatelolco, Axayacatl executed a number of neighboring rulers who had

figure 10.1
Detail of the Codex Borbonicus 10, showing 1 Flint trecena patrons Tonatiuh and Mictlanteuctli. (Drawing by William L. Barnes.)

either pledged support for the defeated city or who had simply not aided the Tenochca in a timely manner.[12] Following the consolidation of the imperial core, Axayacatl in 1474/1475 turned his attentions toward another area of former Tlatelolcan support, western Toluca and the Matlatzinca (Chimalpahin Quauhtlehuanitzin 1997:2:91), where he spent the next three years campaigning.[13] Durán (1994:264) notes that in order to wage legitimate war on the Matlatzinca, Axayacatl had sent requests for stone and wood from the rulers there, building materials needed for his expansion of the ceremonial precinct and, in particular, the structures to hold various sun stones he was having carved. With their refusal to send these goods as pretext for war, Axayacatl and his troops set out for the Toluca

Valley.[14] The Matlatzinca warriors captured in this conflict were then taken back to Tenochtitlan and sacrificed at the militaristic veintena festival of Tlacaxipehualiztli [III], the festival of the flaying of men.

While Durán does not note the year of these events in the campaign, it may well have been at the definitive conclusion of the Matlatzinca campaigns in 12 Rabbit 1478. As noted previously, this year was significant in relation to the cult of Huitzilopochtli in that his name day fell on the tenth day, the midpoint, of both Tlacaxipehualiztli [III] and Panquetzaliztli [XVI]. Both of these festivals had, as a central activity, the sacrifice of captives taken in war. As well, both of these veintena came to a close within the 1 Flint trecena (on its eleventh day, 11 Rabbit). As can be seen in Codex Borbonicus 10 (Figure 10.1), this particular trecena was presided over by the solar deity Tonatiuh along with Mictlanteuctli, the lord of the underworld. Potent divine embodiments of life and death, they were surrounded by the Borbonicus artists with lordly paraphernalia: turquoise diadems, nose plugs, and ear spools as well as turquoise dog pectorals. Between these figures they also placed a large object that closely resembles the Xocotl pole, centerpiece of the Xocotlhuetzi [XI] veintena festival. I will return to this detail in a moment.

Tizoc, upon his own accession four years later, was faced with renewed revolts among the Matlatzinca, particularly, it seems, after his lackluster coronation campaign against Meztitlan (Hassig 1988:189–193). It would not be surprising then, that after these campaigns he would plan to sacrifice his own prisoners in a manner similar to his immediate predecessor, as noted previously. Tizoc's prisoner sacrifices then corresponded with his inauguration of a new Templo Mayor renovation campaign in 4 Reed 1483. What, then, was Tizoc's chosen festival for these sacrifices? The answers to this question are, I believe, encoded in the imagery of the Tizoc Stone (Figures 10.2a–b).

First, the Tizoc Stone is, at its base, a sun stone, a circular monument with a stylized image of the sun atop it. Such monuments play a key role in the celebration of Nahui Ollin (4 Movement; Durán 1971:187–190; Sahagún 1950–1982:2:35, 216–217), the most significant feast dedicated to the sun in the Mexica ritual calendar.[15] As well, in the twelfth of the fifteen roughly chronological scenes on the stone, we see Tizoc, identified by his "leg" name glyph, dressed as a victorious Toltec warrior and wearing the elaborate headdress of Huitzilopochtli and accoutrements of the fire god Xiuhteuctli (Barnes 2009:131–258; Umberger 1981:136a, 1998:242–243, 2010 [1988]; Wicke 1976). Is there, then, a portion of the year 4 Reed (Table 10.3) wherein one can find rites that involve the solar cult, Huitzilopochtli, and the fire god Xiuhteuctli and celebrate or commemorate military victories? The answer is yes.

As always happens in Reed years, the Nahui Ollin (4 Movement) feast takes place on the fourth day of its hosting veintena. As Sahagún (1950–1982:2:216) tells us, this feast is preceded by four days of fasting that would, again, in Reed years, serve as the first four days of any veintena that hosted this principal solar feast. In 4 Reed, these rites occur only once, at the start of Xocotlhuetzi [XI], a major feast dedicated to Huitzilopochtli and Xiuhteuctli (Brown 1988; Caso 1971).

Durán (1971:189) writes that during the 4 Movement moveable feast, a messenger to the sun climbed the steps of the temple of the sun and mounted a sacrificial receptacle he identifies as a *cuauhxicalli*. As most scholars agree, the Tizoc Stone is just such a cuauhxicalli. At the time of this messenger's sacrifice, there was universal bloodletting throughout the land, among nobles and commoners alike (Sahagún 1950–1982:2:216).

When we look to the 365-day xiuhtlapohualli count, we can see that in the year 4 Reed the festival of Xocotlhuetzi [XI] also concluded on the day 7 Reed, a day, as noted previously, of considerable importance to the cult of Quetzalcoatl. Recalling Sahagún's comment that, at times, a moveable feast superseded the regular veintena festival or, as I have suggested, colored it anew, we see that in this year, 4 Reed, the festival would have served to honor a significant cross section of the Mexica "pantheon" of deities, most of whom were, in some way, relevant to Mexica royalty and imperial ideology (see Klein 1987). It also serves to answer the question posed

figure 10.2a
Rollout drawing of the Tizoc Stone conquest scenes marked with the cardinal directions (N = Tlatelolco; E = Chalco; S = Culhuacan; W = Matlatzinca). (Drawing courtesy of Emily Umberger.)

figure 10.2b
Detail of Tizoc Stone conquest scenes: (left) Scene 12, Tizoc conquering the Matlatzinca/Tolocan; and (right) Scene 13, the conquest of Tochpan. (Drawing courtesy of Emily Umberger.)

earlier as to which veintena hosted events of relevance to the sun, Huitzilopochtli, and Xiuhteuctli. What, then, of the military connection?

Historically, in 1483 the *day* 7 Reed would have served well to prompt those with a knowledge of history to recall the *year* 7 Reed 1447, some thirty-six years earlier. In this year, the first Moteuczoma celebrated his own great expansion of the Templo Mayor. The choice of such a date for a grand imperial celebration would have surely been meant to strengthen Tizoc's ties to his uncle Moteuczoma I, an important effort in light of the recent Tlatelolcan conflict.

In addition to serving as a veintena that honored Xiuhteuctli and Huitzilopochtli in Tenochtitlan, Xocotlhuetzi [XI] was also a celebration of the war dead (Brown 1988). Betty Ann Brown (1988:178–179) also argues that the festival had some additional historical significance in Tenochtitlan, where it served to commemorate the Tenochca-Mexica defeat of both the Otomi and the Tepanecs of Azcapotzalco in 1428 (1 Flint), victories that gained a fledgling

Tenochtitlan its independence. All this significance, combined with the fact that the year 4 Reed is often pointed to as the year of the subsequent foundation of the so-called Triple Alliance in 1431 (Umberger 1981:275–276), would have greatly magnified the importance of the celebration of Xocotlhuetzi in 4 Reed 1483, exactly fifty-two years later.

But the Tizoc Stone is mute in regard to all of these dates, and as no pre-Hispanic veintena glyph for Xocotlhuetzi [XI] is known, what then suggests the 1483 celebration of this festival as one of its principal referents? Sahagún (1950–1982:2:114) relates that at the culmination of this midyear festival, Paynal, an enigmatic figure called "the envoy of Huitzilopochtli," grasped sacrificial victims by the hair atop their heads and delivered them to the Templo Mayor, where the priest of Xiuhteuctli, the god of fire, sacrificed them. Those viewing the Tizoc Stone would have immediately recognized that the Huitzilopochtli-clad Tizoc and his victorious analogs sculpted around the circumference of the sculpture were performing this selfsame activity.

Moving from the symbolic to the actual, one can look to Emily Umberger's (2002:100) reminder that captives taken in Tizoc's campaigns against the Matlatzinca are specifically listed as those sacrificed at the inauguration of Tizoc's building campaign, particularly, as Umberger (2002:99–100, figs. 6–7) points out, in the Codex Telleriano-Remensis, folios 38v and 39r. As I (Barnes 2009:136–137) have argued elsewhere, the Tizoc Stone's artists gave the work directional significance, using the solar rays atop the stone (and the flinty mouths of the earth monster below) to highlight four significant conquests that roughly correspond to the cardinal directions (Tlatelolco to the north, Chalco to the east, Culhuacan to the south, and Matlatzinca to the west). When, in 1483, Matlatzinca captives were escorted by the hummingbird-costumed figure of Paynal up the western stairs of the Temple of the Sun, they would have been confronted first and foremost with a depiction of their own conquest at the hands of Tizoc, himself shown on the monument as an "envoy of Huitzilopochtli."

Move ahead four years, and we are back in 8 Reed 1487, the fortieth anniversary of the Templo Mayor renovations celebrated by Moteuczoma I in the year 7 Reed 1447. As noted previously, in 8 Reed the main feast days of both Tlacaxipehualiztli [III] and Panquetzaliztli [XVI] fell on the day 7 Reed. It had been ten years since Axayacatl sacrificed his own Matlatzinca prisoners from the Toluca Valley in the festival of Tlacaxipehualiztli and four years since Tizoc had likely done the same during Xocotlhuetzi [XI] on the day 7 Reed.[16] Both these rulers were dead, and their successor and brother Ahuitzotl was desirous of broadcasting a message of continuity. His completion of the renovations of Tenochtitlan's Templo Mayor would perform just this function, recalling the previous efforts of his brothers and his uncle, the great Moteuczoma I. It is my assertion, then, that Ahuitzotl's celebration of his Templo Mayor completion was timed to have calendrical links to the rituals performed by his ancestors (see Table 10.2).

At the start of the 8 Reed Panquetzaliztli [XVI] veintena, as Sahagún (1950–1982:2:141) tells us, priests would have been concluding an eighty-day

table 10.3
The year 4 Reed in tabular format

4 REED (30/52)	VEINTENA DAY	IZCALLI I	ATLCAHUALO II	TLACAXIPEHUALIZTLI III	TOZOZTONTLI IV	HUEITOZOZTLI V	TOXCATL VI	ETZALCUALIZTLI VII
Jaguar / Ocelotl (14)	*1*	9	3	10	4	11	5	12
Eagle / Cuauhtli (15)	*2*	10	4	11	5	12	6	13
Buzzard / Cozcacuauhtli (16)	*3*	11	5	12	6	13	7	1
Movement / Ollin (17)	*4*	12	6	13	7	1	8	2
Flint / Tecpatl (18)	*5*	13	7	1	8	2	9	3
Rain / Quiahuitl (19)	*6*	1	8	2	9	3	10	4
Flower / Xochitl (20)	*7*	2	9	3	10	4	11	5
Alligator / Cipactli (1)	*8*	3	10	4	11	5	12	6
Wind / Ehecatl (2)	*9*	4	11	5	12	6	13	7
House / Calli (3)	*10*	5	12	6	13	7	1	8
Lizard / Cuetzpallin (4)	*11*	6	13	7	1	8	2	9
Serpent / Coatl (5)	*12*	7	1	8	2	9	3	10
Death / Miquiztli (6)	*13*	8	2	9	3	10	4	11
Deer / Mazatl (7)	*14*	9	3	10	4	11	5	12
Rabbit / Tochtli (8)	*15*	10	4	11	5	12	6	13
Water / Atl (9)	*16*	11	5	12	6	13	7	1
Dog / Itzcuintli (10)	*17*	12	6	13	7	1	8	2
Monkey / Ozomatli (11)	*18*	13	7	1	8	2	9	3
Grass / Malinalli (12)	*19*	1	8	2	9	3	10	4
Reed / Acatl (13)	*20*	2	9	3	10	4	11	5

fast begun immediately after the conclusion of Xocotlhuetzi [XI] earlier in the year. The solar festival of Nahui Ollin here fell in both Tlacaxipehualiztli [III] and Panquetzaliztli [XVI]. And much as he did in Xocotlhuetzi [XI] (and had done quite spectacularly four years earlier in 4 Reed), in the festival of Panquetzaliztli [XVI], Paynal again led sacrificial prisoners in a procession to their sacrifice atop the Templo Mayor (Sahagún 1950–1982:2:142). Accounts of the 8 Reed Templo Mayor dedication claim that some twenty thousand prisoners were sacrificed, so the bellicose image of Tizoc leading rebellious subjects to their sacrifice on the Tizoc Stone would have sent an unambiguous message to any polities thinking of rebelling during the transitional moment following Ahuitzotl's coronation.

That Ahuitzotl saw these activities as a lynchpin in the recent dynastic history of Tenochtitlan

TECUILHUITONTLI	HUEITECUILHUITL	TLAXOCHIMACO	XOCOTLHUETZI	OCHPANIZTLI	TEOTLECO	TEPEILHUITL	QUECHOLLI	PANQUETZALIZTLI	ATEMOZTLI	TITITL	NEMONTEMI
VIII	IX	X	XI	XII	XIII	XIV	XV	XVI	XVII	XVIII	Ø
6	13	7	1	8	2	9	3	10	4	11	5
7	1	8	2	9	3	10	4	11	5	12	6
8	2	9	3	10	4	11	5	12	6	13	7
9	3	10	4	11	5	12	6	13	7	1	8
10	4	11	5	12	6	13	7	1	8	2	9
11	5	12	6	13	7	1	8	2	9	3	
12	6	13	7	1	8	2	9	3	10	4	
13	7	1	8	2	9	3	10	4	11	5	
1	8	2	9	3	10	4	11	5	12	6	
2	9	3	10	4	11	5	12	6	13	7	
3	10	4	11	5	12	6	13	7	1	8	
4	11	5	12	6	13	7	1	8	2	9	
5	12	6	13	7	1	8	2	9	3	10	
6	13	7	1	8	2	9	3	10	4	11	
7	1	8	2	9	3	10	4	11	5	12	
8	2	9	3	10	4	11	5	12	6	13	
9	3	10	4	11	5	12	6	13	7	1	
10	4	11	5	12	6	13	7	1	8	2	
11	5	12	6	13	7	1	8	2	9	3	
12	6	13	7	1	8	2	9	3	10	4	

can then be seen in the so-called Dedication Stone (Figure 10.3), on which he essentially commissioned a visual equation showing himself facing his deceased brother Tizoc across a sacrificial ball, both performing the autosacrificial rites universally required during the Nahui Ollin festival (Barnes 2009:293–296; Sahagún 1950–1982:4:6–7, 35). At the bottom of the stone is the clearly rendered year-bearer date 8 Reed, and above the two figures is the significant day of 7 Reed, the festival date of both Tlacaxipehualiztli [III] and Panquetzaliztli [XVI] in that year and a reminder of the fortieth anniversary of Moteuczoma I's Templo Mayor renovations. The sculpture, carved in a stone similar to that of the Tizoc Stone, makes explicit what the larger work's sculptors had left implicit, relying upon context to communicate their message.

figure 10.3
The Dedication Stone. (Drawing courtesy of Emily Umberger.)

The Aztec Calendar Stone

If such calendrical rhetoric can be winnowed from a monument like the Tizoc Stone, which bears no calendrical glyphs, what, then, can we take away from a monument rife with them? To answer such a question, it makes sense to turn to the Mexica monument with the most extensive set of calendrical glyphs yet recovered: the famous Aztec Calendar Stone, also known as the Stone of the Sun (Figure 10.4).

My intention here is not to review the centuries of extensive iconographic investigations of this well-known monument.[17] It is, instead, to delve into the meaning behind the less well-known date glyphs that are included around the center of the work. Nine date glyphs can be easily identified on the stone. Five of these reference the previous four suns and the fifth, or current, sun, Nahui Ollin (4 Movement). These five stylized glyphs are complemented by twenty reversed day name glyphs carved in a ring around them. All these have been discussed at length in the literature on the stone. There is, though, a group of five glyphs (four date glyphs and an apparent name glyph) on the work that do not seem to form an easily identifiable group and, hence, defy the more straightforward interpretations of their counterparts on the work. It is my assertion that they tell the tale of a different cycle of time, one that has a more human and, in this imperial monument, dynastic relevance.

The most broadly significant of these dates is the year-bearer 13 Reed that artists placed atop the central solar figure's diadem.[18] In most academic accounts, this date is addressed as the year in which the current sun was born in Aztec myth, and at times since the nineteenth century, it has been suggested that this was the date of the monument's creation. The last occurrence of the year 13 Reed before the conquest was 1479, near the end of the reign of Axayacatl. Most specialists now agree that this date is, in fact, too early for the monument's late imperial style.[19] The discrepancy between the sophisticated late style of the monument and this seemingly anachronistic date can be solved, I believe, by viewing the monument as a historical monument as much as a mythical one addressing cyclic completion and renewal. I believe it likely that the stone's primary agents did intend for this glyph to reference 13 Reed 1479, but that they also intended it to do so retrospectively, in the same manner that it references the mythic content that garners most of the attention paid to the stone (the destruction of the previous eras and the birth of the current one [e.g., Matos Moctezuma 2009:264–274]). The question would then be why Moteuczoma II (r. ca. 1503–1519) would have commissioned such a hybrid historic/mythic monument. As is the case with most of the enigmatic work produced during the reign of Moteuczoma II, the answer to this question is

figure 10.4
The Aztec Calendar Stone or Piedra del Sol.
(Drawing courtesy of Emily Umberger.)

complex. I believe the key to understanding it lies in the other four glyphs on the stone that form part of the smaller, recessed, and somewhat irregular central group.

These glyphs are arranged around the central face of the monument and occupy an interstitial space between the glyphs of the four suns and the central arrow of the movement glyph. While some authors have argued these to be directional glyphs (Seler 1992:129–131) or glyphs related to deities and seasons (Fradcourt 2010 [1993]:287), I would expand upon Umberger's (2010 [1988]:252–253) assertion that they reference significant feast days and ritual periods in the year 13 Reed.[20] At the bottom of the group is the day 1 Rain and across from it the day 7 Monkey. Directly above the 7 Monkey glyph is the date 1 Flint, and across from this date glyph is a complex iconographic element that, in the main, serves as the name glyph of Moteuczoma II (as conclusively shown by Umberger [1981:66–71, 2010 (1988):244–250]). And while I do believe this name glyph was meant to reference this Tenochca ruler, its uniqueness in relation to the other glyphs on the stone and its whiff of ambiguity—absent in the others on the stone (and which has contributed to what is now well over a century of variant interpretations)—suggest to me that its agents intended it to stand apart from the more straightforward visual message of the monument. I would even argue that, in operating on a polysemic level, it serves as the visual and referential lynchpin of the monument.

table 10.4
Year 13 Reed in tabular format

13 REED (26/52)	VEINTENA DAY	IZCALLI	ATLCAHUALO	TLACAXIPEHUALIZTLI	TOZOZTONTLI	HUEITOZOZTLI	TOXCATL	ETZALCUALIZTLI
		I	II	III	IV	V	VI	VII
Jaguar / Ocelotl (14)	1	5	12	6	13	7	1	8
Eagle / Cuauhtli (15)	2	6	13	7	1	8	2	9
Buzzard / Cozcacuauhtli (16)	3	7	1	8	2	9	3	10
Movement / Ollin (17)	4	8	2	9	3	10	4	11
Flint / Tecpatl (18)	5	9	3	10	4	11	5	12
Rain / Quiahuitl (19)	6	10	4	11	5	12	6	13
Flower / Xochitl (20)	7	11	5	12	6	13	7	1
Alligator / Cipactli (1)	8	12	6	13	7	1	8	2
Wind / Ehecatl (2)	9	13	7	1	8	2	9	3
House / Calli (3)	10	1	8	2	9	3	10	4
Lizard / Cuetzpallin (4)	11	2	9	3	10	4	11	5
Serpent / Coatl (5)	12	3	10	4	11	5	12	6
Death / Miquiztli (6)	13	4	11	5	12	6	13	7
Deer / Mazatl (7)	14	5	12	6	13	7	1	8
Rabbit / Tochtli (8)	15	6	13	7	1	8	2	9
Water / Atl (9)	16	7	1	8	2	9	3	10
Dog / Itzcuintli (10)	17	8	2	9	3	10	4	11
Monkey / Ozomatli (11)	18	9	3	10	4	11	5	12
Grass / Malinalli (12)	19	10	4	11	5	12	6	13
Reed / Acatl (13)	20	11	5	12	6	13	7	1

In the literary conventions of Mesoamerica, *difrasismo*, or diphrasism (essentially, the use of diphrastic couplets), is quite common both in visual and written forms (see Alcina Franch 1995; Garibay K. 1971 [1953–1954]; Montes de Oca 2000, 2004, 2008; Navarrete 2011). The four glyphs under consideration can be seen to form such a diphrastic couplet. While it is clear that the diagonally opposed 1 Rain and the 1 Flint glyphs could stand for one half of a couplet, such a pairing is not immediately discernable in the lower 7 Monkey glyph and Moteuczoma's name glyph. So where might we turn to discern some possible second-order signification of the glyph paired with 7 Monkey? It is here that the above 13 Reed year sign gives some interpretive direction.

As noted, the year date 13 Reed is often cited as the birth year of the sun. But it also serves as a

TECUILHUITONTLI	HUEITECUILHUITL	TLAXOCHIMACO	XOCOTLHUETZI	OCHPANIZTLI	TEOTLECO	TEPEILHUITL	QUECHOLLI	PANQUETZALIZTLI	ATEMOZTLI	TITITL	NEMONTEMI
VIII	IX	X	XI	XII	XIII	XIV	XV	XVI	XVII	XVIII	Ø
2	9	3	10	4	11	5	12	6	13	7	1
3	10	4	11	5	12	6	13	7	1	8	2
4	11	5	12	6	13	7	1	8	2	9	3
5	12	6	13	7	1	8	2	9	3	10	4
6	13	7	1	8	2	9	3	10	4	11	5
7	1	8	2	9	3	10	4	11	5	12	
8	2	9	3	10	4	11	5	12	6	13	
9	3	10	4	11	5	12	6	13	7	1	
10	4	11	5	12	6	13	7	1	8	2	
11	5	12	6	13	7	1	8	2	9	3	
12	6	13	7	1	8	2	9	3	10	4	
13	7	1	8	2	9	3	10	4	11	5	
1	8	2	9	3	10	4	11	5	12	6	
2	9	3	10	4	11	5	12	6	13	7	
3	10	4	11	5	12	6	13	7	1	8	
4	11	5	12	6	13	7	1	8	2	9	
5	12	6	13	7	1	8	2	9	3	10	
6	13	7	1	8	2	9	3	10	4	11	
7	1	8	2	9	3	10	4	11	5	12	
8	2	9	3	10	4	11	5	12	6	13	

midpoint in the fifty-two-year cycle, a point of passage from the first to the second half of the count. As Umberger (1987) has shown, this transition was traditionally seen as symbolizing the movement from the more mythic to the more historic, from a period of darkness to a period of light. When we view the xiuhtlapohualli count for 13 Reed, we can also see that aside from being one of the few years in the Calendar Round to host an uninterrupted tonalpohualli count, the year itself hosts some interesting features that speak to liminality and transition (Table 10.4).

Following Alfonso Caso's (1967) calendar round scheme (as opposed to that of Tena [1987]), for example, we see that in the year 13 Reed are two occurrences of the Nahui Ollin (4 Movement) festival. The first occurrence of the day 4 Movement in this year falls in one of the only twenty integral tonalpohualli in the fifty-two-year count and likely here serves as the current sun's name day.[21] As well, this moveable feast day falls in the important veintena festival of Toxcatl [VI], the principal veintena dedicated to the cult of Tezcatlipoca. And as we have seen, any Reed-year veintena that hosts the Nahui Ollin festival will then conclude on the day 7 Reed, one of the main name-day feasts of Quetzalcoatl, as noted previously.[22] This fact is of interest here as Quetzalcoatl is, in mythology, cast as both Tezcatlipoca's legendary antagonist and occasional collaborator (Nicholson 1971:398–400, 428–429). It may be that this particular veintena feast, then, had some relevance to the epic creation story conflicts between these two deities and the story of their collaboration in the creation of the current era.

If this Toxcatl [VI] occurrence of 4 Movement may be seen as the apex of the sun's name-day occurrences, then its second appearance in this year can easily be interpreted as its nadir. On the 364th day of the xiuhtlapohualli, 4 Movement sits squarely within the Nemontemi.[23] These were the "unlucky days" that closed the year and during which one was not supposed to move or even go outside of one's house (Sahagún 1950–1982:2:35). One certainly would not want to be born on one of them. The Nemontemi, though, would be a great time for an apocalypse. And so, I argue, this second occurrence of 4 Movement in the year 13 Reed was likely cast as the prophesied death day of the current (fifth) era.[24]

This crossing of the boundary between birth/life and completion/death can then point us to other such events that occur in 13 Reed and that would have been quite significant to the stone's likely patron, Moteuczoma II. At the midpoint of the year is the feast of Hueitecuilhuitl [IX], which can be translated as "Great Feast of the Lords." One of the pivotal events in this festival involved the rulers opening their personal reserves of food and feeding the people during eight days of feasting, music, and song. These feasting activities, according to Sahagún (1950–1982:2:14), began on the twelfth day of the veintena. In 13 Reed years, this day is 7 Serpent. Subsequently, on the twentieth day of Hueitecuilhuitl [IX], the great feast culminated with the dispensing of justice by the ruler and the public execution of criminals. As a festival, then, Hueitecuilhuitl [IX] presents a graphic spectacle contrasting life and death (and the ruler's control of both).

In 13 Reed years, the concluding day of the festival, its principal day of celebration, falls quite prophetically on the day 2 Reed. This day is significant as it will serve as the year-bearer of 2 Reed, which, half of a cycle later, would host the great New Fire ceremony and binding of the years. In 1507 (2 Reed), then, Moteuczoma II would preside over these principal celebrations of the fifty-two-year calendar round that in central Mexico marked the birth of fire and the creation of the calendar.

Thus, I would argue that one of the principal references for this subset of glyphs on the Calendar Stone is an interesting period of time in the year 13 Reed, the birth/death year of the sun, that through its ceremonies contrasting birth and demise simultaneously warned of the death of the current sun and promised the renewal of time and the calendar at the close of the next half of the fifty-two-year cycle.

As such a scheme was fixed and would occur every fifty-two-years, one might ask why Moteuczoma II would reference it in the Calendar Stone. This past period was being highlighted on the monument due to the fact that the last ruler to preside over these important veintena festivals was Axayacatl (r. ca. 1469–1481). In the year 13 Reed 1479, the thirteen-year-old Moteuczoma II, himself on the cusp of majority, would have watched his father, the huei tlatoani Axayacatl, preside over that great festival, much as he would, as a forty-year-old man, preside over the New Fire ceremony referenced by its feast day some twenty-seven years later.[25]

But as Axayacatl would have presided over the entire ritual year, what, then, might lead one to think that Hueitecuilhuitl [IX] is being privileged here? I believe the date glyph 7 Monkey is the key. In 13 Reed, the day 7 Monkey serves as the penultimate day in the veintena immediately following Hueitecuilhuitl [IX], that of Tlaxochimaco [X]. This event was an important feast dedicated to Huitzilopochtli in Tenochtitlan. As related by Sahagún (1950–1982:2:108), the events for Hueitecuilhuitl began two days before the terminal festival day, when all would gather flowers for garlands and begin the preparation of tamales for the great feast. In the year 13 Reed, this day was 7 Monkey.

Additionally, the day 7 Monkey was an important ceremonial date in its own right, as Sahagún (1950–1982:4:74) tells us that any day sign in the seventh position is good, as is any Monkey day sign. He continues, "Seven Monkey they regarded as very good, because, as hath been said in many places, it is a day sign of the seventh place. It was said that it was the time of ritual of Chicomecoatl, the likeness of our sustenance, our wealth and prosperity" (1950–1982:4:73). Chicomecoatl was the calendrically named corn goddess (see Nicholson 1971) whose feast was, of course, also celebrated on the day 7 Serpent (Chicome Coatl). So on the day 7 Monkey, the Aztecs celebrated the feast of 7 Serpent. Furthermore, to strengthen this Monkey/Serpent association even more, the day 7 Monkey falls within the 1 Serpent trecena.[26]

So what of this reference to the maize goddess, Chicomecoatl? Her name is, of course, calendrical, and I do not believe it a coincidence that it happens to fall within the same forty-day period delimited by the dates 1 Rain and 1 Flint (as does 7 Monkey), the only two trecena-initial dates included on the Calendar Stone. Indeed, as noted earlier, in 13 Reed it serves as the very day (the twelfth day) of Hueitecuilhuitl [IX] when the royal-led feasting rituals began.

Returning to the idea of diphrastic couplets, then, and keeping in mind the relationship of Huitzilopochtli and the festival of Tlaxochimaco [X], we see that the Moteuczoma II headdress glyph on the Calendar Stone was also likely meant to reference the veintena of Hueitecuilhuitl [IX] and the day 7 Serpent that falls on its twelfth day.[27] As such, it provides a type of symmetry to the four central subsidiary glyphs on the Calendar Stone. The two glyphs that frame the central solar deity's right hand (our left) can then be related to the festival of Hueitecuilhuitl [IX] as it was celebrated in 13 Reed. The trecena of 1 Rain hosted the start of the feasting celebrations, which fell on its seventh day (the twelfth day of the twenty-day veintena), 7 Serpent. The glyphs that then frame the left hand of the solar deity reference subsequent veintenas in 13 Reed. The first is Tlaxochimaco [X], whose celebrations began on the day 7 Monkey and whose patron Huitzilopochtli's day-name glyph sits directly above it on the stone. The second is Xocotlhuetzi [XI], as 1 Flint falls on the fifth day of this subsequent veintena and closes a forty-day period begun on 1 Rain (begun in Hueitecuilhuitl [IX]).

As discussed in relation to the Tizoc Stone, Xocotlhuetzi [XI] was yet another Huitzilopochtli-related festival that was particularly important to the people of Tenochtitlan. A reference to it here would be appropriate, considering its alternate name and additional significance as Huey Miccailhuitl, "The Great Festival of the Dead" (Sahagún 1997:61). As elements of the feast honored familial dead and those fallen in war (Brown 1988:175; Durán 1971:444–446; Quiñones-Keber 1995:142–143 [Telleriano Remensis fol. 2v]), its being invoked on the Calendar Stone by Moteuczoma II would again point to the 13 Reed festival cycle celebrated by his "fallen" father, Axayacatl.[28]

Again returning to the idea of diphrastic couplets, we can also look at the structure of these glyphs as significant. They not only echo the two curving bands of the central Ollin or Movement glyph that identifies the Fifth Sun, Nahui Ollin, in the center of the stone, but if we understand that the Moteuczoma II headdress represents the day 7 Serpent, among other things, the two dates with a coefficient of seven were then carved diagonally across from each other, as were the two dates with a coefficient of one.

Mathematically, these dates also seem to echo one another. The two glyphs on the right side of the central deity, 7 Serpent (?) and 1 Rain, mark a period of seven days. The other two, 7 Monkey and 1 Flint, are separated by seven days, giving the two groups of glyphs an approximate symmetry. As well, the two trecena initials, 1 Rain and 1 Flint, serve to demarcate the previously mentioned forty-day period. Such sacred periods had yearly analogs, as Umberger (1981:102) has pointed out. In the year 2 Reed 1506/1507, the likely date for this stone, Moteuczoma II would have been celebrating his fortieth birthday if he had been born in 1 Reed (14/52), as evidence suggests. So the significant forty-day ritual period celebrated in 13 Reed 1479 would, in retrospect, be seen to have had prophetic relevance for the then thirteen-year-old future ruler.[29] Adding to this sense would be the previously mentioned fact that in 13 Reed Hueitecuilhuitl [IX] concluded on the day 2 Reed. Twenty-seven years later, this day served as the year-bearer for the 1507 (2 Reed) New Fire ceremony presided over by Moteuczoma II, the stone's patron.

What these four interstitial glyphs do is meld the mythic and the historical more tightly in this monument. The fortuitous concatenation of these events, when viewed from a future date (much as we are doing), would be a strong cosmic endorsement of Axayacatl's son's assumption of power and his taking on the grand responsibility of the New Fire ceremony.

Conclusion

So what we see with the Calendar Stone is not altogether unlike the famous Dedication Stone (see Figure 10.3), on which the living ruler confronts the dead. But where the protagonists of the Dedication Stone were brothers, the Calendar Stone marks a generational shift in rulership and, I believe, the passage of the throne from a father to a son. Moteuczoma II's father was no longer a *teuctli* under the light of the sun but one in the darkness of the underworld, a dark lord, a night lord, or *yohualteuctli*, as Cecelia Klein (2010 [1976]) has suggested. His symbolic likeness in the center of the stone was then set upon the ground as a dark reflection of the *new* era, the new sun, Moteuczoma II. For Moteuczoma II's 1507 New Fire ceremony not only marked the completion of the fifty-two-year count, but it also marked the completion of a meaningful number of *huehuetiliztli* (Sahagún 1950–1982:4:143) cycles, these being periods of 104 years and 65 Venus cycles (Caso 1971:348). In 1507, four huehuetiliztli, a period of 416 years and 260 Venus cycles, would have, effectively, taken one back to 2 Reed 1091, the date of the last New Fire ceremony that the Tenochca-Mexica would have performed while still living in their semi-mythical homeland of Aztlan. Additionally, in 1507 two huehuetiliztli would take one back to 1299, the start of the cycle that hosted the foundation of Tenochtitlan in 2 House 1325.[30]

In the monuments addressed here, Aztec rulers relied upon the divine serendipity of reckoned time in the central Mexican calendar to highlight festivals that, in their day, served them well in communicating messages of stability, continuity, and divine favor. With the Tizoc Stone, then, we can see how an object to be used in calendrically linked rituals—and that also depicts the activities of similarly significant rituals—might have been employed to both link together the various permutations of the Tenochca ruling dynasty and highlight its members' military exploits and pious acts. Ahuitzotl, in 8 Acatl 1487, surely sought to communicate a sense of divine guidance when he used this monument in his Templo Mayor rededication ceremonies, two propitious Nahui Ollin festivals, and, as Klein (1987) suggests, his own coronation events. As a calendrical monument, the Tizoc Stone served to illustrate how the Tenochca dynasty was operating harmoniously within the confines of the sacred calendar.

A similar sense was clearly sought after by Moteuczoma II with his Calendar Stone, itself crafted to be used during the New Fire ceremony and, likely, the two Nahui Ollin festivals of 1507. The monument relates concerns about cosmic death and rebirth as most authors suggest, yet through its more specific calendrical associations it adroitly casts this epic cycle as a dynastic metaphor. As a calendrical monument, the Calendar Stone then conveys through subtle allusions that under Moteuczoma II the royal dynasty of Tenochtitlan no longer existed within the divine confines of time; it was the living embodiment of it.

Acknowledgments

I would like to thank Emily Umberger, Elizabeth Boone, Cecelia Klein, and Angélica Afanador Pujol for their comments on portions of the arguments developed in this essay. I would also like to thank

Anthony Aveni, Colin McEwan, Stella Nair, Victoria Bricker, Leonardo López Luján, Virginia Miller, Mary Miller, Rex Koontz, and other symposium participants for their comments and suggestions on the presentation distilled from the arguments presented here. Some of the ideas presented here were initially presented for an Archaeological Institute of America, Minnesota Chapter, talk delivered in the fall of 2009, and thanks go to Vanessa Rousseau for the opportunity to speak there. Some of the Calendar Stone arguments were further developed at a talk given at the 2012 Cecelia Klein Homenaje at the University of California, Los Angeles, and thanks go to Eulogio Guzman and the organizing committee for the invitation to speak there as well. I am also grateful for a University of St. Thomas Faculty Research Grant in the fall of 2012 that aided in the preparation of this essay. Last, I am thankful for the support of my department and colleagues at University of St. Thomas during the preparation of this work, particularly Victoria Young, Craig Eliason, and Mark Stansbury-O'Donnell.

NOTES

1 Throughout this work, I follow the convention of underlining dates that reference years within the central Mexican fifty-two-year calendar to differentiate them from days that are named using the same convention (8 Reed "year" vs. 8 Reed "day"). In each case, the date consists of a numerical coefficient ranging from one to thirteen and one of the twenty day names. I use the English equivalents of the Nahuatl day names. These day names are Cipactli/Alligator, Ehecatl/Wind, Calli/House, Cuetzpallin/Lizard, Coatl/Serpent, Miquiztli/Death, Mazatl/Deer, Tochtli/Rabbit, Atl/Water, Itzcuintli/Dog, Ozomatli/Monkey, Malinalli/Grass, Acatl/Reed, Ocelotl/Jaguar, Cuauhtli/Eagle, Cozcacuauhtli/Buzzard, Ollin/Movement, Tecpatl/Flint (i.e., Knife), Quiahuitl/Rain, and Xochitl/Flower.

2 For a discussion of the Ex-Archbishop's Stone (also known as the Temalacatl of Moteuczoma I), see Graulich 1992; Matos Moctezuma and López Luján 2009; Padilla et al. 1989.

3 For a brief overview of the Mesoamerican calendar, see Caso 1971 and Tena 1987.

4 When listing veintenas, I follow the practice of including the veintena order in brackets after the name in order to distinguish its order in the year (i.e., Tlacaxipehualiztli [III], indicating that it is the third veintena in the year). Throughout, I follow Alfonso Caso's (1971) veintena order. When a year's placement in the count of fifty-two is significant, it is indicated by the fractional number in parentheses (i.e., 30/52 indicates the thirtieth year in the count). The Maya Long Count was not used in central Mexico, and only dates from within the fifty-two-year Calendar Round appear on monuments.

5 Whereas monuments with a 4 Reed date are fairly common due to the historical importance of this year (see Umberger 1981:275–276), those with 2 Flint, 12 Rabbit, and 7 House are not. It may well be that the unusual gladiatorial stone from Tehuacan (Nicholson 1956), which bears a 4 Reed date glyph, was commissioned to note the significance of Huitzilopochtli's name day falling in the veintena of Tlacaxipehualiztli in the year 4 Reed.

6 It is interesting that these years form a discrete group with multiples of five (15, 20, 25, and 30 out of 52) and that the day's position falls within Panquetzaliztli [XVI] in inverse relation to the year's position in the fifty-two-year cycle (year 15/52, day 20; year 20/52, day 15; year 25/52, day 10; year 30/52, day 5).

7 Pointing to a passage in Sahagún's (1993:252r) *Primeros memoriales*, López Austin and López Luján (2009:253) remind us that in this veintena the Mexica celebrated the birth of Huitzilopochtli. The Nahuatl reads, "Panquetzaliztli, iquac tlacatia in vitzilopochtli" (Panquetzaliztli, when Huitzilopochtli was born). Sullivan's (Sahagún 1997:64) translation of this passage suggests that it was not the god being born, but rather his image being created: "Panquetzaliztli, 'Raising of Banners,' was when [the figure of] Huitzilopochtli

was fashioned." The Nahuatl tlācat(i) is the verb form of "to be born" (Karttunen 1992:253).

8 Some additional calendrical coincidences may be at work here. Umberger (1987:438) interestingly noted Caso's (1967:138) identification of 1 Flint as Huitzilopochtli's birthday (Alvarado Tezozomoc 1998:36) and 1 Death as the date of his death (based on two mentions in Cristobal del Castillo 1908:69, 92). Umberger also points out that these dates are reversed in the tonalpohualli and suggests that the fifty-two days that separate them are, in some way, related to the fifty-two years of the Calendar Round and the concept of rebirth.

9 See Tena 1987 and Umberger 2002 for the argument that the dedication itself likely took place during Panquetzaliztli [XVI]. As I (Barnes 2009:282, 282n19) have argued, if the infamous 8 Acatl Templo Mayor dedication ceremonies were inaugurated during the important imperial/militaristic festival of Tlacaxipehualiztli [III] (Barnes 1997; Broda 1970) and came to a crescendo during Huitzilopochtli's principal festival of Panquetzaliztli [XVI] (as they share the same 7 Acatl principal feast day), this confluence would present one with a 260-day period in which the twenty thousand plus sacrifices reported to have taken place for this event (see López Austin and López Luján 2009:205–206) could have been completed. This fact would change the 8 Reed dedication of the Templo Mayor from being a superhuman event wherein thousands of individuals were sacrificed and disposed of within seconds of one another to one in which, through this 260-day period, eighty prisoners on average were sacrificed each day throughout the city (likely grouped around principal festival days).

10 The 1 Jaguar trecena that begins this veintena on the day 1 Jaguar is the only veintena-initial trecena in any Reed year. In each of the 13 Reed years in the count of fifty-two it falls in a different veintena. Similar relationships exist between other trecena/veintena combinations in each year, varying only by year-bearer, meaning that such conjunctions would happen in a regular, if periodic, fashion.

11 See Boone (1992:153–154) for a list of the various dates given for the accession/death of the various Tenochca huehuei tlatoque in the ethnohistoric record. The dates listed in the Codex Mendoza seem to serve as an acceptable baseline for rulers' accession and death dates.

12 Hassig (1988:183) lists Xochimilco, Cuitlahuac, Mizquic, Mexicatzingo, Huitzilopochco, Colhuacan, and Cuitlahuac.

13 My interpretation of these events here is at some odds with that of Hassig (1988), who sees these campaigns as an attempt by the Aztecs to forestall Tarrascan expansion into this region.

14 As one of his first foreign campaigns, this would have been a real test for Axayacatl, and while his armies were, in the end, victorious, he suffered a grievous wound to his leg during an apparent assassination attempt at Xiquipilco (Durán 1994:266).

15 For more information on Mexica sun stones, see Umberger 1987 and Matos Moctezuma and Solís 2004.

16 As Umberger (2002:100) points out, additional sacrificial ceremonies also took place in the following year, 6 House. In this year, the 1 Death trecena fell in both Tlacaxipehualiztli and Panquetzaliztli, likely marking these veintenas as the dates of the additional Templo Mayor sacrifices.

17 As Villela and colleagues (2010:15) note, "Writing about the *Calendar Stone* often entails a rehearsal of the monument's interpretation that can be so long and exhausting that it can threaten to overwhelm the argument or new materials in books and articles"; hence, I would steer the interested reader toward Villela and Miller's (2010) thorough treatment of the sculpture and its previous interpretations rather than repeat them here. See also Graulich 1997.

18 This feature is actually like a solar ray made of the upper tip of the central "arrow" motif used to create the complex 4 Movement name glyph that dominates the center of the monument. I find it unlikely that the artists did not plan for this element to recall the pointed turquoise diadem worn by central Mexican *tlatoque*. See Neurath 1992 for a discussion of the *xiuhuitzolli* and solar rays.

19 More recently, Matos Moctezuma and Solís (2004:66) have suggested a 13 Reed 1479 date for the monument; however, earlier in the same work they acknowledge that 13 Reed may have been a symbolic date rather than an actual one (Moctezuma and Solís 2004:37). Matos Moctezuma (Matos Moctezuma and López Luján 2009:279) still attributes the stone's patronage to Axayacatl, whereas Olivier and López Luján (2009:78–92) place it firmly within the reign of Moteuczoma II.

20 Umberger addresses 1 Flint and 1 Rain as ceremonially significant days, though she is not sure about the ultimate significance of 7 Monkey. The diadem glyph has not been treated calendrically. In the main, authors generally avoid addressing these glyphs as a cohesive group, instead offering interpretations for each glyph in relative isolation (e.g., Graulich 2010 [1992]:260–261; Umberger 2010 [1988]:241–255).

21 By "integral," I mean that the 260-day tonalpohualli is not interrupted by the year-ending Nemontemi "nondays" and, hence, retains its integrity as a complete count. These integral counts occur an even twenty times in the fifty-two-year count (in years 2, 4, 7, 9, 14, 16, 19, 21, 24, 26, 29, 31, 34, 36, 39, 41, 44, 46, 49, and 51) using the Caso correlation. Much as is the case in the fifty-two-year count, here 13 Reed (as year 26) again serves to close out the first half of the group. Interestingly, the most ritually significant Reed years all fall into this group (2 Reed, 1 Reed, 13 Reed, 8 Reed, 7 Reed), and here 13 Reed is the fulcrum year. Such is not the case with the other yearbearers. Unfortunately, sixteenth-century sources are mute on these complete tonalpohualli-count years' significance to day keepers.

22 The day 7 Reed serves as a veintena feast day eighteen times in the fifty-two-year cycle, and does so in thirteen different years. It falls in Toxcatl only four times in the fifty-two-year calendar round, and aside from this 13 Reed occurrence, the others fall in the second half of the count (year 26, day 20; year 31, day 15; year 36, day 10; year 41, day 5). Only in 13 Reed, the first time it falls in Toxcatl, does it serve as the veintena feast day.

23 As I (Barnes 2009:269n6) have noted, if one follows Caso's correlation scheme that argues for an Izcalli [I] initial veintena, seventy-two occurrences of the Nahui Ollin festival fall within the fifty-two-year cycle. This second 13 Acatl Nemontemi occurrence of the feast serves to divide them into two equal halves.

24 Hence, perhaps, strengthening Klein's (2010 [1976]) embattled identification of the face in the center of the 4 Movement glyph as Yohualteuctli and also supporting Fradcourt's (2010 [1993]) argument suggesting the stone shows the sun at midnight.

25 See Hajovsky 2007:211–213 and Barnes 2009:334, 334n1 for a discussion of 1 Reed as Moteuczoma II's birth year. McEwan and López Luján (2009:14) also list this year. I have cited Nicholson 1961:409–411, Davies 1974:208, and Alvarado Tezozomoc 1987:572–573 for supporting evidence of a 1 Acatl birth date for Moteuczoma II. Tezozomoc, being the most specific, claims Moteuczoma was thirty-four when he gained the throne in 1502/1503. This makes a 1 Acatl 1467 birth year quite likely. As well, this would place his birth during the later reign of Moteuczoma I, making the choice of this ruler for his namesake quite logical.

26 Indeed, the ninth trecena of 1 Serpent, in which 7 Monkey falls, is one of the more propitious trecenas in the 260-day count, as Umberger (2010 [1988]:253) points out. Its importance to the long-distance merchants referred to as *pochteca* was such that in his discussion of this trecena, Sahagún (1950–1982:2:59–72) spends four chapters simply describing its first five days, more than any other passage of time discussed in Book 4 of his *Florentine Codex*.

27 See Umberger 1981:66–71, 199 for a thorough review of the variant interpretations of this headdress glyph and a convincing argument for it serving, in the main, as Moteuczoma II's name glyph. In more recent times, scholarly consensus seems to be shifting toward Umberger's assertion (originally put forward by Peñafiel [1990] in 1890, as she points out). See, for example, the entries in McEwan and López Luján 2009, particularly Olivier and López Luján 2009:78–92. Variant early interpretations of this glyph claimed it to be a reference to the fire god Xiuhteuctli (Beyer 1965 [1923]) and the spirit of dead warriors (Seler 1992:96).

28 As well, Sahagún (1950–1982:2:17) notes that preparations for Xocotlhuetzi [XI] began in Tlaxochimaco [X]. Durán (1971:444) also claims that this festival was a continuation of the previous festival.

29 As it was more common for brothers to succeed one another to office in central Mexico, in 13 Reed 1479 it would have seemed unlikely that Moteuczoma II might accede to the throne anytime soon, as his father Axayacatl had at least two brothers who could rule after his passing. Hence, this forty-day period would not have been all that remarkable in that year. Twenty-seven years later it surely would have stood out, however. Along those same lines, Moteuczoma's putative name day of 7 Coatl in his likely birth year, 1 Acatl 1467, falls in the feasting veintena of Etzalcualiztli [VII] much as it falls in the

feasting veintena of Hueitecuilhuitl [IX] in 13 Reed years. As a day sign related to a maize goddess, a goddess of plenty, its placement in these feasting veintena would have had resonance. As well, in the year 1 Reed the day 1 Flint, the day that concludes this forty-day ritual period begun on 1 Rain, falls as the fifth day of Hueitecuilhuitl [IX], the principal feast being referenced by the Calendar Stone.

30 Such cyclical relationships were first suggested by Palacios (1918, 1921).

REFERENCES CITED

Alcina Franch, José

1995 Lenguaje metafórico e iconografía en el arte mexica. *Anales del Instituto de Investigaciones Estéticas* 66:7–44. Mexico City.

Alvarado Tezozomoc, Hernando

1987 *Crónica mexicana.* Edited by Manuel Orozco y Berra. Editorial Porrúa, Mexico City.

1998 *Crónica mexicáyotl.* Translated by Adrián León. Universidad Nacional Autónoma de México, Mexico City.

Anders, Ferdinand, Maarten Jansen, and Luis Reyes García

1991 *El libro del Ciuacoatl: Homenaje para el año del Fuego Nuevo; Libro explicativo del llamado Códice Borbónico.* Fondo de Cultura Económica, Mexico City.

Barlow, Robert H.

1987 *Tlatelolco: Rival de Tenochtitlan; Obras de Robert H. Barlow,* vol. 1. Edited by Jesús Monjarás-Ruis, Elena Limón, and María de la Cruz Paillés H. Instituto Nacional de Antropología e Historia, Mexico City.

Barnes, William L.

1997 *The Great Flayer God of Central Mexico: An Iconographic and Interpretive Study of Xipe Totec in Precolumbian Style Codices.* MA thesis, Department of Art, Tulane University, New Orleans, La.

2009 Icons of Empire: The Art and History of Aztec Royal Presentation. PhD dissertation, Department of Art History and Archaeology, Tulane University, New Orleans, La.

Beyer, Hermann

1965 [1923] Algunos datos nuevos sobre el calendarios azteca. *El México antiguo* 10:257–260.

Boone, Elizabeth Hill

1992 The Aztec Pictorial History of the Codex Mendoza. In *The Codex Mendoza,* edited by Frances Berdan and Patricia Anawalt, vol. 1, pp. 35–54, 152–153. University of California Press, Berkeley.

Broda, Johanna

1970 Tlacaxipehualiztli: A Reconstruction of an Aztec Calendar Festival from Sixteenth-Century Sources. *Revista española antropología americana* 5:197–276.

Brown, Betty Ann

1988 All Around the Xocotl Pole: Reexamination of an Aztec Sacrificial Ceremony. In *Smoke and Mist: Mesoamerican Studies in Memory of Thelma D. Sullivan,* edited by J. Kathryn Josserand and Karen Dakin, vol. 2, pp. 173–189. BAR International Series 402. British Archaeological Reports, Oxford.

Caso, Alfonso

1959 Nombres calendáricos de los dioses. *El México antiguo* 9:77–100.

1967 *Los calendarios prehispánicos.* Intituto de Investigaciones Históricas, Universidad Nacional Autónoma de México, Mexico City.

1971 Calendrical Systems of Central Mexico. In *Archaeology of Northern Mesoamerica,* edited by Gordon F. Ekholm and Ignacio Bernal, pp. 333–348. Vol. 10 of *Handbook of Middle American Indians,* edited by Robert Wauchope. University of Texas Press, Austin.

Castillo, Cristobal del
1908 *Fragmentos de la obra general sobre historia de los mexicanos escrita en lengua náuatl.* Translated by Francisco del Paso y Troncoso. S. Landi, Florence.

Chimalpahin Quauhtlehuanitzin, Domingo De San Antón Muñón
1997 *Codex Chimalpahin: Society and Politics in Mexico Tenochtitlan, Tlatelolco, Texcoco, Culhuacan, and Other Nahua Altepetl in Central Mexico.* 2 vols. Edited and translated by Arthur J. O. Anderson and Susan Schroeder. University of Oklahoma Press, Norman.

Cortés, Hernán
2001 *Letters from Mexico.* Translated by Anthony R. Pagden. Yale University Press, New Haven, Conn.

Davies, Nigel
1974 *The Aztecs: A History.* Putnam, New York.

Durán, Diego
1971 *Book of the Gods and Rites and the Ancient Calendar.* Translated and edited by Fernando Horcasitas and Doris Heyden. University of Oklahoma Press, Norman.

1994 *The History of the Indies of New Spain.* Translated and edited by Doris Heyden. University of Oklahoma Press, Norman.

Fradcourt, Ariane
2010 [1993] New Insights on the Interpretation of the *Aztec Calendar Stone* (with Notes on Skeletonization). In *The Aztec Calendar Stone*, edited by Khristaan D. Villela and Mary Ellen Miller, pp. 276–292. Getty Research Institute, Los Angeles.

Garibay K., Angel M.
1971 [1953–1954] *Historia de la literatura náhuatl.* 2 vols. Editorial Porrúa, Mexico City.

Graulich, Michel
1992 On the So-Called "Cuauhxicalli of Motecuhzoma Ilhuicamina": The Sánchez-Nava Monolith. *Mexicon* 14(1):5–10.

1997 Reflexiones sobre dos obras maestras del arte azteca: La Piedra del Calendario y el Teocalli de la Guerra Sagrada. In *De hombres y dioses*, edited by Xavier Nogues and Alfredo López Austin, pp. 155–208. El Colegio de México and El Colegio de Michoacán, Mexico City.

2010 [1992] The Stone of the Sun. In *The Aztec Calendar Stone*, edited by Khristaan D. Villela and Mary Ellen Miller, pp. 258–262. Getty Research Institute, Los Angeles.

Hajovsky, Patrick
2007 On the Lips of Others: Fame and the Transformation of Moctezuma's Image. PhD dissertation, Department of Art History, University of Chicago, Chicago.

Hassig, Ross
1988 *Aztec Warfare: Imperial Expansion and Political Control.* University of Oklahoma Press, Norman.

Karttunen, Frances
1992 *An Analytical Dictionary of Nahuatl.* University of Texas Press, Austin.

Klein, Cecelia F.
1987 The Ideology of Autosacrifice at the Templo Mayor. In *The Aztec Templo Mayor*, edited by Elizabeth Hill Boone, pp. 293–370. Dumbarton Oaks Research Library and Collection, Washington, D.C.

2010 [1976] The Identity of the Central Deity on the *Aztec Calendar Stone*. In *The Aztec Calendar Stone*, edited by Khristaan D. Villela and Mary Ellen Miller, pp. 199–217. Getty Research Institute, Los Angeles.

López Austin, Alfredo, and Leonardo López Luján
2009 *Monte Sagrado–Templo Mayor.* Universidad Nacional Autónoma de México, Instituto de Investigaciones Antropológicas, Mexico City.

Matos Moctezuma, Eduardo
2009 The Sun Stone or Aztec Calendar. In *Monumental Mexica Sculpture*, edited by Eduardo Matos Moctezuma and Leonardo López Luján, pp. 231–290. Fundación Conmemoraciones 2010, Mexico City.

Matos Moctezuma, Eduardo, and Leonardo López Luján (editors)
2009 *Monumental Mexica Sculpture.* Fundación Conmemoraciones 2010, Mexico City.

Matos Moctezuma, Eduardo, and Felipe Solís
2004 *The Aztec Calendar and Other Solar Monuments.* CONACULTA/INAH and Grupo Azabache, Mexico City.

McEwan, Colin, and Leonardo López Luján (editors)
2009 *Moctezuma: Aztec Ruler.* British Museum Press, London.

Montes de Oca, Mercedes
2000 *Los difrasismos en el náhuatl del siglo XVI.* PhD dissertation, Universidad Nacional Autónoma de México, Mexico City.
2004 Nahuatl Couplets: An Overview. *Sprachtypologie und Universalienforschung: STUF* 51(1):70–84.
2008 Los difrasismos: Un rasgo del lenguaje ritual. *Estudios de cultura náhuatl* 39:225–238.

Navarrete, Federico
2011 Writing, Images, and Time-Space in Aztec Monuments and Books. In *Their Way of Writing: Scripts, Signs, and Pictographies in Pre-Columbian America,* edited by Elizabeth Hill Boone and Gary Urton, pp. 175–195. Dumbarton Oaks Research Library and Collection, Washington, D.C.

Neurath, Johannes
1992 Xiuhuitzolli: Motecuhzoma's Diadem of Turquoise, Fire, and Time; A First Approach to the Problem. *Archiv für Völkerkunde* 46:123–148.

Nicholson, H. B.
1955 Aztec-Style Calendric Inscriptions of Possible Historical Significance: A Survey. Paper prepared for the Mesa Redonda on Calendric Problems of Ancient Central Mexico, Chapultepec.
1956 The Temalacatl of Tehuacan. *El México antiguo* 8:95–134.
1961 The Chapultepec Cliff Sculpture of Motecuhzoma Xocoyotzin. *El México antiguo* 9:379–442.
1971 Religion in Pre-Hispanic Central Mexico. In *Archaeology of Northern Mesoamerica,* edited by Gordon F. Ekholm and Ignacio Bernal, pp. 91–134. Vol. 10 of *Handbook of Middle American Indians,* edited by Robert Wauchope.

Nicholson, H. B., and Eloise Quiñones-Keber
1983 *Art of Aztec Mexico: Treasures of Tenochtitlan.* National Gallery of Art, Washington, D.C.

Olivier, Guilhem, and Leonardo López Luján
2009 Images of Moctezuma and His Symbols of Power. In *Moctezuma: Aztec Ruler,* edited by Colin McEwan and Leonardo López Luján, pp. 78–91. British Museum Press, London.

Padilla, J., P. F. Sánchez-Nava, and Felipe Solis Olguín
1989 The Cuauhxicalli of Motecuhzoma Ilhuicamina. *Mexicon* 11(2):24–25.

Palacios, Enrique Juan
1918 La Piedra del Sol y el primer capítulo de la historia de México. *Memorias y revista de la Sociedad Científica "Antonio Alzate"* 38:1–100.
1921 *The Stone of the Sun and the First Chapter of the History of Mexico.* Translated by Frederick Starr. University of Chicago Press, Chicago.

Peñafiel, Antonio
1990 Destrucción del Templo Mayor de México. In *Trabajos arqueológicos en el centro de la ciudad de México,* edited by Eduardo Matos Moctezuma, pp. 179–268. Instituto Nacional de Antropología e Historia, Mexico City.

Quiñones-Keber, Eloise
1995 *Codex Telleriano-Remensis: Ritual, Divination, and History of a Pictorial Aztec Manuscript.* University of Texas Press, Austin.

Sahagún, Bernadino de
1950–1982 *Florentine Codex: General History of the Things of New Spain.* 13 vols. Edited and translated by Charles E. Dibble and Arthur J. O. Anderson. School of American Research, Santa Fe, N.Mex.
1993 *Primeros Memoriales.* Facsimile photographs by Ferdinand Anders. University of Oklahoma Press, Norman, in cooperation with the Patrimonio Nacional and the Real Academia de la Historia, Madrid.

1997 *Primeros Memoriales*. Translated by Thelma D. Sullivan, edited by H. B. Nicholson. University of Oklahoma Press, Norman.

Seler, Eduard

1992 *Collected Works in Mesoamerican Linguistics and Archaeology*, vol. 3. Edited by J. Eric Thompson and Francis Richardson. Labrynthos, Culver City, Calif.

Tena, Rafael

1987 *El calendario mexica y la cronografía*. Colección Científica 161. Instituto Nacional de Antropología e Historia, Mexico City.

Townsend, Richard F.

1979 *State and Cosmos in the Art of Tenochtitlan*. Studies in Pre-Columbian Art and Archaeology 20. Dumbarton Oaks Research Library and Collection, Washington, D.C.

Umberger, Emily G.

1981 Aztec Sculptures, Hieroglyphs, and History. PhD dissertation, Department of Art History, Columbia University, New York.

1987 Events Commemorated by Date Plaques at the Templo Mayor: Further Thoughts on the Solar Metaphor. In *The Aztec Templo Mayor*, edited by Elizabeth Hill Boone, pp. 411–450. Dumbarton Oaks Research Library and Collection, Washington, D.C.

1998 New Blood from an Old Stone. *Estudios de cultura náhuatl* 28:241–256.

2002 Notions of Aztec History: The Case of the Great Temple Dedication. *RES: Anthropology and Aesthetics* 42:86–108.

2010 [1988] A Reconsideration of Some Hieroglyphs on the Mexica Calendar Stone. In *The Aztec Calendar Stone*, edited by Khristaan D. Villela and Mary Ellen Miller, pp. 238–257. Getty Research Institute, Los Angeles.

Villela, Khristaan D., Matthew Robb, and Mary Ellen Miller

2010 Introduction. In *The Aztec Calendar Stone*, edited by Khristaan D. Villela and Mary Ellen Miller, pp. 1–41. Getty Research Institute, Los Angeles.

Villela, Khristaan D., and Mary Ellen Miller (editors)

2010 *The Aztec Calendar Stone*. Getty Research Institute, Los Angeles.

Wicke, Charles R.

1976 Once More around the Tizoc Stone: A Reconsideration. *Proceedings of the International Congress of Americanists* 2:209–222. Mexico City.

11

Ages of the World in the Andes

JUAN M. OSSIO A.

GIVEN THAT DOUBTS HAVE ARISEN ABOUT the existence of the belief in successive ages in the Andean world, I intend in this essay to present evidence to the contrary. My main support in this endeavor will be *El primer nueva corónica y buen gobierno* of the Indian chronicler Felipe Guaman Poma de Ayala. Through an examination of this document, I will show that these successive ages were part of a conception of time that can be considered as cyclical or static and that they were hidden behind a linear view adopted as part of the Western cultural language in which Guaman Poma expresses himself. Hence, relevant to my approach will be Mircea Eliade's distinction between linear and cyclical time.

Additional evidence in favor of my contention is the fact that the Andean area has many cultural traces in common with the rest of the continent, and if we can attest with certainty the existence of this belief in Mesoamerica, it might also have been present in the Andes. In fact, a solid confirmation of this hypothesis can be seen in the theme, closely related to a conception of ages, of the "revolt of the objects" (see Figure 11.3), which is represented both in the Popol Vuh and in Mochica pottery. Given these resemblances, and to enhance the Andean particularities, I will present a brief description of the conceptions that existed in Mesoamerica.

The division of a cosmological time that accounts for the duration of the world since its beginning to the present day into a fixed number of ages, cycles, stages, or periods, each with its own characteristics, is a belief common to many societies. The fact that these units of time often have a beginning and an end that is succeeded by the emergence of another unit that will follow the same course has resulted in the idea that this conceptualization of time can be represented as cyclical, in opposition to another view, characteristic of Judaism/Christianity, that sees time as nonrepeatable or linear. One of the best contrasts between these two modes of representing cosmological time was developed by the historian of religion Mircea Eliade (1954) in his renowned book *Cosmos and History*.

In a linear conception, each event attains a meaning in itself because it is seen as a manifestation of the will of God. Eliade writes:

> For the first time, we find affirmed, and increasingly accepted, the idea that historical events have a value in themselves, insofar as they are determined by the will of God. This God of the Jewish people is no longer an Oriental divinity, creator of archetypal gestures, but a personality who ceaselessly intervenes in history, who reveals his will through events (invasions, sieges, battles, and so on). Historical facts thus become "situations" of man in respect to God, and as such they acquire a religious value that nothing had previously been able to confer on them. It may, then, be said with truth that the Hebrews were the first to discover the meaning of history as the epiphany of God, and this conception, as we should expect, was taken up and amplified by Christianity. (Eliade 1954:104)

A cyclical conception, on the other hand,

> reveals an ontology uncontaminated by time and becoming. Just as the Greeks, in their myth of eternal return, sought to satisfy their metaphysical thirst for the "ontic" and the static (for, from the point of view of the infinite, the becoming of things that perpetually revert to the same state is, as a result, implicitly annulled and it can even be affirmed that "the world stands still"), even so the primitive, by conferring a cyclic direction upon time, annuls its irreversibility. Everything begins over again at its commencement every instant. The past is but a prefiguration of the future. No event is irreversible and no transformation is final. (Eliade 1954:89)

He later adds: "Almost all these theories of the 'Great Time' are found in conjunction with the myth of successive ages, the 'age of gold' always occurring at the beginning of the cycle, close to the paradigmatic illud tempus. In the two doctrines—that of cyclical time, and that of limited cyclical time—this age of gold is recoverable; in other words, it is repeatable, an infinite number of times in the former doctrine, once only in the latter." (Eliade 1954:112)

As examples of these doctrines, Eliade refers to the "Yugas" from India and other concepts derived from Greece, all of which have in common the idea of succeeding ages that depart from a kind of paradise and that through different stages (generally four or five) follow a decadent process, ending in a period of darkness that is associated with the creation of humans. For Eliade, these conceptions represent cosmological devices developed by humans to face the terror of history. Thus, "history could be tolerated" (Eliade 1954:132).

The Mesoamerican Case

I shall first briefly describe the Aztec and Maya cyclical conceptions of time; then, I shall make a more detailed analysis of the Andean time categories as described by several chroniclers, in particular the Peruvian Indian chronicler Guaman Poma.

In contrast to the degenerative sequence of the ages described previously, those of the Aztec and Maya show no signs of a similar deteriorating or devolving process. The Mexican ages also differ in the nature of the inhabitants of each age and in the way they vanish. But what they have in common is that they are also divided into a fixed number of eras (five), including the present. Each ends according to different cataclysmic causes, and in the case of Mesoamerica, the end occurs systematically via cataclysms that generally are derived from one of the four elements.

Historian of religion Kay Read (1998:70) argues: "No great moral lies behind these destructions. The remains of each sun simply change into the forms of the next according to a schedule. Transformation is naturally described and historically circumscribed, not in any way morally prescribed." For Read, there is no angry god behind the destruction, nor is it the result of terrible transgressions. Rather, destruction takes place because each age is part of a sequence that is organized neither according to a process of decay nor randomly. According to several scholars (e.g., Florescano 1987:122), the idea of progress from one age to the next seems to be behind the sequence, which corresponds to the hierarchical patterns behind the quadripartite division of space.

Mexican historian Enrique Florescano (1987) suggests that although different versions of the five suns myth exist, almost all begin with the existence of a dual supreme divinity (Tonacatecuhtli and Tonacihuatl), known among the Mexica as Ometeotl. This divinity occupies the highest, or thirteenth, level of the sky. From here originate four divinities that perform very important roles in relation to the suns that will be created. These four divinities all have the same name of Tezcatlipoca, but each of them is associated with a different color (red, black, white, or blue) and cardinal direction.[1] Florescano (1987:113) adds that Tonacatecuhtli and Tonacihuatl, after six hundred years of inactivity, began "the creation of the world." Thus, they created fire, an incomplete sun, and a man and a woman. The first was entrusted with agriculture and the second with weaving. From them came the first generation of *macehuales*, or peasants. The underworld, heaven, and water were also created with their respective gods. All were created when time was still not measured—that is, when neither days nor years nor ages existed (Florescano 1987:11).

From this moment, the time of the suns started—that is, the period governed by the divine power that created movement. Once more, the four gods were the principal actors in this creation, each having responsibility for the origin and destruction of the suns. As Florescano notes, Tezcatlipoca, who disguised himself as a feline, was the first god to create the sun, which initiated the beginning of the ages of the world and from which point in time the years began to be counted. Giants who fed from wild plants and animals populated this age. It ended suddenly on the day 4 Ocelot, and its duration was 676 years.[2] Then followed a second age called the Sun of Wind, which was followed by a third one named Sun of Fire, and then a fourth one called Sun of Water.

After the disappearance of these four suns came an intermediate stage in which Tezcatlipoca and Quetzalcoatl accomplished the task of reconstructing the universe. This universe was the final sun, called Nahui Ollin, or Movement Sun (Florescano 1987:14). Then, the gods commanded Quetzalcoatl to create human beings. In order to accomplish this task, he descended to the lower world, the kingdom of Mictlantecuhtli, to obtain the bones and ashes of humans from the earlier cycles. On his return, he fell in a hole placed by Mictlantecuhtli and the bones broke apart. Finally, Quetzalcoatl arrived at Tamoanchan, where all the gods were gathered, and delivered his load. Next, the goddess Cihuacoatl-Quilaztli ground the bones and made dough that she deposited in a container. Quetzalcoatl let blood from his penis and deposited this divine blood into the container. The rest of the gods also made sacrifices. Two years later, the macehuales were created, and Quetzalcoatl provided them with maize after performing some magical actions.

Once people and their main means of subsistence were created, the gods decided to provide light and movement to the world by creating the fifth sun. This creation happened in the year 13 Acatl (Reed). The place they chose was Teotihuacan. Then followed the transformation of the ugly deity Nanahuatzin, who accepted the task of sacrificing himself by jumping into a fire, followed somewhat later by Tecuciztecatl. The first became the sun and the second the moon. But immobility yet remained. And so the self-sacrifice of the gods followed. But even this was not enough. Only with the sacrifice of human beings could the movement of the sun result in originating the idea of war to obtain victims (Florescano 1987:121). Clearly, the concept of ages or suns was vital to Aztec cosmology and to their existence as it was tied to war and sacrifice. From the previous description, we see that each of the first four suns was autonomous, although each was also part of the set of predetermined temporal and spatial positions shown in the great Aztec Sun Stone (Figures 11.1 and 11.2).

In addition to the correspondences between time and space, the Mexica myth of the five ages shares with other American traditions the concept of the development of the cosmos and humanity as a process that moves from darkness to light.

figure 11.1
Stone of the Sun (El zodiaco), from Carl Nebel, *Voyage pittoresque et archéologique dans la partie la plus intéressante du Mexique par planches lithographiées avec texte explicatif*, 1836.

Temporal Sequence	Spatial distribution and associated gods according to several authors and location in the Stone of the Sun
1 Jaguar 4	North, Tezcatlipoca, cold, and upper left side
2 Viento (Wind) 4	West, Quetzalcóatl, magicians, and upper right side
3 Lluvia de fuego (Fire Rain) 4	South, Tlaloc, fire, and lower right side
4 Agua (Water)	East, Chalchiuhtlicue, and lower left side

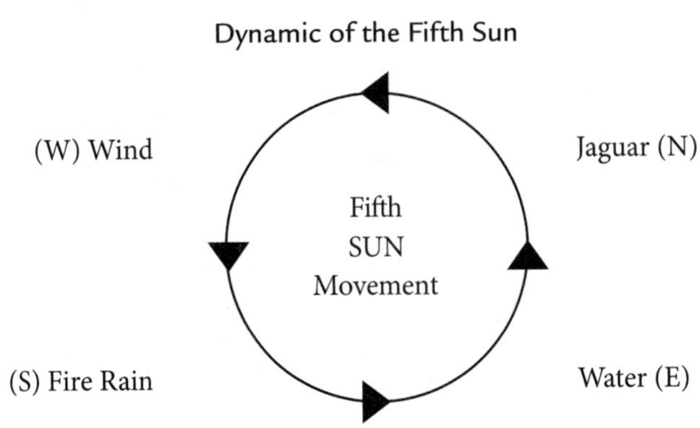

figure 11.2
Mexica time and space. (Illustration by Juan M. Ossio A.)

But there are not many cases in which the origin of human beings unfolds in different ages of durations associated with a fixed number of years. The Popol Vuh of the Quiché Maya, for example, describes the creation of human beings in stages but not in a formal sequence, as was envisioned by the Nahua. According to the text, at the beginning all was silent, only the sky and the sea existed:

> All alone are the Framer and the Shaper, Sovereign and Plumed Serpent, They Who Have Borne Children and They Who Have Begotten Sons. Luminous they are in the water, wrapped in *quetzal* feathers and *cotinga* feathers. Thus they are called Plumed Serpent. In their essence, they are great sages, great possessors of knowledge. Thus surely there is the sky. There is also Heart of Sky, which is said to be the name of the god. (Anonymous 2007:58–59)

All was dark when Heart of the Sky (Thunderbolt Huracan, Youngest Thunderbolt, and Sudden Thunderbolt) spoke with the Sovereign and Plumed Serpent to give birth to humanity.

First, the gods created the earth in the shape of a plate; then, they created the mountains and the valleys. Water divided the sky from the earth, which was covered with a cypress and pine forest. Then animals were conceived, but since they were incapable of talking and worshiping the gods, the gods decided to make a second creation. This creature was made of mud and earth, but it became sodden and mushy. It fell apart and dissolved. These beings were cast aside, and, with the help of other gods, effigies of carved wood were created, but they did not have hearts or minds. They walked without purpose. They had no understanding, especially about how to worship the gods who created them, so they were crushed and killed by a resinous flood and by different animals and objects. They were attacked by their cooking utensils: "Then the grinding stones said this to them: 'We were ground upon by you. Every day, every day, in the evening and at dawn, always you did holi, holi, huki, huki on our faces. This was our service for you who were the first people. But this day you shall feel our strength. We shall grind you like maize. We shall grind up your flesh,' said their grinding stones to them" (Anonymous 2007:76–77). Their descendants are the spider monkeys. The final creation episode in the Popol Vuh deals with the youthful twin creator gods Hunahpu and Xbalanque. They defeat the gods of the death and pestilence in the underworld and are transformed into the sun and the moon. Only then, in the fifth age, comes the final creation of human beings.

While the Maya creation story recounts four stages before the emergence of human beings, this story coincides with that of the Aztec in that each of the ages in the Popol Vuh differs from the previous (although no estimates are given of their duration) and their end is not clearly ascribed to a cataclysm. Only that of the wooden men ends in something similar, combining, on the one hand, a flood of resin and, on the other, destruction by other forces (animals and utensils used in daily life). But the Maya are well known for having organized a complex system of time cycles (see Bricker and Bricker, this volume).

The Andes Case

Apart from those of the Aztec and the Maya, the concepts of world ages in other parts of pre-Hispanic America are not known to be as complex as the versions summarized earlier. For example, the Hopi had a conception of this kind in North America, but it was possibly derived from the Mexican system. In South America, such concepts are found in several societies such as the Pilagá, the Makiritare, the Waiwai, and the Sirionó, as mentioned by Lawrence Sullivan (1988:51–52). These ages, generally two or three in number, were also believed to end in destruction, either by flood, fire, or rain. Since these stories are thought to derive from post-Hispanic versions, however, doubts have been raised about the beliefs being autochthonous.

Such is the case of several descriptions of cosmological beliefs recorded in the Andes. For example, the chronicler Fray Martín de Murúa (2008 [1613]:39v [55v]), who copied information from different chronicles (as was common), reproduced almost exactly the Mexican chronicler López de Gómara's description of the ages of the world. John Rowe (1987:758) notes instances of copying and assumes that a belief in world ages was alien to Andean culture: "The product of Murúa robbery is a trick for the innocent reader who easily accepts as Peruvian several customs and beliefs pertaining to other areas. . . . The five Mexican suns are equally exotic to the Andean area although Murúa was not the only writer who tried to import them. Murúa just made a simple transference; the others manipulated the idea to build something different in its details" (my translation).

Among "the others," Rowe mentions Blas Valera, the anonymous Jesuit, and Fernando de Montesinos. Apart from suggesting that Blas Valera also might have been influenced by López de Gómara because of a reference Blas Valera makes in his description of the Mexican ages, Rowe's (1987:759) contention is that all these sources are late, either from the end of the sixteenth or first half of the seventeenth century, and that the millenarian lengths given to the ages, in contrast to those of the Mexicans, might have been due to influences received from the Western tradition, which since 1570 had stressed biblical chronologies.

That some Andean chroniclers might have described traditions borrowed from Mexico or the West is perfectly possible. Alternatively, one could ask: might it be that if these chroniclers copied from external sources it was because something like these traditions did, in fact, exist in the culture they were describing but in terms that were not so immediately understandable to a Western reader? Otherwise, what particular common interest did these diverse authors possess that would have caused them to invent for the Andes something that did not exist? Additionally, other authors not mentioned by Rowe make references to Andean world ages (e.g., Oliva 1895:71; Salinas y Córdova 1957:13), as do other chronicles (mentioned by Tom Zuidema [1964:229]), including those of Sarmiento de Gamboa, Cabello Valboa, and Vázquez de Espinoza, who estimated the total duration of the Inca kings in symbolic numbers as one thousand or five hundred years. In addition, I suggest that paintings representing Inca and Spanish kings also indicate this belief.

The fact that we have a large corpus of evidence regarding Mexica belief in cosmological ages before the arrival of the Spaniards suggests that this concept might have been ever present throughout Mesoamerica. And if it existed in Mesoamerica as such, might it also have expanded in some form to the south much earlier than the conquest? Perhaps the versions in South America were not reproduced identically; rather, parts of them were selected, for example, the destruction of mankind by water or by fire, as expressed by some Amazonian groups. In other areas, as in the Andes, versions could have been more similar but became accommodated to the numerical patterns that were based on a decimal system.

That echoes of Mesoamerican traditions in Peru arrived before the conquest can be seen both in documents and images. For example, in some Moche representations, we can see scenes similar to those mentioned in the Popol Vuh about the revolt of the objects against the wooden beings created in the fourth Maya epoch. Also, in chapter three of the Huarochirí Manuscript (supposedly written by seventeenth-century chronicler Francisco de Ávila [1873:131]), the only set of Andean myths written in Quechua, we find the following: "They relate that, a long time ago, the sun disappeared and the world was dark for a space of five days; that the stones knocked one against the other; and that the mortars, which they call *mutca*, and the pestles called *marop*, rose against their masters, who were also attacked by their sheep, both those fastened in the houses and those in the fields." Lope de Atienza (1931:1) also mentions a similar belief, but in relation to eclipses. In this case, it was believed that if the animals succeeded in destroying the moon or the sun, then the pots and jars would be transformed into serpents that would threaten mankind. Jeffrey Quilter (1989) provides us with a Moche representation (Figure 11.3) and with an interesting analysis of these parallels.

The tradition of the revolt of utilitarian material objects remains until today. It is not associated

figure 11.3
The Revolt of the Objects in Moche iconography. (Drawing by Donna McClelland.)

with stages of humanity, but rather with the perils of the soul in the afterlife, which is probably a Christian inflection. Ricardo Valderrama and Carmen Escalante mention that, after crossing some rivers, the soul arrives:

> At the cats' village all of them attack us with their claws until washing us in blood. From there our poor soul crosses the hen village. Here they say to us:
>
> You in your house did not feed us.
>
> You in your house have me eating ground and worms.
>
> And they kick and bite us until the flesh of our body hangs like rags from our bones. Running away from this suffering it arrives to the guinea pigs' village here they also say to us:
>
> You cut my throat and put me in boiling water.
>
> You roast me in fire to eat me.
>
> Does it hurt you? It also hurt me as such.
>
> They talk as such and rolled us with their bites until pulling out pieces of our flesh and even of our bones. Thus running away almost as a skeleton arrives to the pots' village because they say they also have a soul. Truly they are made of living earth from the Pacha (Earth). That is the reason they say they have their village in the way to the afterlife. In their village the pots do not produce any suffering to the men. They do not even talk to us but only recriminate the women:
>
> You put me alive upon fire. If I did that to you, you would also feel it.
>
> Being you a woman like I you gave me a bad treatment.
>
> That is what the pots say when women pass through their village. (Valderrama and Escalante 1980:258; my translation)

If these echoes remain in the Andes, why could it not be that those concerning the ages of the world, which today are widely encountered all over Andean communities, are rooted in concepts and beliefs that existed before the arrival of the Spaniards? I regard the fact that a chronicler might have copied descriptions from other cultural areas or provided information years after the earlier moments of the conquest as insufficient evidence that a belief in cosmological ages was alien to the Andes. Instead of basing his argument on Murúa's reproduction of a description made by López de Gómara, Rowe should have been more cautious and perhaps have questioned whether there were not grounds in the Andes to explain these copied passages, especially when other authors, especially Guaman Poma, who was very close to Murúa, talked in depth about this tradition in their chronicles.[3]

The Evidence from Guaman Poma

As Monica Barnes (1994), Victoria Cox (2002), and David Fleming (1994) have demonstrated, Guaman Poma was well acquainted with Western literature, particularly the *Cronografía* written by Hieronimo de Chaves (1580),[4] which divided the past into a fixed number of ages and attempted to estimate their total duration until the present. But does this mean that what he developed in relation to the ages of the world derived only from this tradition and that something similar did not exist in his own culture?

Drawing on my own views about the Andean nature of *El primer nueva corónica y buen gobierno* (Ossio 2008a), I shall here argue that behind Guaman Poma's description of the European and Andean ages lies a logic, although phrased in similar terms to those of Chaves, that is quite different from that of the *Cronografía* and closer to what other chroniclers such as Montesinos refer to regarding Andean ages.

Whereas the chroniclers who describe Andean ages present them as traditions worthy of comparison with those of the Mexica, Guaman Poma presents the ages basically to proclaim that the Indians were Christians before the arrival of the Spaniards, thereby negating the main justification for the conquest, which was the evangelization of the Indians. As we shall see, his argument rests on two different Andean models, one cosmological and the other genealogical. The first one has for Guaman Poma a messianic connotation of an order that could only be restored in the Andes if all of the Spaniards returned to their place of origin (Ossio 2008a); the second is used to legitimize, through principles of descent, the rights of the Indians to their own territory.

It is precisely in the nature of Guaman Poma's argument that we see the emergence of significant evidence in favor of the existence of the idea of five ages in the Andes prior to the arrival of the Spaniards. The reason is that his argument, although incorporating overtones of a linear conception of time, is rooted in a cyclical or static view common to conceptions of time as encapsulated in ages structured by a symbolic number.

At the beginning of *El primer nueva corónica y buen gobierno*, we find Guaman Poma, as if using an ideal pattern, presenting indigenous space and time textually in a temporal scheme divided into two sets of five ages. Thus, his representation of both European and Indian space is divided into four realms, with Cuzco at its center, modeled on Tahuantinsuyo (Figures 11.4 and 11.5). That they are strictly parallel and static is confirmed by the fact that each set has almost the same duration: the European 6,612 and the Andean 6,613 years (Figures 11.6 and 11.7; Ossio 1970, 1973, 2008a). Manipulating them linearly on the basis of a genealogical model, however, he states that the Indians were Christians because they descended from Adam and Eve and worshipped Viracocha, who was seen as Spanish or Christian, but only until the time of the Inca, who occupy a fifth age and whom he regards as idolaters.[5]

We see that, according to Guaman Poma, Andean people were Christians before Spanish attempts to convert them. But he states that the Inca, whom he locates within the fifth age, were idolaters because they did not have a known father and their mother was a witch.[6] Only by playing with the ages and creating one model that is cosmological and

figure 11.4
The Indies and Spain, drawing 42 [42] of Felipe Guaman Poma de Ayala, *El primer nueva corónica y buen gobierno*, 1615. (Photograph courtesy of The Royal Library, Copenhagen.)

figure 11.5
Mapa mundi, drawing 983–984 [1001–1002] of Felipe Guaman Poma de Ayala, *El primer nueva corónica y buen gobierno*, 1615. (Photograph courtesy of The Royal Library, Copenhagen.)

Ages of the Spanish Universe
1) The First World, Adam and Eve
2) The Second World, Noah and the Flood
3) The Third Age of the World, Abraham
4) The Fourth Age of the World, David
5) The Fifth Age of the World, birth of Jesus Christ
} 6,612 years

Popes List
1) Saint Peter
2) Saint Hieronymus and Pope Damazo
3) Pope Juanes (approved by the emperors)
4) Pope Leo
5) Pope Boniface IX of Naples (Discovery of the Indies)

Ages of the Indian Universe
1) Uari Uiracocha Runa 800 years
2) Uari Runa 1,300 years
3) Purun Runa 1,100 years
4) Auca Runa 2,100 years
5) Inca Runa in p. 37 1,515 years
p. 91 1,548 years
p. 118 1,522 years
p. 453 1,500 years
p. 948 1,500 years
6) Pachacuti Runa 32 years
7) Conquest
8) Rebellion against the Spanish king
9) Kings Charles V, Philip II, and Philip III
} 24 years
} 112 years
10) "Considerations"
} 6,613 years

figure 11.6
Guaman Poma's time tables. (Illustration by Juan M. Ossio A.)

another that is genealogical can he accomplish what he wishes to argue. In the cosmological model, the five ages have a beginning and an ending, with cataclysmic destructions proceeding one after another (Pachacuti). In the genealogical model, there are five generations that permit him to both create continuity and allow him to separate the age of the Inca as idolaters from the four proceeding ages of Christian Indians. But since it was so rooted in the cosmological model that the present had to be located in a fifth age, Guaman Poma cannot avoid suggesting that the fourth age of the Indians was indeed a fifth age, because until that age the duration he gives to all of them, starting with the first, called Uari Uiracocharuna, is around five thousand years (fifty-three hundred, to be exact).

The number and duration of Guaman Poma's ages also suggest a link to Andean pre-Hispanic concepts of time. If he had clearly understood Chaves's and other historians' biblical concept of ages, he would have identified six,[7] following the number of days it took God to create the world. For example, Chaves (1580:56v; my translation) states that "this division was made thus according to the six days in which the world was created and this is the common division of Eusebio (de Cesarea) and of all the historians." He (Chaves [1580:58]) estimates the time since Noah to be 2,952 (not 6,612)

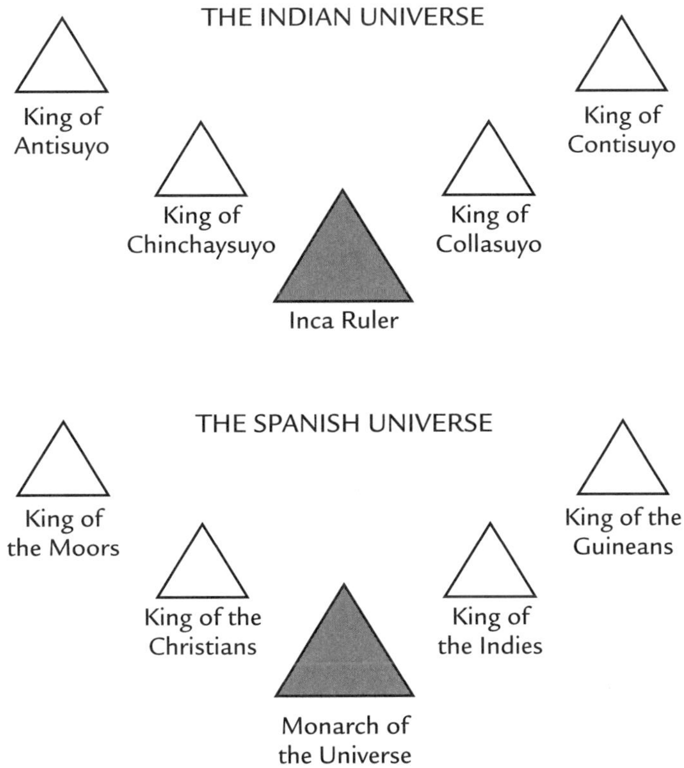

figure 11.7
Two parallel sets of spaces and ages: European and Indian. (Illustration by Juan M. Ossio A.)

years, which none of the other historians contemporaneous with Chaves state.

Guaman Poma's estimation of the length of the two parallel series of ages also suggests the Andean roots of his time sequence. It would seem that—faced with the duration estimates made in the *cronografías* he was following, but unaware of the grounds that underlay them—he felt compelled to make his own estimates. Lacking other alternatives, he used a method that, years later, Fernando de Montesinos would attribute to pre-Hispanic time reckoning: to create a length of five thousand years, he gives each of the five ages a length of one thousand years (Capac Huatan), which ran up to the Christian years in which he was writing, 1612 and 1613 (see Figure 11.6).[8]

Another similarity to the traditions that Montesinos (2010) recounts in his *Memorias antiguas, historiales y políticas* is the use of the notion of "Pachacuti" as a cataclysmic breaking point between periods of either five hundred or one thousand years within a sequence of five ages of one thousand years each known as Capac Huatan. Apart from the different occasions on which Guaman Poma uses lengths of five thousand, one thousand, and five hundred years (see Figure 11.6), this pattern can also be seen clearly when, on page 911 [925] of the *Nueva corónica*, his regular sequence of five ages is extended to a sixth one and the name he gives to it is "Pachacuti Runa." But the coincidences go even further, because both authors locate the last ruling Incas in the second half of a fifth age, which corresponds to a period of five hundred years marked between 4500 and 5000 CE.

This last point brings to light yet another similarity that consists of Guaman Poma placing in a ninth position within the ages of the last ten Incas or the last Inca dynasty the principle of order represented by the Inca called Pachacuti. Hence, according to Montesinos, who lists more than one hundred Inca rulers from the first to the fifth age, the Inca Pachacuti of this last stage was not only the ninth Inca of the list of the ten last Incas, but the ninth Pachacuti within the list of one hundred Incas (Table 11.1). In Guaman Poma's work, on the other hand, it is in the ninth age that follows

table 11.1
Fernando de Montesinos's five mythical ages (Imbelloni 1946:109).

600 of America	Pachacuti I
1000 First Sun (end of age)	Pachacuti II
1500	Pachacuti III
2000 Second Sun (end of age)	Pachacuti IV
2500	Pachacuti V
3000 Third Sun (end of age)	Pachacuti VI
3500	Pachacuti VII
4000 Fourth Sun (end of age)	Pachacuti VIII
4500 middle point of the Fifth Sun corresponding to the Inca period	Pachacuti IX

"Pachacuti Runa" that he places as ordering principles the three Spanish kings who governed Spain during his life-span: Charles V, Philip II, and Philip III (see Figure 11.6).[9]

This view of the conquest as a Pachacuti, or cataclysmic breaking point, is not exclusive to Guaman Poma. Zuidema notes that the chronicler Antonio Vázquez de Espinoza (1969 [1629]:ch. CIII, para. 1585, p. 391), writing about the conflict between Huascar and Atahuallpa, describes the last years of the Inca empire as "ending the extremely wealthy pagan monarchy of the incas that had started in the year 1031[10] since the first Manco Capac and lasted 500 years in its greatness and wealth in which those kings conquered so many nations who ordered with their laws . . . in order to facilitate the reception of the Evangelical one" (my translation).

As mentioned by Zuidema (1964:10), other chroniclers (e.g., Pedro Sarmiento de Gamboa [1960 (1572)] and Miguel Cabello de Valboa [1951 (1586)]) also give similar estimates about the length of Inca history. In all of them, the encounter with the Spaniards is seen as the end of a five-hundred- or one-thousand-year period, and the duration of an Inca's rule is accommodated to fit to this number; for example, in Sarmiento de Gamboa's work, the recurrent number is one hundred years, which seems to be at the base of Guaman Poma's calculations of the life-span of each of the Incas, whereas in Cabello de Valboa's it is around fifty years.[11]

The Term *Pachacuti* as Renovation and Destruction

Because the term *Pachacuti* refers to a breaking or liminal point between two structured periods, it connotes both destruction and renovation. Nevertheless, and perhaps due to Western apocalyptic beliefs introduced during the colonial period, the first part of its meaning has been expanded upon the most. Accordingly, Inca Garcilaso de la Vega (1960:bk. V, ch. XXVIII, 304; my translation) writes: "By way of proverb they say pacham cutin; it means the world changes and generally they say it when the big things change from good to bad and rarely is it said when the change is from bad to good."

Perhaps also under the influence of the *Tercer catecismo limense* of 1584, the Quechua vocabularies published after this work also list *Pachacuti* and translate it as the "end of the world, big destruction, pestilence, lost ruin, common harm" (González Holguín 1952:270; my translation).

For Guaman Poma, Pachacuti means "God's punishment," but he also uses it to mean the "universal flood," which he calls "Uno Yaku Pachacuti."[12] In addition, when he describes the government of the ninth Inca Pachacuti, Guaman Poma (2009:109 [109]) says, "During the time of this Inca many Indians died. There was hunger, thirst and pestilence, punishments given by God. It did not rain for seven years; some say ten years. . . . This Inca was named Pachacuti Inca" (Figure 11.8).[13]

figure 11.8
Pahacuti Inca, drawing 108 [108] of Felipe Guaman Poma de Ayala, *El primer nueva corónica y buen gobierno*, 1615. (Photograph courtesy of The Royal Library, Copenhagen.)

Apart from the pen-and-ink drawing and textual description of this Inca, Guaman Poma also uses the term *Pachacuti* to refer to the first in a series of *capitanes*, who appear immediately after the series of the *coyas*, or Inca wives. The first is named Ynca Yupanqui, Pachacutichic Ynga,[14] who is described as the first son of Manco Capac Inca. This captain is the same as another described by Murúa in different terms (and who will be mentioned subsequently). According to Guaman Poma's (2009:146) text and drawing, this individual/character was very lazy, slept until the sun was high in the sky, and "made no conquests, nor did they do anything. They just slept, ate, went whoring, enjoyed themselves, had parties and banquets and strolled through the city of Cuzco with the other nobles, auquicona and Incacona [Incas]. They ended their days in Cuzco during the time of Manco Capac." Guaman Poma's drawing depicts this figure lying asleep in an European-style bed next to a window through which the sun shines brightly (Figure 11.9).

Ludovico Bertonio (1984 [1612]:141), as he defines the term in his Aymara vocabulary, regards the Quechua connotation of Pachacuti as a dissipated or chaotic person: "Pachacuti haque, vel.

figure 11.9
Captain Pachacutichic, drawing 145 [145] of Felipe Guaman Poma de Ayala, *El primer nueva corónica y buen gobierno*, 1615. (Photograph courtesy of The Royal Library, Copenhagen.)

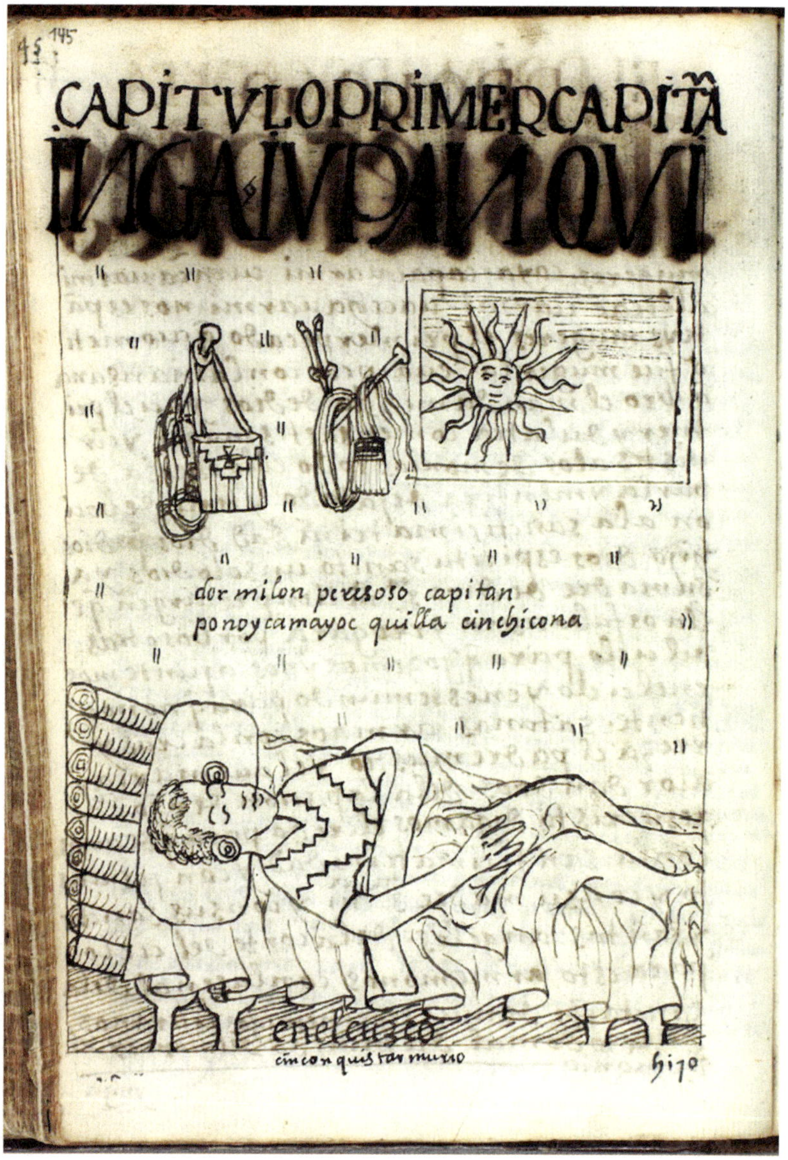

Kanchilla aroni haque: Braggart who says that will do and say wonders and at the end does not make anything" (my translation).[15] In other words, a Pachacuti is someone who is a braggart, who claims to be capable of doing wonderful things, but who in the end does nothing.

Fray Martín de Murúa, although he worked very closely with Guaman Poma (Cummins and Ossio 2013; Ossio 2004, 2008b), includes nothing like this connotation when talking about Pachacuti in his two manuscripts. But Pachacuti's wife, Mama Anahuarque or Ipahuaco, is described as somebody who was remembered for appeasing with sacrifices a kind of unruly Pachacuti, represented by the eruption of a volcano in Arequipa. In his two versions, this Mercedarian friar also mentions the same *capitán* who is mentioned by Guaman Poma, but very differently. In common with the Pachacuti capitán of the *Nueva corónica*, he also is identified as the son of the first Inca, Manco Capac, but the context that surrounds him is more messianic, as revealed by the presence in Guaman Poma's drawing of an apocalyptic angel shown descending from the sky and blowing a horn. This figure is clearly derived from the angel Gabriel, who announces the beginning of the Last Judgment. Below the figure

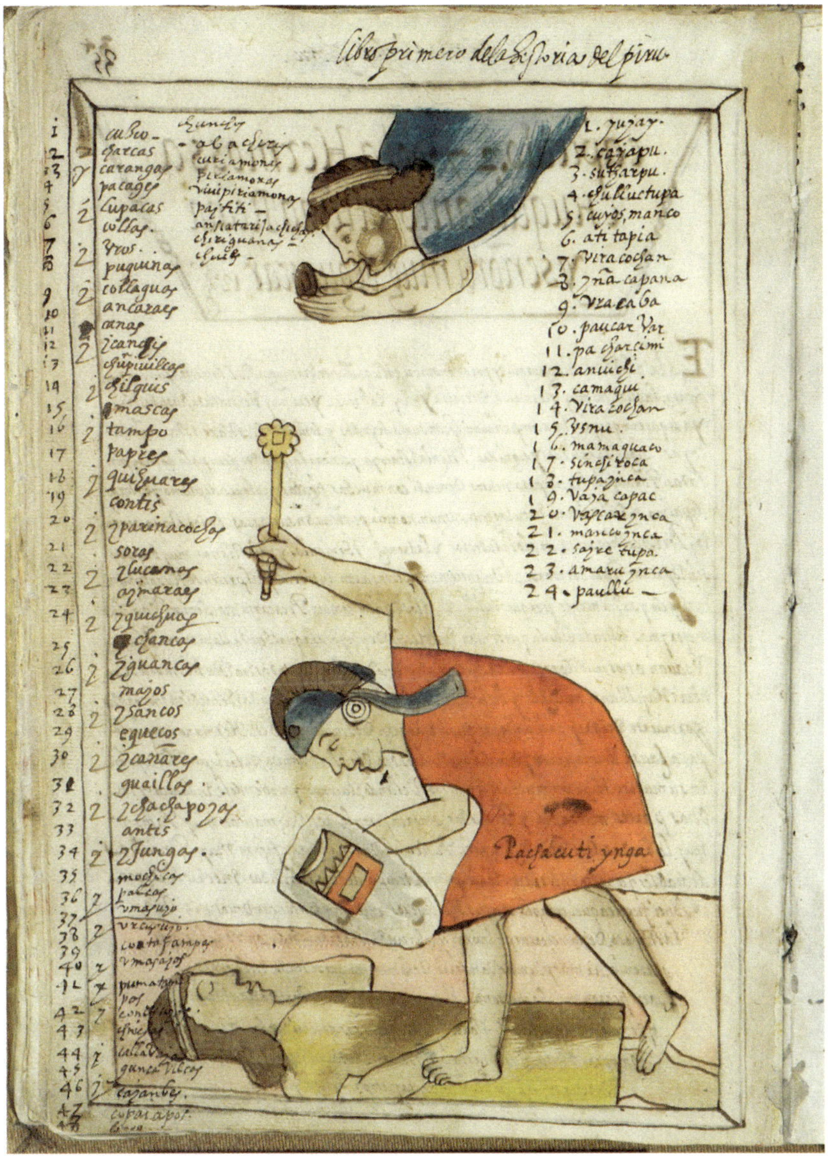

figure 11.10
Captain Pachacuti, from Martín de Murúa, *Códice Murúa*, ca. 1590. Private collection.

of Pachacuti is depicted, in an orange knee-length tunic, a figure who is described as a giant and shown killing an enemy (Figure 11.10). The text is as follows:

> They say he was not so brave as cruel because he was of rough condition and the first to command worship of the guacas and order how they should be sacrificed. And distributed them and ordered to be worshipped throughout the kingdom. And some want to say, although as a fable, that that was the cause that in the time of this brave captain and infant Pachacuti on the top of this city that they call Chetacaca and by another name Sapi a person dressed in red, as shown in this painting, and with a trumpet in one hand and a baton on the other showed up, but before he appeared for a whole month it had been raining day and night continuously, and they became afraid that the earth wanted to turn upside down, something they call Pachacuti. And that this person had come over the water and to Pisac, four leagues from Cuzco, the infant encountered him and requested him not to play his trumpet because they feared that if he played it the earth might turn upside

> down and they would become brothers. And he did not play it and after some days he turned into stone. And for this reason he was called Pachacuti, which means the earth upside down and also withdrawn or disinherited from what belongs to him. (Murúa 2004 [ca. 1590]:fol. 36r, 2008 [1613]:fol. 203v; my translation)

Are the two versions of Guaman Poma and Murúa as different as they appear to be? At the moment, I do not have a clear answer, but given the similar nature and position of the two captains of the same name, it deserves further careful analysis. Perhaps the negative side that Guaman Poma emphasizes in his portrayal of this captain might be related to his negative view of the Inca, starting with Manco Capac, whom he sees as an idolater. In the case of Murúa, his portrayal might be connected to a less ideological antipathy toward the Inca and a great interest in Inca myth and history as it was told to him in which this figure, as suggested by Zuidema (2010:588), is, in calendrical terms, related to the alternation of a rainy with a dry season.

For the majority of the chroniclers, the name Pachacuti was given to this Inca, whose original name was Inca Yupanqui, because after defeating the Chancas, the greatest enemy threat to the Inca, he became responsible for the reconstruction of Cuzco. Pedro Cieza de León (1967 [1553]:121), for example, introduces this term in the following context: "While crowning him on his head Viracocha Inca said to him 'I name you from today . . . Pachacutec Yupanqui Capac Indichuri that means change of time, Rey Yupanqui, Son of the Sun'" (my translation).

One other colonial context in which the idea of radical change associated with the term *Pachacuti* occurs is related to the messianic and nativistic movement called Taqui Onqoy, which took place between 1560 and 1570. According to Fray Cristóbal de Molina (el Cusqueño), the followers of this movement believed that

> all the huacas which the Christians had burnt and destroyed had been brought to life again; and that they had been divided into two parts, one of which was united with the huaca *Pachacama*, and the other with the huaca *Titicaca*. The story went on that they had formed in the air, in order of battle against God, and that they had conquered Him. But when the Marquise entered this land, it was held that God had conquered the huacas, as the Spaniards had overcome the Indians. Now, however, it was believed that things were changed, that God and the Spaniards were conquered, all the Spaniards killed, and their cities destroyed, and that the sea would rise to drink them up, that they might be remembered no more. (Molina 1873:60)[16]

In the *Información de servicios* of Cristóbal de Albornoz, who, along with Luis de Olivera, conducted an extirpation campaign against the Taqui Onqoy, many of the witnesses speak of a change of *mita* (turn) and, in one case, an image of an upside-down transformation, a proper or chaotic condition of a Pachacuti. It is expressed in the following terms: "And if they did not worship the guacas and made such ceremonies and sacrifices being preached to them they would die and walk upside down their heads next to the ground and their feet above them, and others would turn into guanacos, venados (deer) and vicuñas and other animals, and mentally disturbed they would jump into abysses, and that their guacas would make a new world and other people" (Millones 2007:156 [43v]; my translation).

This evidence makes it clear that in both Quechua and Aymara languages there existed a term for referring to the breaking point between two ages. This was *Pachacuti* or *Pachaticra*, which, under the influence of Christian millenarian preaching, became associated with the Last Judgment. In addition to this cosmological connotation, it seems to have been related to the idea of "dissipation" linked with youth in the process of being initiated, thus suggesting a further association with a breaking point in relation to the life cycle or the transition between age classes, whose organization, as we shall see, was similar to that of the cosmological ages.

Further Issues with Guaman Poma's Five Ages

One additional issue that requires comment is Guaman Poma's description of the names and content of the five Indian ages. As mentioned, his interest in the ages is based on his assertion that, through descent, the Andeans became Christians long before the arrival of the Spaniards. Assuming that this status could be attained by means of descent, he presents the sequential nature of the ages according to a genealogical model not alien to Andean society. Such is the impact of this model that on some occasions he describes the ages as if they were individuals. Thus, talking about a list of kings he says:

> The generation and lineage of people from very ancient times that God brought to this kingdom of pagan Indians who descended from Adam and Eve and the offspring of Noah of the Flood—these were Vari Viracocha Runa, Yarovilca, which is to say he is a much more important lord than all of the nations. From Vari Viracocha Runa descended Vari Runa, Yarovilca, as did Purun Runa, Yarovilca, and Auca Runa, Yarovilca, which refers to the legitimate king who was a descendant of Adam and Noah. Yarovilca means king; Yarovilca also means Inca. (Guaman Poma de Ayala 2009:75)

This manner of presenting the ages seems to have had an influence on some followers of

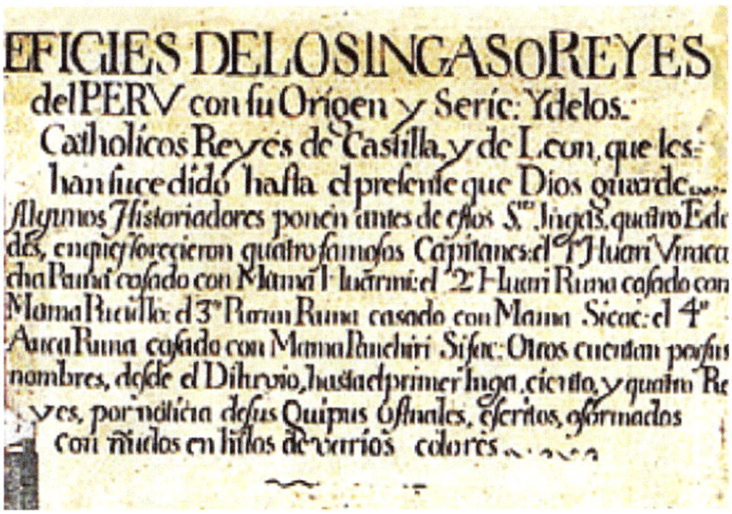
EFIGIES DELOSINCASOREYES
delPERV con fu Origen y Seríe: Ydelos:
Cathólicos Reyes de Castilla, y de Leon, que les:
han fucedido hafta el prefente que Dios guarde...
Algunos Historiadores ponen antes de eftos S.res Jngas quatro Edades, en que ſe forecieron quatro famofos Capitanes: el 1.° Huari Viracocha Pacha cafado con Mama I huarmi: el 2 Huari Runa cafado con Mama Pucillo: el 3.° Purun Runa casado con Mama Sicac: el 4.° Auca Runa cafado con Mama Panchiri Sifac: Otros cuentan por fus nombres, defde el Diluvio, hasta el primer Inga, ciento, y quatro Reyes, por noticia de fus Quipus, ó Anales, efcritos, ó formados con ñudos en hilos de varios colores...

figure 11.11
Detail of the *Genealogy of the Incas and Their Spanish successors*, ca. 1750, Nuestra Señora de Copacabana, Lima.

figure 11.12
Perez Bocanegra model, from Juan Pérez Bocanegra, *Ritual, formulario, e institución de curas, para administrar a los naturales de este reyno los Sanctos Sacramentos*, 1631.

Guaman Poma, for example, Fray Buenaventura de Salinas y Córdova (1957:ch. 1), and some paintings that represent the dynasty of the Inca and Spanish kings where, in the upper left corners, these ages are mentioned as "Capitanes" and, as in the case of Fray Buenaventura, a wife is assigned to each of them (Figure 11.11). Moreover, the use Guaman Poma makes of a genealogical model matches perfectly with the cosmological ages because, as suggested by Zuidema (1989:77–82), this corresponded to the numbers of generations within an "ayllu" as evidenced by a drawing made by Juan Pérez Bocanegra (Figure 11.12) associated with marriage regulations and kinship hierarchical positions.[17]

Apart from this genealogical model, there is an additional linear pattern that has to do with the nature of the populations associated with each of the ages. Seeking, at a cosmological level, that the ages might present continuity between them, Guaman Poma adopts an evolutionary pattern that unfolds from nature to culture. Thus, before the beginning of the ages the world is described as having been inhabited by "serpents, amaros; savages, zacha

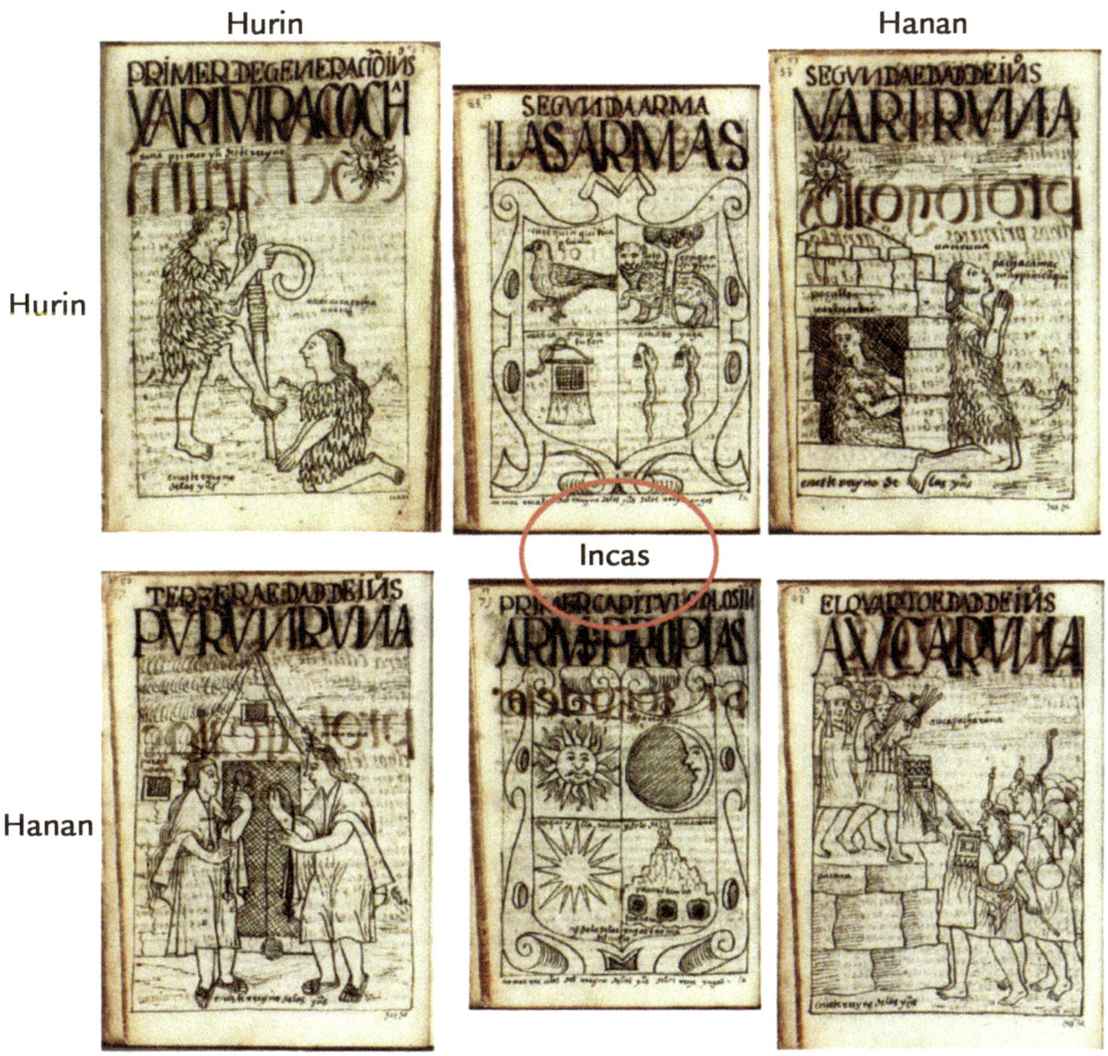

figure 11.13
Inhabitants of the ages, drawings 48 [48], 83 [83], 53 [53], 57 [57], 79 [79], and 63 [63] of Felipe Guaman Poma de Ayala, *El primer nueva corónica y buen gobierno*, 1615. (Photograph courtesy of The Royal Library, Copenhagen.)

runa-uchuc ullcos; jaguars, otoroncos; goblins, hapinunos; mountain lions, pomas; foxes, atocs; bears, ucumaris; deer, luychos. These first Indians, Vari Viracocha, killed them, conquered the land and ruled over them, entering this kingdom by the command of God" (Guaman Poma 2009:50). Then follow the Uariuiracocharuna, who lived in caves and appear to have been dressed with tree leaves; then the Uariruna, who lived in *pucullos* (tombs) and were dressed in skins of animals; then the Purunruna, who moved to houses and learned to weave; then those of the fourth age, the Aucaruna, who reached a high stage of culture that included kingship; and finally come the Incaruna, who are divided into two: the descendants of the first four ages and the idolaters, each of whom are associated with two different coats of arms (Figure 11.13).

There being a sequence of five ages with links established between them, Guaman Poma cannot avoid seeing a correspondence with the organization of groups in space, wherein the first four are divided into two sets that are framed under genealogical and cosmological considerations. Whereas the first two have in common the term

Uari, which connotes the idea of autochthonous people within a realm or owners of the land, those that follow include terms such as *Purun* and *Auca*, whose meanings are more of a cosmological kind because they refer to humans who lived in the first case in a barren environment and in the second in perpetual warfare. On the other hand, the fifth age, as in the case of space, is associated with the principle of order that, for the pre-Hispanic past, represented the Inca who, in the colonial present, became the monarch of the world (see Figures 11.6 and 11.7). Since this age is not represented in a single drawing, because it corresponds to the twelve Incas described in the successive chapters, I am representing it with the coats of arms that Guaman Poma associates with the legitimate Incas Tocai Capac and Pinahua Capac and the idolater Manco Capac (Figure 11.13 and Table 11.2)

The structure of Table 11.2 illustrates the conceptual connection between space and time. The first part depicts the division of the social and political space of Tahuantinsuyu, the empire of the Inca as they conceived it. The second part demonstrates the time scheme of the five ages as they relate to each other in a social as well as temporal model. If we also look at the organization of the *acllas*, or virgins of the suns, dedicated to the service of the main Inca divinities, arranged according to age grades, we find another example. This model of service by gradations of five years is by analogy further

table 11.2
The organization of space and time.

SPACE		
	Hanan	**Urin**
Hanan	Chinchaysuyu	Antisuyu
	Inca or Cuzco	
Urin	Collasuyu	Contisuyu
TIME		
	Hanan	**Urin**
Hanan	Aucaruna	Purunruna
	Inca runa	
Urin	Huariruni	Huari Viracocharuna

evidence in favor of the existence of a concept of the ages of the world. Again, Guaman Poma provides us with the most accurate description:

> The first virgins, uayrur aclla [high priestesses]. They were consecrated to the cult of the Sun, the Moon, stars, chasca cuyllor [Venus] and chuqui ylla [lightning]. These virgins did not speak to any man in their lifetime. They started this service at twenty years of age.
>
> There were virgins of the huaca of Huanacauri, a sacred shrine of the Incas. These ladies were called sumac aclla [beautiful chosen women]. They did not sin nor have any dealings with men. They were thirty years old.
>
> There were also virgins of the major huacas. These women were called uayror aclla sumac [most important chosen beauties]. They were twenty-five years old and remained chaste until death.
>
> The virgins of the secondary huacas were called sumac acllap catiquin. They were thirty-five years old. These ladies spin and weave the garments of delicate cloth for the huacas of this kingdom.
>
> There were virgins of the minor huacas. These women were called aclla chaupi catiquin sumac aclla. They were forty years old and worked in the fields and made garments.
>
> There were virgins called pampa acllacona [peasant priestesses] who served the Moon, the stars and other common huaca vilca. These women were weavers of chumpi, uincha, chuspa huatu [cords for hanging a pouch], chuspa ystalla [women's pouches] and other fineries. They were fifty years old. They never sinned and were daughters of auquicona, Inca princes. (Guaman Poma 2009:299 [301])

The similarity between Guaman Poma's description of the five ages of the Andean world and the organization of the acllas is clearly suggested by the fact that this time, instead of two doubled periods of time plus a fifth age, there are two main categories of acllas. Each has secondary ones and a sixth age that seems to have a marginal condition, as can be seen by a comparison of Table 11.2 and Table 11.3.

table 11.3
The organization of religious acllas

	HANAN	HURIN
HANAN	Guayrur Aclla Twenty years old; served the sun, moon, stars, and chuquiylla.	Sumac Aclla Thirty years old; uanacauri.
HURIN	Guayrur Aclla Sumac Twenty-five years old; main guacas.	Sumac Aclla Catequen Thirty-five years old; secondary guacas
		Chaupi Catequen Sumac Forty years old; lower guacas served in agriculture and to make cloth.
	Pampa Acllaconas Fifty years old; served the common gods; were daughters of Auquiconas and wove *chumpis* (waist bands) and *chuspas* (bags).	

This being the case, is it not possible then that the age classes from twenty-five to fifty mentioned for the tributaries by several chroniclers[18] may have also acted as a significant model for the ages of the world and as further evidence of its existence in the Andes?

Conclusion

I have attempted to describe the evidence in favor of the existence of a belief in five ages in precontact Andean culture. The strongest argument is provided by Felipe Guaman Poma de Ayala's *El primer nueva corónica y buen gobierno*. The support it provides does not lie at the surface of his text, but rather in the interpretive reconstruction of spatial and temporal categories corresponding to a cultural tradition that is very different from that of the language and form in which he expresses himself. Accordingly, behind this reconstruction lies a conflict between a linear view of time intimately associated with the Western historical terms with which he expresses himself and an Andean cyclical or static view deeply rooted in the categories of thought he inherited from his ancestors, which were part of the Andean conception of five ages.

The ages of the world are not for Guaman Poma only a curiosity or a cosmological achievement worthy of people who had attained a high level of civilization. This idea is the main instrument he uses to contradict the strongest argument of the Spaniards used to legitimize their invasion of the Andean territory, namely the Christianization of the Indians.

Guaman Poma's way of arguing pertains to different cultural standards. It is not one wherein "contract" predominates over "status," as Sumner Maine (1972) would say. Guaman Poma argues that one can attain the status of Christian through descent, rather than through the acceptance of transmitted doctrine. For this purpose, the Indian author accommodates the ages to a genealogical model, which was part of his own cultural tradition, to argue that the Indians were already Christians before the Spaniards arrived because they descended from Christian Spaniards, as did Noah or Viracocha; however, the link with Noah is not very consistent because this patriarch is associated with a set of ages that correspond to the European world, which is clearly presented as independent of the Indian ages. It is precisely at this point that Guaman Poma confirms the validity of all those chroniclers, starting with Montesinos, who suggest in one way or another the existence of cosmological ages as closed cycles in the Andes. The independence of the two sets of ages is confirmed by the fact that both receive similar measurements of time. But what leaves no doubt about the Indian writer's involvement in the Andean cosmological conception of the ages as described by Montesinos and the other authors is his estimate of the duration of the age gradations. As we have seen, he calculates the duration by adding the year that he wrote his chronicle (1612 or 1613) to the symbolic figure of five thousand years that is equivalent to the number of years of five ages, each of one thousand years.

Further support for Guaman Poma's involvement with this cosmological tradition can also be seen in his attempt to extend these cycles into the present historical time—that is, he also divides the history of European contact up to his present into five periods or stages. Once again, the pattern consists of five ages but, as we see in Figure 11.6, it departs from a sixth age that, in consonance with the breaking points described by Montesinos, is called "Pachacutiruna." This age is followed by a seventh age he calls "Conquista Cristiano runa" and then by an eighth age that deals with the Spanish civil wars. This eighth age completes a trilogy of ages that have in common chaotic circumstances. Finally, the sequence ends with a ninth age that, as in the case of the ninth position in the Inca royal dynasties, was occupied by "Pachacuti" as "Reformer of the World"; here, he places high Spanish personages in the same role. These are the Spanish kings Charles V, Philip II, and Philip III, who governed Spain from the European encounter until his own present. After this age what follows, in a tenth position, are his moral considerations for a good government or for the restoration of order, which he sees as having been disrupted by the conquest.

Additional evidence for temporal organization rooted in an Andean tradition of *pacha* as time can be seen in the Andean tradition of the organization of pacha as space (see Nair, this volume)—that is, the organization of time and space are expressed through the single term *pacha*, which strongly suggests their intimate conceptual relationship just as it existed in Mexico, as outlined in the introduction to this essay. A clear confirmation exists that Guaman Poma organized things in space and time according to a predetermined divisionary scheme of five, which can be seen in his attempt to organize both the European space that he depicts below the Indies (see Figure 11.4) and the parallel sequence of ages of these two worlds (see Figure 11.6). Moreover, if we look at the organization of the first four Indian ages, we may notice that the same system of oppositions behind the *suyus* of the Tahuantinsuyu is also present in them, and if we go even further, it might also be seen in the system of age classes as expressed in the description that Guaman Poma makes of the different acllas or virgins of the suns, as they were called by the Spaniards.

Finally, it is Guaman Poma de Ayala, a late chronicler, who, although an Indian, provides us with the strongest evidence in favor of the existence of cosmological ages in the Andean world. He achieves it all via Andean temporal, spatial, and social categories that underlie the European organization of his document.

Acknowledgments

My special gratitude is extended to Tom Cummins for his generous patience in correcting the style of my English in this essay.

NOTES

1 Florescano suggests that the one associated with white is Quetzalcoatl and the one with blue, Huitzilopochtli. Jacques Soustelle (1968:129), on the other hand, states that the red Tezcatlipoca is also known as Xipe-Toltec, the god of the east; the second is the black god of the north, night, cold, and the nocturnal sky; the third, Quetzalcoatl, is the "feather serpent," the white god of the west; and the fourth, Huitzilopochtli, the warrior god painted blue, the triumphant sun of noon, the eponymous god of the Aztec capital.

2 It is equivalent to thirteen days (*trecena* cycle) × fifty-two-year cycle = 676 years.

3 Murúa took not only from authors who dealt with Mexico but those who wrote about Peru, such as Juan Polo de Ondegardo, Diego Fernández el Palentino, Cristóbal de Molina el Cusqueño (perhaps through Cabello de Valboa), Jerónimo de Oré, and many others, but this was also true of almost all the chroniclers.

4 According to Cox (2002), Guaman Poma might have consulted Andrés de Li (1999 [1546]) and Rodrigo Zamorano (1594), who also were authors of chronographies. Gonzáles Díaz (2012) adds a "Repertorio de los tiempos" written by an anonymous (1554) author.

5 For Guaman Poma, the term *Viracocha* is synonymous with "Spaniard," "Christian," and "outsider," as well as with the Andean divinity associated with creation. In his description of the first Indian age, he writes: "Vari Viracocha Runa, First generation of Indians, descendants of the Spaniards whom God brought to this kingdom of the Indies, the ones who came from Noah's ark after the Flood. After the people from Noah's ark multiplied by God's order, they spread throughout the world. This first generation lasted and multiplied a few years, 830 years in this New World called the Indies, where God sent them. These Indians were called Vari Viracocha Runa because they were descendants of Spaniards and for that reason the Spaniards were called Viracocha" (Guaman Poma 2009:49; my translation). The Inca, considered by Guaman Poma as idolaters, are separated from this line of descent. Hence, the first four generations are treated as if they were five and that of the Inca is doubled into two kinds of Inca: Tocai Capac and Pinau Capac as descendants of the previous ages, and Manco Capac as separated for having an unknown father and being the son of the witch Mama Huaco.

6 Guaman Poma does not fully identify with the Inca and echoes Viceroy Toledo's criticisms against them. Guaman Poma (although arguing that an ancestor married the Inca ruler Tupac Yupanqui) states that Manco Capac, the first Inca ruler, had an unknown father and therefore was not the descendant of the first four ages. Incapable of avoiding that the Inca occupy a fifth position, he acknowledges Tocay and Pinahua Capac, two *curacas* of pre-Inca ethnic groups of Cuzco, as legitimate Incas.

7 Zamorano (1594:87v) gives seven ages, and their total duration is 3,967 years.

8 To determine the duration of the Age of the Inca, or fifth age, he performs something similar when he states: "One hundred and fifty years from the discovery and one hundred and twelve years from the conquest and from this moment this count is made from 1613 since the birth of the Lord" (Guaman Poma 2001 [1615]:435 [437]; my translation).

9 To enhance the ordering position of these three kings, the sixth, seventh, and eighth ages are characterized by unordered events, such as the war between Huascar and Atahualpa within the sixth, the conquest within the seventh, and the Spanish civil wars within the eighth.

10 This is the date given for the foundation of Cuzco by Manco Capac.

11 Although not associated with millenarian dates, we have further evidence that the conquest was seen as a Pachacuti. It comes from Juan de Betanzos (1996:128), who mentions that the ninth Inca Pachacuti forecast that "after the days of his grandson Huayna Capac there would be *pachacuti*, which means 'change of the world.' Those lords asked him if that change of the world would be from floods, fire, or pestilence. He told them it would not be for any of those reasons but, rather, because white, bearded, and very tall men would come."

12 Sarmiento de Gamboa and Anello Oliva also use this term for a similar purpose.

13 To translate Guaman Poma's texts, I have used the edition by Roman Hamilton, whose translation

into English is very accurate. But since it is always very difficult to convey in another language what Guaman Poma says, I recommend the digitized version on the website organized by the Royal Library of Copenhagen.

14 Guaman Poma is the only author who writes the epithet of this character in this way, according to Imbelloni (1946:117–130), who dedicates ample text to the discussion of this term and regards it as the proper Quechua expression for referring to somebody who performs the action of transforming the world.

15 In addition, he also mentions the word *Pachacuti*, which Bertonio (1984 [1612]:241) translates as "Tiempo de Guerra. Y tambien agora lo toman para significar el juyzio final."

16 In a document from the community of Guancaraylla published by Abdón Yaranga (1978:167), the authenticity of which has been questioned because it does not include a copy of the original version, this turn of the world is mentioned as "Pachaticramuni" or "Pachacutimuni."

17 As Zuidema suggests when talking about the third representation in *The Ceque System of Cuzco* (1964) and, later, in the Spanish edition of his book (1995) as well as in *El calendario inca* (2010), the list of Inca kings and their *panacas* associated with the dual division of Hanan and Hurin could be seen as two timeless, parallel genealogical sequences.

18 The number of age grades frequently mentioned is either ten or twelve, but almost all coincide in specifying that the working obligations corresponded to individuals between the ages of twenty or twenty-five until the age of fifty. Significant to my argument is that this cluster of ages, considered as tributaries, is, in turn, divided into five formal stages with a span of five years each. Bandera suggests this when describing the collective marriages legitimized by an Inca representative. He says that when the villages became aware that a *tucuyricuc* of the Inca (literally "he who sees everything," a royal visitor who represented the Inca) was about to arrive, they all adopted the following organization: "[They] placed by their order and in the main square the Indians who did not have women from fifteen to twenty years old, from twenty-five to thirty, from thirty-five to forty, each age by itself; and likewise the single women by their ages. The men facing the women and from here first he gave women to the 'caciques and principales' who did not have or needed more, and afterward to the rest of the Indians by their ages with their equivalent" (Bandera 1965:179; my translation). I am very grateful to Tom Zuidema for having suggested to me this line of research, which expands the possible ground for the ages of the world to other Ameridian groups with age classes, such as the Bororo, Ge, and Tukanos.

REFERENCES CITED

Anonymous

1554 *Repertorio de los tiempos*. Printed by Francisco Fernández de Córdova, Valladolid. Electronic document, http://gredos.usal. es/jspui/handle/10366/82575, accessed March 7, 2011.

Anonymous

2007 *Popol Vuh*. Translation and commentary by Allen J. Christenson. Electronic version of original 2003 publication, www.mesoweb.com/publications/Christenson/PopolVuh.pdf, date accessed June 26, 2014.

Atienza, Lope de

1931 Compendio historial del estado de los indios del Perú. In *Religión del imperio de los incas*, vol. 1, edited by Jacinto Jijón y Caamaño. Escuela Tipográfica Salesiana, Quito.

Ávila, Francisco de

1873 A Narrative of the Errors, False Gods, and Other Superstitions and Diabolical Rites in which the Indians of the Provinces of Huarochirí, Mama, and Chaclla Lived in Ancient Times, and in which They Even Now Live, to the Great

Perdition of Their Souls. In *Narratives of the Rites and Laws of the Yncas*, edited and translated by Clements R. Markham, pp. 121–147. Ashgate, London.

Bandera, Damián de la
1965 Relación general de la disposición y calidad de la provincia de Guamanga, llamada San Joan de la Frontera, y de la vivienda y costumbres de los naturales della (1557). In *Relaciones geográficas de indias*, vol. 1, pp. 176–180. Atlas, Madrid.

Barnes, Monica
1994 Las edades del hombre y del mundo según Hierónimo Chaves, de Sevilla, y Guaman Poma de Ayala, del Perú. In *Humanismo siglo XX: Estudios dedicados al Dr. Juan Adolfo Vázquez*, edited by Juan Schobinger, pp. 291–297. Universidad Nacional de San Juan, San Juan.

Bertonio, Ludovico
1984 [1612] *Vocabulario de la lengua aymara*. Talleres Gráficos El Buitre, La Paz.

Betanzos, Juan de
1996 *Narrative of the Incas*. University of Texas Press, Austin.

Cabello de Valboa, Miguel
1951 [1586] *Miscelánea antártica*. Universidad Nacional Mayor de San Marcos, Lima.

Chaves, Hieronimo de
1580 *Chronografía o repertorio de los tiempos: El mas copioso que hasta ahora ha salido a luz*. Casa de Fernando Díaz, Seville.

Cieza de León, Pedro
1967 [1553] *El señorío de los incas*. Instituto de Estudios Peruanos, Lima.

Cox, Victoria
2002 *Guaman Poma de Ayala: Entre los conceptos andino y europeo de tiempo*. Centro Bartolomé de las Casas, Cuzco.

Cummins, Tom, and Juan M. Ossio A.
2013 "Muchas veces dudé Real Mag. açeptar esta dicha ympressa": La tarea de hacer "La Famossa Historia de los Reyes Incas" de Fray Martín de Murúa. In *Au miroir de l'anthropologie historique: Mélanges offerts a Nathan Wachtel*, edited by Juan Carlos Garavaglia, Jacques Poloni-Simard, and Gilles Riviére, pp. 151–170. Rennes University Press, Rennes.

Eliade, Mircea
1954 *Cosmos and History: The Myth of the Eternal Return*. Harper Torchbooks, New York.

Fleming, David
1994 Guaman Poma, Hieronymo de Chaves, and the Kings of Persia. *Latin American Indian Literatures Journal* 10(1):46–60.

Florescano, Enrique
1987 *Memoria mexicana: Ensayo sobre la reconstrucción del pasado época prehispánica-1821*. Joaquín Mortiz, Mexico City.

Garcilaso de la Vega, El Inca
1960 Los comentarios reales de los incas. In *Obras completas del Inca Garcilaso de la Vega*, vol. 2. Atlas, Madrid.

Gonzalez Díaz, Soledad
2012 Guaman Poma y el repertorio anónimo (1554): Una nueva fuente para las edades del mundo en la *Nueva corónica y buen gobierno*. *Chungara: Revista de antropología chilena* 44(3):377–388.

Gonzalez Holguín, Fray Diego
1952 *Vocabulario de la lengua general de todo el Perú llamada lengua quichua o del inca*. Universidad Nacional Mayor de San Marcos, Lima.

Guaman Poma de Ayala, Felipe
2001 [1615] *El primer nueva corónica y buen gobierno*. Edited by Rolena Adorno. Royal Library of Denmark, Copenhagen. Complete digital facsimile edition of the manuscript, with a corrected online version, available at www.kb.dk/elib/mss/poma/, accessed 2001.
2009 *The First New Chronicle and Good Government: On the History of the World and the Incas up to 1615*. Edited and translated by Ronald Hamilton. University of Texas Press, Austin.

Imbelloni, José
1946 *Pachakuti IX*. Editorial Nova, Buenos Aires.

Li, Andrés de
1999 [1546] *Repertorio de los tiempos*. Tamesis, London.

Millones S. G., Luis

2007 Taki Oncoy: De la enfermedad del canto a la epidemia. In *Fuentes para el estúdio de la colonia IV.* Dirección de Bibliotecas, Archivos y Museos, Centro de Investigaciones Barros Arana, Santiago.

Molina, Cristóbal de

1873 The Fables and Rites of the Inca. In *Narratives of the Rites and Laws of the Yncas*, edited by Clements R. Markham. Electronic document, http://www.sacred-texts.com/nam/inca/rly/index.htm, accessed 2003.

Montesinos, Fernando de

2010 Memorias antiguas historiales y políticas. In *The Quito Manuscript*, edited by Sabine Hyland, pp. 105–154. Yale Peabody Museum, New Haven, Conn.

Murúa, Fray Martín de

2004 [ca. 1590] *Códice Murúa* (Galvin manuscript). Edited by Juan M. Ossio A. Testimonio, Madrid.

2008 [1613] *Historia general del Perú.* Facsimile ed. Getty Research Institute, Los Angeles.

Oliva, Anello

1895 *Historia del reino y provincias del Perú.* Imprenta y Librería de San Pedro, Lima.

Ossio A., Juan M.

1970 The Idea of History in Felipe Guaman Poma de Ayala. B. Litt thesis, Oxford University. Electronic document, http://base.kb.dk/manus_pub/cv/manus/ManusIntro.xsql?nnoc=manus_pub&p_ManusId=65&p_Lang=main, accessed February 16, 2014.

1973 Guaman Poma: Nueva corónica o carta al rey; Un intento de aproximación a las categorías del pensamiento del mundo andino. In *Ideología mesiánica del mundo andino*, edited by Juan M. Ossio A., pp. 153–213. Ignacio Prado Pastor, Lima.

2004 *Códice Murúa.* Testimonio, Madrid.

2008a *En busca del orden perdido.* Fondo Editorial de la Pontificia Universidad Católica del Perú, Lima.

2008b Murúa's Two Manuscripts: A Comparison. In *The Getty Murúa: Essays on the Making of Martin de Murúa's "Historia General del Peru," J. Paul Getty Museum Ms. Ludwig XIII 16*, edited by Tom Cummins and Barbara Anderson, pp. 77–94. Getty Research Institute, Los Angeles.

Pérez Bocanegra, Juan

1631 *Ritual, formulario, e institución de curas, para administrar a los naturales de este reyno los Sanctos Sacramentos.* Geronymo de Contreras, Lima.

Quilter, Jeffrey

1989 The Moche Revolt of the Objects. *Latin American Antiquity* 1(1):42–65.

Read, Kay Almere

1998 *Time and Sacrifice in the Aztec Cosmos.* Indiana University Press, Bloomington.

Rowe, John

1987 La mentira literaria en la obra de Martín de Murúa. In *Homenaje a Aurelio Miró Quesada Sosa*, edited by Héctor López Martínez, pp. 753–761. Villanueva, Lima.

Salinas y Cordova, Fray Buenaventura

1957 *Memorial de las historias del Nuevo Mundo.* Universidad Nacional Mayor de San Marcos, Lima.

Sarmiento de Gamboa, Pedro

1960 [1572] Historia indica. In *Obras completas del Inca Garcilaso de la Vega*, vol. 4, pp. 189–279. Atlas, Madrid.

Sullivan, Lawrence

1988 *Incanchu's Drum.* Macmillan, New York.

Sumner Maine, Henry

1972 *Ancient Law.* Everyman's Library, New York.

Valderrama, Ricardo, and Carmen Escalante

1980 Apu Qoropuna (Visión del mundo de los muertos en la comunidad de Awkimarca). *Debates en antropología* 5:233–264.

Vásquez de Espinoza, Antonio

1969 [1629] *Compendio y descripcion de las Indias Occidentales.* Atlas, Madrid.

Yaranga, Abdón

1978 Taki Onqo ou la vision des vaincus au XVIe siècle. In *Les mentalités dans la Péninsule Ibérique et en Amérique Latine aux XVIe et XVIIe siècles: Histoire et problematique*, pp. 119–179. Serie Etudes Hispaniques 1. Publications de L'Université de Tours, Tours.

Zamorano, Rodrigo

1594 *Chronología y repertorio de la razón de los tiempos*. Seville.

Zuidema, R. Tom

1964 *The Ceque System of Cuzco: The Social Organization of the Capital of the Incas*. Brill, Leiden.

1989 El Parentesco Inca. In *Reyes y guerreros*, edited by R. Tom Zuidema and Manuel Burga, pp. 54–116. FOMCIENCIAS, Lima.

1995 *El sistema de los ceques del Cuzco: La organización social de la capital de los incas*. Fondo Editorial, Pontificia Universidad Católica del Perú, Lima.

2010 *El calendario inca*. Fondo Editorial, Pontificia Universidad Católica del Perú, Lima.

12

Refounding the House

Time, Politics, and Metallogenesis in a Colonial Aymara Coat of Arms

TRISTAN PLATT

We know how a taste event can surprise us and then persist in memory while also diminishing in intensity until renewed or replaced by another bite. In Potosí Quechua, this initial surprise is expressed by the so-called suffix of sudden discovery, *-sqa*, as in "¡Chay uchu haya ka-sqa!" or "That red pepper is/was hot!"[1] The suffix situates the time of speaking as coming just after the experience of an "event" whose impact was like a train that rushed through a station before tapering off into the distance. Our being at any time is weighted by past experiences, the memories of which—conscious and unconscious—interact with new "events" to create new present states, layered pasts, and future possibilities.[2] Living memories are selected, "tweaked," and transformed according to the circumstances surrounding recall.

In this essay, I examine the early colonial re-membering[3] of his lineage's historical experience by a *mallku*, or condor chief, of the Aymara-speaking federation and Inca Province of Qaraqara, don Fernando Ayra de Ariutu, who lived in what today is the Department of Potosí, Bolivia. This dynastic history was compiled to support a petition for an *encomienda* and a coat of arms sent by the mallku to the Spanish king Philip IV in 1635 (Platt et al. 2011 [2006]:docs. 16–17).[4] It is contained in don Fernando's *Probanza de méritos y servicios* (Proof of merits and services) and is of particular interest for comparing with the Cuzco-centered accounts dominant in Andean ethnohistory.

How was the passage of time conceptualized retrospectively by the Qaraqara? I shall not argue here for "exact" time measurement or for the static structure of a fixed ethnoperiodization. I am interested in the re-imagination of earlier historical periods from the perspective of later conjunctures, which give rise to successive ways of reframing remembered traces from the past. Analysis of a coat of arms can offer us, I believe, a fresh perspective on the mining and religious history of South Andean societies under Inca and Spanish rule and on the workings of collective memory under the pressures of colonialism.

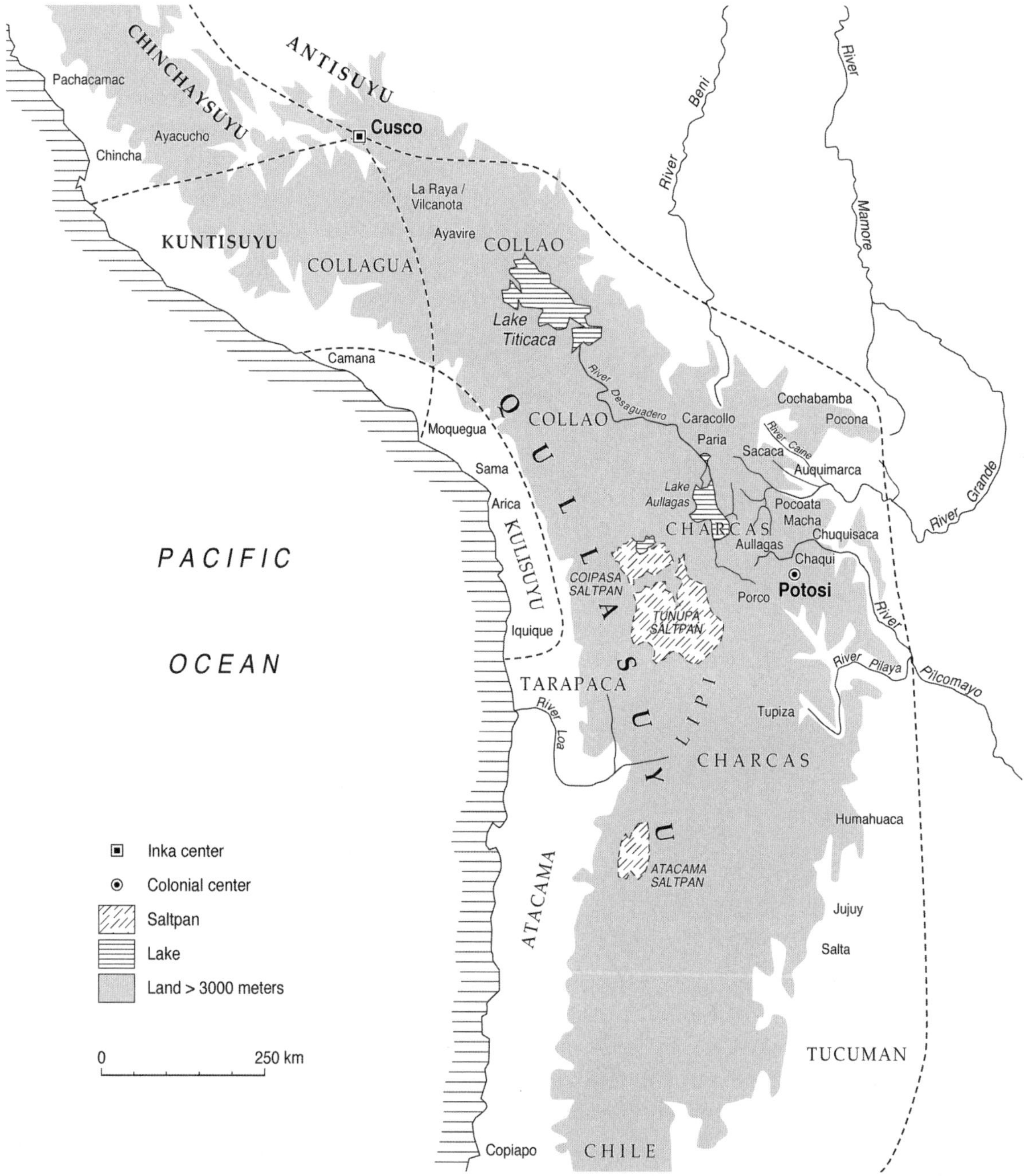

figure 12.1

Collao and Charcas within the Inca's Qullasuyu. (Drawing by Graeme Sandeman.)

At the time of the Spanish occupation of the Inca's Qullasuyu in 1538 (Figure 12.1),[5] Qaraqara elites were linked politically and ceremonially with those of the neighboring "Province of Charka" (see Platt et al. 2011 [2006]:docs. 20–22).[6] "Qaraqara," according to don Fernando's *Probanza*, was, in fact, an honorific title meaning "dawn" (*alba*) given to the federation by Inca Wayna Qhapaq. Previously, they were probably known as the White Charka, linked to the neighboring Province of the (Red) Charka, but with Wayna Qhapaq's gift, both became what we have called the Qaraqara-Charka confederation.[7] They were two of several "nations" making up the larger Inca "Province of Charcas" when the Spanish arrived, and on the frontier between the larger "provinces" of Charcas and Collao, in the Sura federation's territory, lay the Inca administrative town of Paria, whose ruins and storehouses have recently been rediscovered near modern Oruro (Condarco et al. 2002; Pärssinen 2010; Del Río 2005; see Figures 12.1 and 12.4).

To interpret don Fernando's arms, we need to know the conditions in which they were produced. Blazons were issued by the Royal College of Heralds and granted by the king in Madrid, but they were formulated on the basis of proposals made by petitioners drawing on their own iconographic repertoires (Estenssoro Fuchs 2010). Colonial Andean and Mexican blazons were therefore polysemic, combining European and Amerindian meanings copresent and available to different readerships. Rather than constituting an ex nihilo "invention of tradition" following colonial rupture (as is sometimes argued), they could draw on pre-Hispanic traditions, even if in European guise. They provide rich material for interpreting the refoundation of pre-Hispanic collective identities in early colonial circumstances.

Probanzas from several Qaraqara, Charka, and other South Andean dynasties have been found, and others may yet turn up.[8] I focus here on the questionnaire, testimonies, and accompanying documentation sent to Philip IV of Spain in 1638 by the *corregidor* of Potosí, don Juan de Lizarazu, in answer to the king's demand for evidence and advice with which to respond to don Fernando Ayra de Ariutu's request. I also compare a sketch or template (*tarjeta*) of the coat of arms, sent from Potosí to Madrid to guide the College of Heralds, with the official blazon returned from Madrid to Potosí.

I approach the documents, therefore, without treating their use of alphabetic and notarial Spanish, or even of Spanish heraldic formats, as sufficient proof of social identities cut adrift from their pre-Hispanic past. Andean self-identifications and dynastic memories could persist even while transformed under the pressures of colonization, land expropriation, and demographic decline. The deepest memories appear to recall "events" from around two hundred years before—quite a modest lapse of time, in comparative terms. Don Fernando presented himself as a member of a Christian Aymara nobility, with privileges granted by the Inca that he hoped would be matched by the king. His version of the history of his house, and the ancestral symbols he proposed for his blazon, are not simply "fictitious." By probing their meanings, we can understand better the redeployment of Andean temporal and symbolic ideas under early colonial rule.

The Dynastic History of the Qaraqara

In his questionnaire, don Fernando Ayra de Ariutu presented his antecedents in nineteen questions, which were then supported and enriched by the testimony of thirteen witnesses between the ages of sixty-five and eighty-five. Questions two to seven concern the lineage of his father, don Fernando Chinchi, while questions eight to thirteen concern the lineage of his mother, doña Úrsula Anco Tutumpi Ayra Kanchi. These two lineages apparently descended from two male founding ancestors, Qhapaq Lukalarama of Pocoata and Tata Ayra Kanchi Anco Tutumpi of Macha (Figure 12.2).[9] Both were from their respective group's lower moiety (Urinsaya), although their authority could extend over the upper moiety (Anansaya) as well. Macha and Pocoata were (and are) neighboring ethnic groups (ayllus), in which these ancestors had lived as supreme lords of the White Charka. Both are said to have been the contemporaries of Inca Yupanki [Pachacuti] of Cuzco, probably in the mid-fifteenth century. The spectacular marriage between don Fernando Chinchi and doña Úrsula around 1595 created an alliance between the two ancient "houses," and don Fernando Ayra de Ariutu as their offspring.

In the *Probanza*, don Fernando and his witnesses chronicle three broad periods in the history of Qaraqara (or White Charka), each initiated by certain key "events" and separated from the next

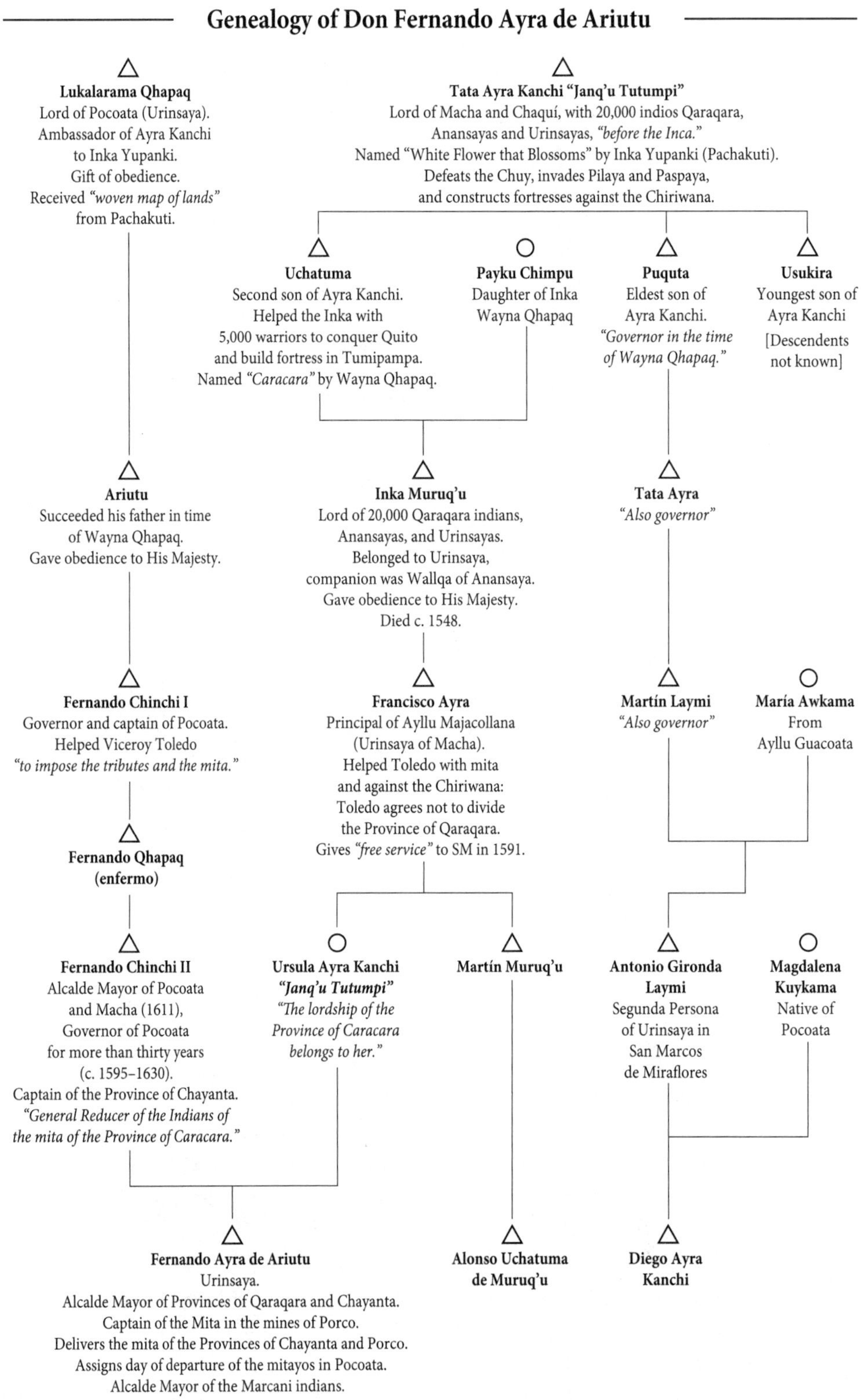

figure 12.2
Genealogy of don Fernando Ayra de Ariutu. (Drawing by Graeme Sandeman, after Platt, Bouysse-Cassagne, and Harris 2011 [2006]:fig.5.1.)

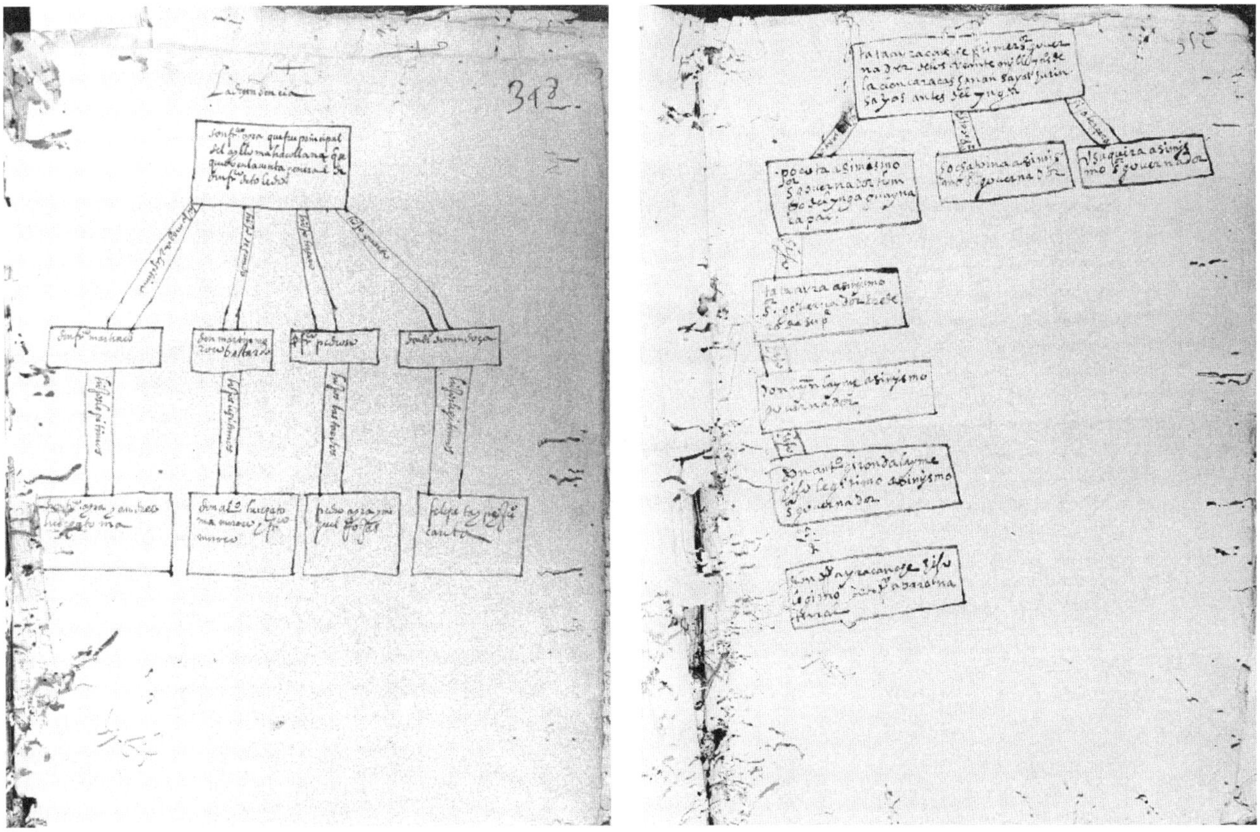

a b

figure 12.3
Linked genealogies of (a) Alonso Uchatuma de Muruq'u; and (b) Diego Ayra Kanchi (Platt et al. 2011 [2006]:799).

by a period of turmoil. These three "events" are as follows:

1. An exchange of gifts between Inca Yupanki [Pachacuti] and Ayra de Ariutu's maternal ancestor Tata Ayra Kanchi, who was "first lord governor of the 20,000 indians of the Caracas [na]tion [*sic* = *nación Caracaras*], Hanansayas Hurinsayas before the Inca" (Platt et al. 2011 [2006]:767–781; Figure 12.3b).[10] Ayra Kanchi was the lord of two Qaraqara *provincias* or *parcialidades*, Macha and Chaquí.[11] A cacique of Pocoata, don Juan Vissaya, said: "He was so courageous and feared a captain that no one could excel him: he subjected as far as the Chuy, and scoured the lands of Pilaya and Paspaya where he built some fortresses,[12] and this witness has seen some *memorias* [possibly a *khipu*; Aym. = *chinu*], and he was someone who travelled on a golden litter from town to town" (Platt et al. 2011 [2006]:737).

Ayra Kanchi sent his ambassador, Lukalarama of Pocoata, to Cuzco, where he "gave" Pachacuti the "news" (*noticia*) of the four quarters of the Tawantinsuyu[13] receiving in exchange a "woven map" of Qaraqara lands (*carpatira*; a copy probably stayed in Cuzco).[14] Inca Yupanki, in turn, sent an ambassador to Ayra Kanchi to bestow upon him the name of "Anco Tutumpi," translated in the text as "flor blanca que brota," "White Flower That Blossoms" (Platt et al. 2011 [2006]:728, 730–731, 734–735, 737).[15]

Lukalarama's embassy to Pachacuti in Cuzco was recalled by another witness summoned to give evidence in 1637, don Fernando Callapay, the governor and *cacique principal* of the Province of Asanaki on the Altiplanic shores of Lake Aullagas (today Poopó) and "Mita Captain of this Town [Potosí]":

> The said [Localarama] from whom [don Fernando Ayra de Ariutu] descends was the first ambassador who gave the obedience to the Inca for Anco Tutumpi Ayra Canchi, who was lord of more than 20,000 indians, and they say that he gave the Inca the news of the four provinces of Collasuio, Antesuio, Condesuio, Chinchaisuio, and of the quality of the lands, and that the Inca gave him as a reward a map woven [fol. 16v] in *cumbe* cloth, which the Indians call Carpatira, and he gave him the surname of Capax. (Platt et al. 2011 [2006]:741)

Obviously, don Fernando Callapay was not himself part of the embassy, which is placed some two hundred years previously; he is giving his version of what had become a shared "collective memory" among at least some Charcas lords.

The "map woven in *cumbe* cloth" exemplifies the diversity of political arrangements devised by the Inca for the different provinces of the Tawantinsuyu (Burger et al. 2008; Malpass and Alconini 2010).[16] It may have born a formal representation of the lands, groups, and, possibly, calendrical responsibilities of the province. It is presented as an Inca antecedent to the desired grant of a Spanish coat of arms and is cited in support of don Fernando's claim to the lordship of the "Province of Qaraqara."[17]

2. Several decades pass before new "gifts" are made by Inca Wayna Qhapaq to Uchatuma, the son of Tata Ayra Kanchi, who personally visits the Inca in Cuzco "to give him his obedience, as to his King and natural lord": for the mallku, a "shirt of gold disks" ("una camiseta de roeles de oro"; the term *roel* is heraldic), the title of Qhapaq, and the hand of the Inca's daughter Payku Chimpu; for his Indians, the name Caracara, meaning "dawn" (alba). These gifts reward Uchatuma's military support with five thousand crack troops during the Inca's campaigns in Quito and Pasto, as well as against the "Condesuios and Andesuios"; he also helped construct a fortress at Tumipampa (modern Ecuador; Platt et al. 2011 [2006]:730–731, 737).[18]

3. Three more decades pass, and then, in 1538, another "gift" takes place, this time to the Hapsburg emperor Charles V through the mediation of the invaders, Hernando and Gonzalo Pizarro. It is made by several Aymara-speaking federations, led by the lords of the Qaraqara and the Charka: Muruq'u (son of Uchatuma) and his Charka companion Kuysara.[19] In the *Probanza*, Uchatuma's son Muruq'u succeeds his father in the time of Inca Waskar, rival and half-brother of Atawallpa, and is said to have "given the obedience" to the emperor before being baptized don Juan and supporting the royal forces during the rebellion of the encomenderos against Charles V (1545–1548; question nine). Another source tells of the mallku's "gift" to the emperor at the end of 1538 of the silver mine that had belonged to Wayna Qhapaq in the silver mountain of Porco (Anónimo 1934 [1539]; Platt et al. 2011 [2006]:125).[20] This gift constitutes a transfer of obedience (*traslatio imperii*) that is confirmed by the dis-covery of Potosí to the Spanish in 1545, just as the encomenderos under Gonzalo Pizarro were preparing their rebellion against the crown (Platt and Quisbert 2008).[21]

But the profile of the new period is confused. Gonzalo Pizarro led the encomenderos to defeat at Xaquixaguana in 1548, when many Andean federations fought for the emperor. To Andean dismay, however, new encomiendas were granted to his majesty's supporters by the "pacifier" La Gasca, which began the division of the Qaraqara.[22] In the 1550s, some managed to get themselves placed under the direct administration of crown officials ("en cabeza de Su Magestad"), even helping the colonial authorities put down further rebellions, but Lascasian hopes and debate gave way in the 1560s to royal doubts and debts, culminating in the repressive colonial government of Viceroy Francisco de Toledo (1569–1581).

The situation now becomes clearer: the Spanish are there to stay, and around 1570, following Lope García de Castro's Ordenanzas of 1565, the Qaraqara are administratively divided between two new colonial provinces, Chayanta and Porco, with the pre-Hispanic *provincia* of Macha assigned to Chayanta and most of the old provincia of Chaqui (including the mines of Porco and Potosí) falling within Porco Province (Figure 12.4).

The division of Qaraqara motivates the following question:

> Is it true that the said don Juan Inca Moroco had as his legitimate son don Francisco Ayra, grandfather of the said don Fernando Ayra de Ariutu, which said don Francisco succeeded his father don Juan Inca Moroco in the said government and lordship of the said province of Caracara, although in the Visita General made by the said Viceroy [Toledo] the towns were divided, not because of any failing of don Francisco Ayra, but so that he should be less overburdened and have less work in the delivery and government of his said indians, and having petitioned for his province not to be divided, the said Viceroy, seeing his quality, capacity and understanding, repeated the *reducción*, and [ordered] that he govern it as his fathers and ancestors had done? (Platt et al. 2011 [2006]:731)

The question is supported by a certificate given to don Francisco Ayra (for whom see Figure 12.2) by Viceroy Toledo in 1575, which returned to the mallku his authority over the Indians of Porco Province; a copy of the certificate was carefully included by don Fernando Ayra de Ariutu in his *Probanza*. There can be little doubt that this concession followed negotiations between Toledo and don Francisco Ayra over the delivery of Qaraqara *mitayos* to Potosí[23] and also recognized the mallku's support for Toledo's unsuccessful campaign against the Chiriwana in 1573.[24]

But the new colonial provinces persisted, each governed by a different Spanish corregidor and *justicia mayor*. With the departure of Toledo, therefore, the Qaraqara lords tried to strengthen their case by *refounding* the house (casa) of Tata Ayra Kanchi, reaffirming the unity of the Qaraqara through an interdynastic marriage of around 1595 (see Figure 12.2). The bride, doña Úrsula Ayra Kanchi, took the name and title that Inca Pachacuti had granted to her ancestor—Anco Tutumpi, "White Flower That Blossoms"—and she claimed the same extended jurisdiction ("le pertenece el señorío de la Provincia de Caracara"; Platt et al. 2011 [2006]:732, 738).[25] Her husband, don Fernando Chinchi II, governed until 1630, and in 1629 (according to a certificate signed by a priest of Pocoata, Padre Manuel Salvanes) their son don Fernando married a lady from Porco Province, doña Inés de Amcoma, daughter of the cacique of Tacobamba, thereby creating a new bond between the colonial Provinces of Chayanta and Porco (Platt et al. 2011 [2006]:754).

In 1630, don Fernando Chinchi II died, and his lordship passed to don Fernando Ayra de Ariutu, who sent his request to King Philip IV of Spain in 1635. Clearly, he wished to officialize with a coat of arms his parents' efforts to "refound" the ancestral jurisdiction of their two lineages. He called witnesses to validate the dynastic history transcribed and presented in his questionnaire. It is this seventeenth-century "incapsulation" of a Qaraqara historical narrative that has left us the main genealogy (complemented with additional sources in Figure 12.2), together with the chronicle of key "events" and also the sketch or template (tarjeta) of the coat of arms (Figure 12.5).[26] I shall interpret the arms, taking into account both the long historical trajectory that lies behind them and the political conjuncture of their grant.

By the mid-seventeenth century, then, we find a retrospective sequence of three periods in the relations of this Aymara-speaking nobility with the Inca-Hapsburg state between about 1450 and 1640. The second and third periods had been separated from their predecessors by interstitial periods of invasion, warfare, and social upheaval: the second by conflict with Pachacuti's son, Tupaq Yupanki (events that are repressed in the dynastic narrative for reasons we shall see but that can be partially recovered from other sources) and the third by the Inca civil war between Atawallpa and

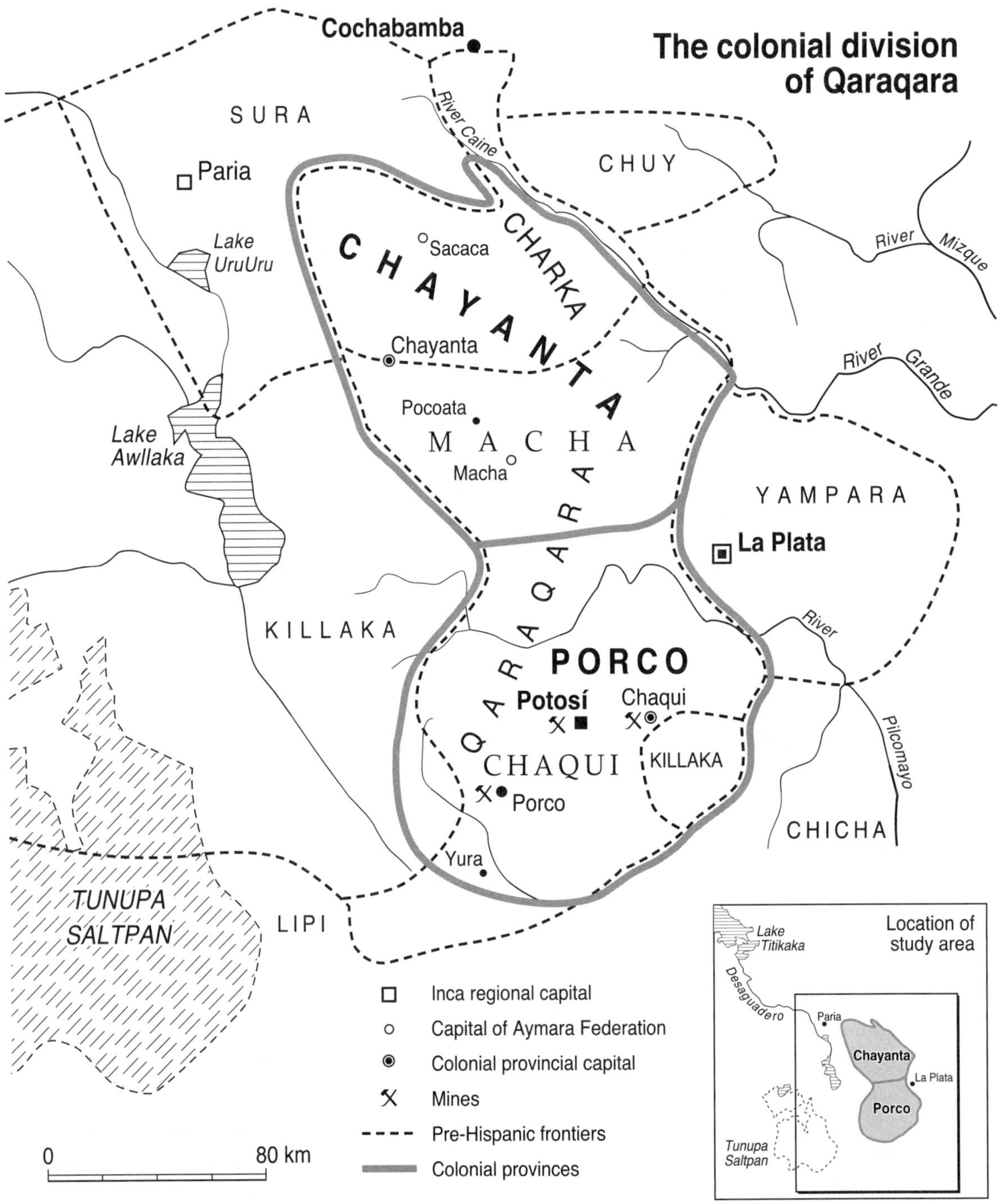

figure 12.4

The colonial division of Qaraqara. (Drawing by Graeme Sandeman.)

Waskar, followed by the chaotic consequences of the Spanish invasion after the early gift of Wayna Qhapaq's mine in Porco (consequences that are briefly mentioned in the *Probanza* but can also be recovered in more detail). We may note the efforts of these Aymara-speaking lords to negotiate with succeeding sovereigns, both Inca and Spanish, transforming their relations with the state and between themselves under each regime and, in the process, reframing their memories of earlier periods. This shows us an early colonial periodization in the making and the active intervention of colonial lords in the process of re-membering and reframing "events" from the past.

Moreover, these periods seem to have been symbolized for the Qaraqara by reference to aspects of Venus, the evening and morning star. The planet appears in 1640 as the *lucero* in the "first quarter" of the blazon granted by Philip IV to Tata Ayra Kanchi's great-great-great-grandson, don Fernando Ayra de Ariutu, and the emblems in the blazon are said to come from his ancestors. The pre-Hispanic significance of the morning star is confirmed by the new name of Qaraqara, "Dawn," given to the White Charka by Wayna Qhapaq. But the aspect of Venus could change in the passage from one period to the next.

Venus, celestial marker of the twilight and the dawn (Urton 1981),[27] was important for many Amerindian groups. The dusky threshold when the evening and morning star appears is a transition directly perceptible to human senses and can be used metaphorically to symbolize larger historical changes not so evident to the senses. Venus appears as the evening star for 263 earth days before disappearing in front of the sun's glare for 8 days and then reappearing as the morning star for another 263 days, on average, before again disappearing behind the sun for some 50 days, making a total of 584 earth days. This cycle, which was well understood by the Maya, may not have been observed in the same detail by the Qaraqara.[28] I argue, however, that the passage from evening to morning star, with a transitional period of obscurity, could be used to symbolize successive periods in Qaraqara relations with the Inca solar state. We can then ask how the lucero of the blazon might have been understood, many years later, in 1640.

The pre-Hispanic provincia of Chaquí (approximately equivalent to the post-1570 Province of Porco) was found by Europeans to be literally *sown* with silver; we even hear of silver "potatoes" clinging to the roots of unearthed plants (Benino 1966 [1573]).[29] In the midst of this silver-impregnated land lay the Porco and Potosí silver mountains, whose metals were exploited by Incas Pachacuti, Tupaq Yupanki, and Wayna Qhapaq as well as by the lords and commons of the Aymara federations.[30]

Porco was an ancestral site of sacred power, a huaca, since before the Spanish invasion. In a late sixteenth-century extirpation of idolatry document, the huaca of Porco had been removed for safekeeping to Caltama near Toropalca on the frontiers with the Chicha federation: a triple lightning deity, patron of mining, warfare, rainwater, and health (see Platt et al. 2011 [2006]:doc. 1).[31] Access to Porco had been controlled at the mine's mouth by a "porter" (*punkukamayu*) or shaman (*hechicero*), and the existence of an Inca cult center on top of the mountain of Porco has been confirmed archaeologically. Porco may have provided a seasonal focus for devotees of the deity and his precious metals; in the late sixteenth century, we hear of pilgrims from all the old "seven nations" of Charcas still congregating secretly in Caltama to pay their devotions to "Tata Porco" in exile.

Potosí, by contrast, had a cult center devoted to the sun on its peak, probably connected with the solar sanctuary at Copacabana; it, too, was known and venerated before the Spanish arrived, although it appears that the Rich Seam (Veta Rica) had not yet been exploited.[32] A third silver deposit was the mountain of Chaquí itself, whose fame circulated in the sixteenth and seventeenth centuries (Alonso Barba 1967 [1640]; Cruz 2009; Platt et al. 2011 [2006]; Platt and Quisbert 2008; Del Río 2005).

With the Indians' dis-covery of Potosí to the Spaniards in 1545, a new industrial hub arose in Qaraqara lands at the foot of the mountain. The largest city in the Spanish empire, Potosí generated flows of goods, money, and labor throughout the Andes and beyond, initially on the basis of pre-Hispanic technologies and labor relations (Assadourian 1982, 1989). After Toledo's introduction of amalgamation with quicksilver in 1573, South Andean lords competed for power and prestige as mita captains and lord mayors (*alcaldes mayores*), while providing subsidies of labor and subsistence to Potosi's early capitalist mining economy.[33] The transformation of Andean space, demography and politics through invasion, missionization, and silver production, together with the demand of the mining market, provides the backdrop to don Fernando's request to be recognized and honored as the descendent of Tata Ayra Kanchi, lord of "Macha and Chaquí . . . before the Incas" (Platt et al. 2011 [2006]:798–799; see Figure 12.3b).

By 1630, when he succeeded his father, don Fernando Chinchi II, don Fernando Ayra de Ariutu often resided in Potosí, where, along with other Andean mita captains and lord mayors, he governed the mitayos and Indians of La Plata, Potosí, and the provinces. But already by the end of the sixteenth century, his parents had probably established their houses[34] in the new *reducción* towns of San Pedro de Macha and San Juan Bautista de Pocoata (Chayanta Province).

The *cédula real* granting the coat of arms to don Fernando, signed by Philip IV in 1640, was sent from Madrid to Potosí with the blazon describing the contents of each quarter of the shield (Platt et al. 2011 [2006]:761). Before analyzing the blazon, we must consider the status of the information in the questionnaire and testimonies. The deeds of don Fernando's ancestors are described in the *Probanza* generation by generation, as a form of res gestae; they are summarized (with some mistakes) by the king and Lizarazu in their cedulas and *pareceres*. But how were these narrative fragments constructed?

In the *Probanza*, each fragment is attached to a genealogical ascendent's name. Do these fragments reflect the work of an individual memorizer, with or without a khipu (Aym. = chinu)? Or are they the result of collective negotiations, such as those Maurice Halbwachs (1992) took to be at the roots of a group's collective memory? We know that the lords and principals took part in "meetings" (juntas) where different versions of the past were discussed and confronted, and agreement was reached on which was "true." An example is the debate after the Spanish occupation about the importance of Kuysara, lord of the (Red) Charka and commander of all Charcas troops in Cochabamba in 1538. In 1584, a Spanish witness for the *Probanza* of Kuysara's grandson, don Fernando Ayawiri, said that "the past things of old were discussed in a certain meeting [junta] that took place, of caciques from this province, and it was debated and this witness heard it said, that the said Cumsara [Kuysara] had spoken of the fault he had had in the said resistance" (Platt et al. 2011 [2006]:921).[35]

Here the memory work was carried out through collective negotiations in which the views of several lords and authorities were discussed at a meeting, corrected, agreed upon, and assigned as mental attachments to corresponding ancestors. The oral tradition was broken down into "bits" of information, codified "memory traces" associated with specific genealogical figures, whose names served to elicit recall.

This leads us to qualify Estenssoro's (2010:173) Orwellian assertion that the Spaniards learned that "it was more efficient to change the past by [changing] memory [*por la memoria*] than to change the future by deeds [*por los hechos*]." Andean memory was not putty in the hands of the Spanish to remodel as they chose; its mechanisms were managed through conclaves of Aymara elders and memory workers who discussed and agreed on their preferred versions. We have access to some of their conclusions, caught in the notarial format of the probanzas.

Now, several Charcas lords found their privileges curtailed when Viceroy Toledo included them as tributaries in the new *tasa* (tribute lists), with loss of salary, lands, and retainers, and in some cases they were replaced by new authorities chosen by Toledo's inspectors. Together with the Potosí mita, the new tribute was applied to Indians who were evading their resettlement in reducción towns. Subversive meetings (juntas) were held in the early 1570s by discontented Aymara lords in the old Inca Province of Charcas, with encouragement from the priest of the (Red) Charka capital, Sacaca, Padre Joan de Padilla; though when discovered, the lords wrote a fawning letter to the viceroy apologizing unreservedly and disassociating themselves from Padilla (Platt et al. 2011 [2006]:855–857).[36]

We have seen that Muruq'u's son, don Francisco Ayra, raised with Toledo his loss of jurisdiction, and Toledo heard his complaint, granting him the right—as "Captain of the Qaraqara Mita"—to be attended by two halberdiers[37] and to exercise authority over tributaries and mitayos living in the southern colonial Province of Porco. This grant meant the recovery of his jurisdiction, but after Toledo's departure (1581) the Charcas lords worked with Dr. Manuel Barros de San Millán to write their *Memorial* of 1582, in which they drew parallels between Inca practice and the privileges they claimed under Philip II, while denouncing Toledo's arbitrariness. Still, their policy of loyalty to the king was maintained. In 1591, Charka and Qaraqara lords, along with others from Charcas and the Collao, contributed to the "free service" (*servicio gracioso*) in money requested by the king from his American vassals to help him fight the English and Turkish "heretics" (Platt et al. 2011 [2006]:974–995). To understand the coat of arms we must recognize the loyalty to the state (whether pre- or post-Spanish invasion) professed by these Qaraqara and Charka lords.[38]

But the tension between Inca and colonial ideas of "province" persisted throughout the seventeenth century. The Aymara nobility of Macha and Pocoata sometimes received authority as alcaldes mayores (Indian lord mayors) over the whole Province of Qaraqara, at others times being confined to those Qaraqara towns and groups situated within the Province of Chayanta. As late as 1673, don Fernando was having new copies made of his *Probanza*, the blazon grant, and the accompanying documents (Platt et al. 2011 [2006]:763). Probably, the eclipse of the pre-Hispanic Province of Qaraqara was not complete until the early eighteenth century.

Don Fernando's request was, therefore, part of a view of the colonial order that made its legitimacy dependent on pre-Hispanic precedent. It was also a way of enunciating the fresh colonial identity of the beneficiary (Estenssoro Fuchs 2010). Don Fernando adopted a style of self-presentation that combined key names from both his ascendant lineages: Ayra, the founder of his mother's patrilineage, who had exchanged gifts with Inca Pachacuti, and Ariutu from his father's line, the son of Lukalarama who gave obedience to the Spanish. This combination was appropriate, for the names showed his descent from two key ancestors who had collaborated, at different stages, in the construction of the Andean state.[39]

The Tarjeta and the Blazon

Don Fernando's initial letter to the king is missing from the file. The first surviving document in the series leading to the grant of 1640 is a cédula real sent by the king to the corregidor of Potosí, don Juan de Lizarazu, dated April 4, 1636. The cedula states that don Fernando had requested both the encomienda and the coat of arms. It repeats some of the contents of don Fernando's missing letter, summarizing his merits and services and those of his ancestors, and asks Lizarazu to collect information to assess don Fernando's claims.

Don Fernando, therefore, gathered together further material—*provisiones*, *títulos*, *testimonios*, and *recaudos*—that was received by Lizarazu on February 29, 1638. It is significant that included with the information was our only surviving copy of the *Cédula de encomienda* by which the whole Province of Qaraqara was granted in 1540 by Francisco Pizarro to his brother Gonzalo. Clearly, if don Fernando could secure this same encomienda for himself, his political objective would be assured (Platt et al. 2011 [2006]:289–301, doc. 3). Copies of the documentation were taken before the originals were returned to don Fernando. These included Toledo's "Title of Cacique to don Francisco Ayra" of April 25, 1575; an earlier copy had already been taken of this key document on June 26, 1635, and may have been included with don Fernando's missing letter to the king.

Lizarazu summarized the testimony and gave his recommendation (*parecer*) on March 3, 1638:

figure 12.5
The tarjeta, from the Archivo General de Indias, MP-Escudos 75. Photograph courtesy of the Ministerio de Educación, Cultura y Deporte, Archivo General de Indias, Spain.

> Your Majesty can grant him the *coat-of-arms, whose tarjeta I send with the papers*, which [arms] are *those that don Fernando himself chooses as belonging to his ancestors*, allowing him to put them on the door of his house, in the church where he has his tomb, and to enjoy this honor, he, his sons and descendents, like a *noble man descended from the Incas* [through Payku Chimpu], and you can also *grant him the post of Alcalde Mayor de Indios in the Corregimiento of Chayanta*, with the privileges that brings, for all the days of his life. (Platt et al. 2011 [2006]:725–726; emphasis added)

Note that Lizarazu recommended that don Fernando be made lord mayor (alcalde mayor) of the Province of Chayanta but pointedly omitted all reference to the encomienda or to the Province of Qaraqara. Clearly, the corregidor considered the two colonial Provinces of Chayanta and Porco there to stay. On March 1, 1639, he sent the file to Madrid for the king to make his decision.

The tarjeta mentioned by Lizarazu (Figure 12.5) was prepared in Charcas to help the Spanish College of Heralds design the blazon and was sent to Madrid with the *Probanza*. Lizarazu says it was based on ancestral emblems chosen by don Fernando himself, and the mallku may have resorted to the services of his own draftsman, for we know from a denunciation by some of his Indians that he kept "an examined painter" (*un pintor examinado*) in his household, who "adorned his stews [*potajes*] with gold" (Platt et al. 2011 [2006]:757, doc. 16, appendix, paragraph 7).[40] Pocoata continues to be known for its placer gold deposits, and, according to the blazon, the shield also contained gold.

The blazon sent from Madrid in 1640 can be imagined as the top layer of a damaged three-leveled palimpsest. It includes two emblems absent from the top-left quarter of the painted tarjeta sent in 1639, which forms the second layer. Indeed, were it not for the blazon we would not realize what is missing, for the quarter is empty except for a six-rayed silver star. The third layer is the missing letter sent by don Fernando to King Philip IV in 1635, which gave rise to the cédula real of 1636. Although this layer has disappeared, we can infer some of its content from the blazon.

The first quarter of the coat-of-arms is of particular interest, for it appears to evoke Andean ideas about the cosmic processes of metallogenesis. Recent archaeological research has discovered evidence of advanced metallurgical techniques in Qaraqara territory (Pulacayo, Chaqui Province) that date back to the Middle Horizon (Lechtman et al. 2011). The Incas seem to have learned many mining and metallurgical skills, with their associated religious knowledge, from the miners of the south. They will have valued their relationship with Tata Ayra Kanchi and his descendents because Porco, Potosí, and other mines lay in their territory, and they possessed the shamanic and metallurgical capacity to communicate with their silver huacas. I will show that the emblems given in the first quarter of the blazon represent Qaraqara ideas about the organic generation of silver by lightning and falling stars, a process involving the planet Venus.

The First Quarter

"In the first [quarter] in the upper half [the top left of the shield for the viewer, heraldic right], in a blue field, a green plant with three white lilies [*tres azucenas blancas*] in a green meadow, and in the upper part of the quarter a splendor [*resplandor*], and in the midst of it a star [lucero] of gold with eight rays."[41] This quarter is the most complex, with three elements placed in vertical relation to each other. But the first two are missing in the tarjeta, while the third, the star or lucero, is shown at the top of the quarter as a silver star with six rays instead of in the middle as a gold star with eight rays. I will take each element in turn.

The Three White Lilies or *Amancaya*

One reason for the omission from the tarjeta of the three white lilies may be that their combination with the Star of David against a blue ground could smack of heresy to an inquisitorial mind, suspicious of any unorthodox representation of the Trinity. Perhaps the painter of the tarjeta thought it prudent to omit the three lilies. But they also had peninsular heraldic associations with the Order of the Azucenas in Madrid, which would have allayed susceptibilities when it came to designing the blazon. The star-of-Bethlehem would naturally be placed in the highest position, although the planet Venus was placed in the middle position in the blazon for reasons we shall see.

The emblems can also be read from an Andean perspective. First, the "White Flower That Blossoms," Hanq'u Tutumpi (sent as an honorific by Pachacuti to Ayra Kanchi in the mid-fifteenth century and later adopted as her title by doña Úrsula, mother of don Fernando Ayra de Ariutu), was an Andean species of "lily" as the blazon tells us ("tres azucenas blancas"), and the lexicographer of Aymara, Bertonio (1956 [1612]), translates *azucena* as "amancaya" (*Ismene narcissiflora*), here in its white variety.[42]

The amancaya is a beautiful Andean flower that usually blossoms, according to Cobo (1956 [1653]:1:179–181), in groups of three or four: "From each stalk, of the many that are produced, two cubits [*codos*] in height and as thick as a thumb, are born ten or twelve flowers, not all together, but successively in *groups of three or four*, and as some wither, *others blossom*" (emphasis added). The "three white liles" were clearly a group of white *amancayas* (Figure 12.6). They may have grown at the dawning place (*paqarisqa*) of the descent group, commonly linked with a tree, shrub, or roots (Que./Aym. = mallki, *tunu*).

Cobo (1956 [1653]:1:179–181) also describes a single flower of the white amancaya: "It has *six white petals* similar to those of the lily, and within them a beautiful little white calix, four fingers from the center to the rim, and the mouth is three fingers

figure 12.6
"Three White Lilies": the Amancai, estampa 152 from Baltasar Jaime Martínez Compañón, *Trujillo del Perú*, 1781–1789, vol. 5, Biblioteca del Palacio Real, Madrid. © Patrimonio Nacional, Madrid.

figure 12.7a
Ismene Calathina, "Six Green Veins, Six Yellow Buds," from *Flore des serres et des jardins de l'Europe: Annales générales d'horticulture* (Gand, Belgium: L. van Houtte, 1845–1880), vol. 5, pl. 440.

wide, and it ends in *six beaks or points*, and *within it there issue from the center six green veins, which form ridges within the bell itself*, and from the end of each springs a little yellow bud the size of a grain of wheat" (emphasis added). Figures 12.7a–b shows these "six green veins" forming a six-rayed star that radiates from the center between the petals.

Invoking now a concept well known in Andean studies, I suggest that the amancaya was the *kama-sqa* of the six-rayed silver lucero shown in the tarjeta. *Kama-y* indicates the animation and multiplication of a life-form by a powerful archetype; *kama-sqa* is the concrete living manifestation thus animated (Duviols 1978; Taylor 1974). The idea that stars can represent vivifying forces that bring life and increase to their organic counterparts on earth

figure 12.7b
Hymenocallis Sp. (Photograph by Hibert Huaylla.)

figure 12.8
(a) Church of Curaguara de Carangas, 1608, showing (b) four amancayas with angels and stars on the sacristy ceiling; (c) roof of the nave with six-petaled lilies; and (d) presbytery roof with stars. (Photographs courtesy of La Iglesia de Curaguara de Carangas.)

is attributed to the Inca by Polo and, following Polo, by the Jesuits Acosta (1987 [1590]), Pachacuti Yamqui Salcamaygua (1993 [1613?]), and Cobo (1956 [1653]). Similar astrological beliefs existed in contemporary Europe. The close association of lilies and stars may also be observed in the paintings of the church of Curaguara de Carangas (1608; see Figure 12.8). We can see in the "White Flower That Blossoms" an organic manifestation of the six-rayed evening star, which is the kama-q (with agentive *-q*) of the white flowers of the amancaya and the seed of the white metal within the earth.[43]

The Slinger of Lightning

The second emblem, "at the top" of the quarter, is the resplandor, a word often used in contemporary sources for lightning (Ziólkowski 1997). Polo Ondegardo (1585 [1559?]:ch. 1.3, fol. 7v) does *not* use the word *resplandor* when he writes: "After Viracocha and the Sun, the third Huaca and the most venerated was the thunder . . . which they called by three names Chuqui illa, Catu illa, Intillapa, pretending that he is *a man in the Sky with a sling and a mace, and that he is able to make it rain and hail and thunder*, and everything else belonging to the region of air where clouds are formed" (emphasis added). One hundred years later, however, Cobo (1956 [1653]:2:160–161), who in other respects follows Polo closely, adds important details: "They imagined him as a man in the sky *formed of stars* with *a mace in the left hand and a sling in the right, dressed in shining robes, which caused the flash of the lightning* [resplandor] *when he turned to loose his sling; and that its crack caused the thunders*, which he made when he wanted water to fall" (emphasis added). According to Cobo, the lightning is the shimmer of the warrior's clothes when he moves to sling the bolt. The thunder is the crack of the sling as it is released. But where is the bolt (*rayo*)?

An answer to this question is proposed by Mariusz Ziólkowski.[44] The lucero (the third item mentioned in don Fernando's blazon) was called in Quechua "Chasca Cuyllur" (González Holguín 1952 [1608]:98, 570). It was part of the initiation ceremony for Inca youth: "Another temple of the Lucero Chasca Cuyllor, Chuqui Ylla, uaca billcacona. Where the lords and ladies, princes, entered to sacrifice; they were gods of these, of the younger ones" (Guaman Poma 1980 [1612]:236). Here "Chasca Cuyllor" is associated with "Chuqui Ylla," and Zuidema (1974–1976:222) has suggested a close relation between these two divinities, for both are Andean names for the lucero (Guaman Poma 1980 [1612]:272). In Guaman Poma's (1980 [1612]:79 [79]) representation of the Incas' "own coat of arms" (*armas propias*), besides the sun, moon, and the three-fold Wana Kawri/Pacaritambo/Tambo Toco, there appears in the lower left quarter a brilliant star called Choqui ylla uillca, identified as lucero in the adjacent text (Guaman Poma 1980 [1612]:80 [80]; Figure 12.9). Again, in the drawing of the "idols" (Guaman Poma 1980 [1612]:263 [265]), the star in the text corresponds indistinctly to the names Chasca Cuyllor and Chuqui Ylla (Guaman Poma 1980 [1612]:239). Ziólkowski (1997:61, 131), therefore, proposes that these two names correspond to the two aspects of Venus as morning and evening star, and "the *Evening Star* [Choque Illa, also the name of a form of lightning] would represent a projectile perfectly, since it moves across the constellations and also performs a movement like a 'fall' from above, for it sets after the Sun to disappear below the horizon."[45]

Where in the sky might we find the celestial slinger? Ziólkowski's account can be complemented with ethnographic data from Maragua (Chuquisaca) and the Salar de Uyuni (Potosí) in southern Bolivia, where the slinger (*hondero*) is equated with the llama herder (*llamero*), in association with the sling (*warak'a*). According to the ethnoastronomer Giuseppe Ciancia,

> The Slinger is formed by a region of dark interstellar filaments and clouds located in the southern Milky Way, nearby the western constellation of Scorpio. It is a well-recognised figure in the sky, as it stands out against the stellar background, and appears as a half human-like figure with an elongated arm which holds a sling. The Sling itself is described as a stellar asterism formed by a few stars of the western Scorpio. In particular, it is formed by α, β, δ, π, σ, τ Scorpii. The combination of the Slinger and the Sling is

figure 12.9
The Inca's Arms, drawing 79 [79] of Felipe Guaman Poma de Ayala, *El primer nueva corónica y buen gobierno*, 1615. (Photograph courtesy of The Royal Library, Copenhagen.)

> therefore formed by a mixture of a "dark" constellation and a stellar one. Antares, the alpha and brightest star in Scorpio, stands where the hand of the Slinger holds the Sling. . . . In the Andean region of southern Bolivia the Slinger (*Hondero*, or *Llamero*) is always associated with the Sling (*Honda*, or *Warak'a*). They are indicated and described together. (Giuseppe Ciancia, personal communication 2013; Figure 12.10 a–d)[46]

The "fall of stars" to earth, particularly the setting of Venus as evening star, was associated with the sowing and germination of metals and minerals underground. In twentieth-century Misminay (Cuzco), Gary Urton (1981) also found references to the sowing of silver in the earth by lightning, although the figure of the slinger is not present; for that, we have to turn to the ethnoastronomy of the southern mining region. But lightning, thunderbolts, and shooting stars were all part of the family of "atmospheric phenomena" that left their ore-shoots, silver "potatoes," and, perhaps, amancaya tubers (*kamasqa*) within the earth.[47]

Just as the slinger and lightning are associated with warfare, so the Inca modeled himself on the divine slinger when he went to war, on foot or on his red-feathered war litter (*pillco rampa*), dressed in glittering clothes and adornments and slinging golden projectiles against his enemies (Figure 12.11; Bouysse-Cassagne 1993, 1997; Ziółkowski 1997). The same practice was probably followed by Tata Ayra Kanchi, lord of the White Charka, when he traveled on his "golden litter" and expanded his domain toward Cochabamba (Chuy territory and, probably,

a

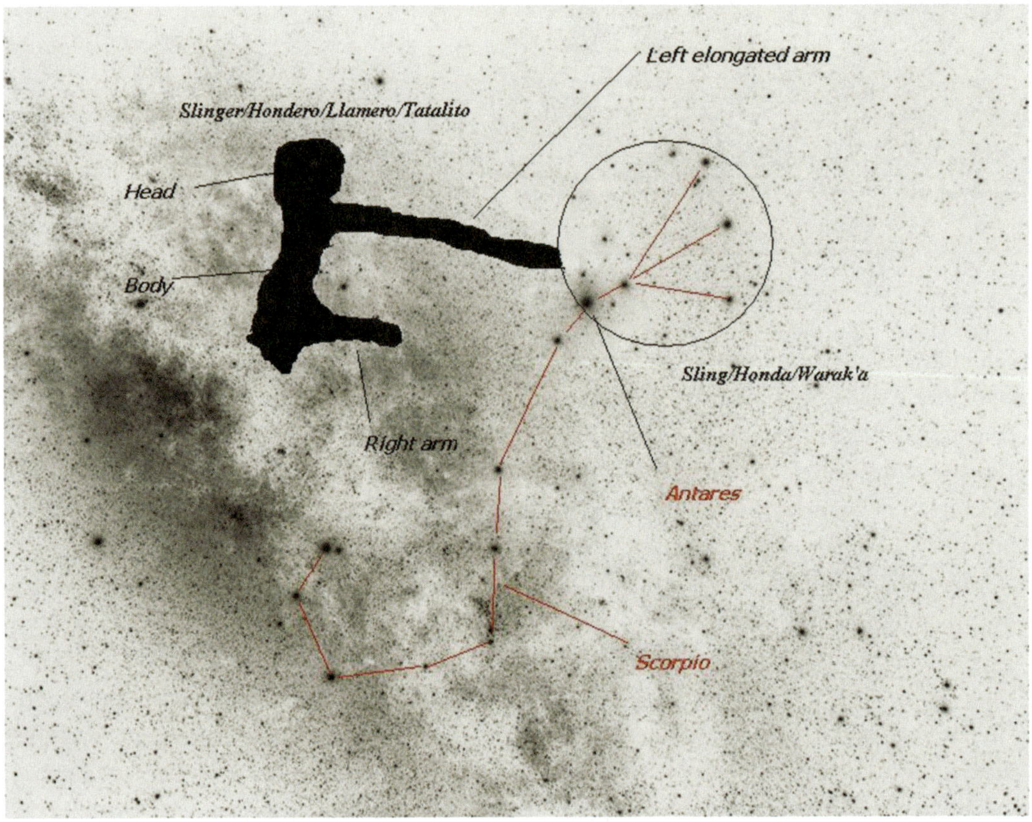

b

figure 12.10
(a) The constellation of the celestial slinger; (b) the slinger and the sling with Scorpio; (c) the slinger and the sling with Venus; and (d) the slinger in context with projectile. (Drawings by Giuseppe Ciancia.)

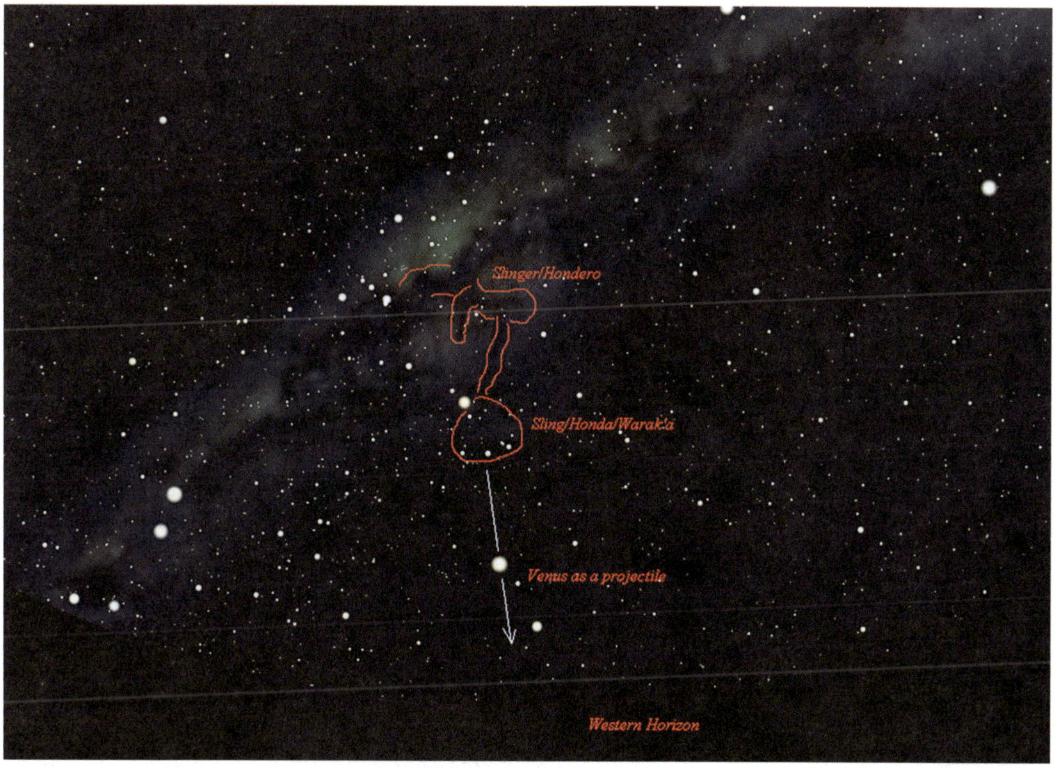
Slinger/Hondero
Sling/Honda/Warak'a
Venus as a projectile
Western Horizon

c

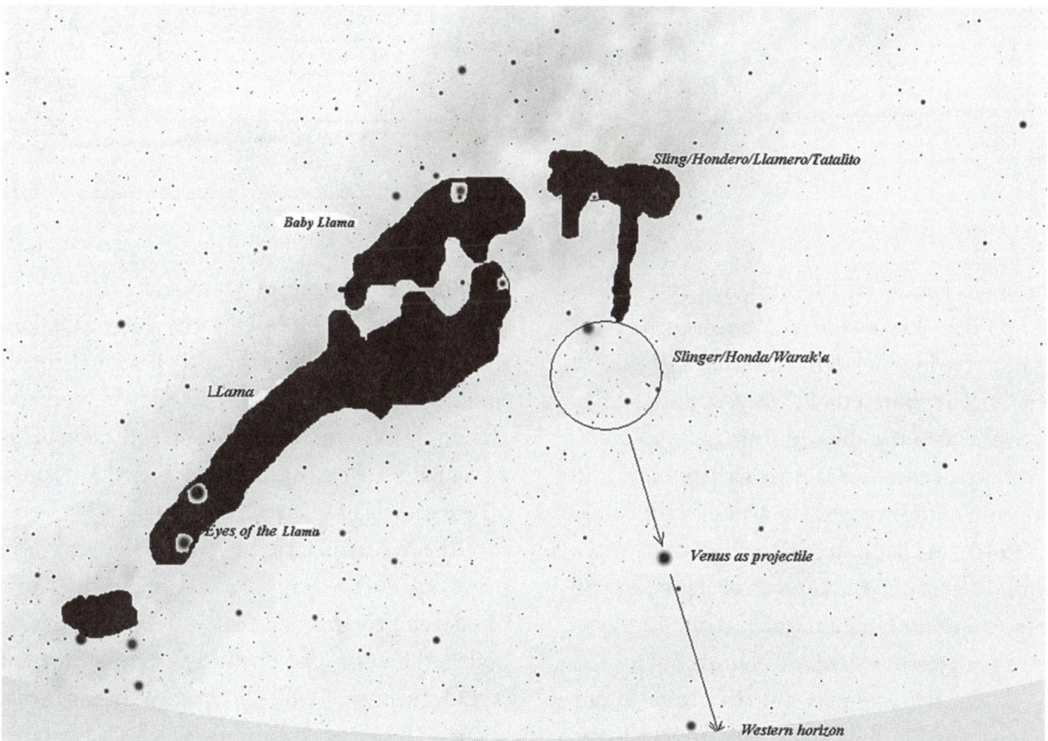
Sling/Hondero/Llamero/Tatalito
Baby Llama
Slinger/Honda/Warak'a
LLama
Eyes of the Llama
Venus as projectile
Western horizon

d

figure 12.11
The Inca's War Litter or Pillco Rampa, drawing 333 [335] of Felipe Guaman Poma de Ayala, *El primer nueva corónica y buen gobierno*, 1615. "Guayna capac ynga va a la conquista de los Cayambes, Guancabilcas, Cañaris, Ciccho, Chachapoya, Quito, Lataconga. Llevan los yndios Andamarcas y Soras Lucanas Parinacochas a la guerra y batalla de prisa lo llevan. Batalla del ynga." On page 332 [334], we read: "Cómo el Ynga pelea con su enemigo de encima de las andas. Tira con piedras de oro fino de su pillco ranpa [andas con plumas rojas] a su contrario al Apo Pinto, Guayna Pinto, y conquista la prouincia de Quito." (Photograph courtesy of The Royal Library, Copenhagen.)

the *cocales* of Tiraque), on the one hand, and toward Porco and Pilaya-Paspaya (Cinti), on the other. The images of glittering metal, the celestial slinger and sling, the stellar "thunderbolts" (*rayos*), and, finally, the silver potatoes and the white amancaya, kamasqa of Venus, all express Qaraqara astronomic and metallogenetic knowledge. No wonder the silver mine of Porco was dedicated to the lightning huaca of war, mining, curing, and agropastural production.

Now, González Holguín (1989 [1608]:59) translates *yuraq* not only as "white" (blanco) but also as "dawn" (alba). This suggests a further step in our analysis, for the white lily *yuraq amancae* can also be translated as the "flower of the dawn." It is animated by the six-pointed lucero of the evening star (Chuqui Illa) that sows the silver, but it blossoms under Venus's alternative aspect as morning star (Chasca Coyllur). The name of Qaraqara (also "Dawn"), given to all the White Charka by Wayna Qhapaq, is thus an expansion of Pachacuti's personal gift of the name Hanco Tutumpi to Ayra Kanchi. Perhaps we should think of this change as an extension of "solar citizenship" from the lord and ally Ayra Kanchi to all the vassals of his son Uchatuma.

The White Charka, in whose territory lay the mines of Porco, Potosí, and Chaquí, had probably developed this complex of religious ideas well before the arrival of Pachacuti.[48] This Inca called Ayra Kanchi the "white amancaya" because his silver deposits were exceptionally rich, and he enjoyed shamanic communication with the celestial slinger of the night. He offered Pachacuti a channel of communication with the lightning god from whom the Inca received his own initiation as shamanic warrior in Cuzco at the pool of Susurpuquio.[49] But he was also the "White Flower that Blossoms" at the rising of the sun. The title (which may reiterate one already assumed by Ayra Kanchi) recognized the mallku's intimate relation with the amancaya as paqarisqa (Que. = "dawning place") of his people or lineage and *mallki* (Aym./Que. = "plant of origin") of the White Charka.[50]

We have other evidence of the relationship between Inca and South Andean ideas of metallogenesis. Zuidema and Urton (1976) have shown the association of thermal springs with the setting of a star or planet. One of the shrines in Cobo's *Relación de las huacas* is the spring in the square of Aucaypata (Cuzco) in which "the priests of Chuqui Illa said that the Thunder bathed," suggesting a place of union between earth and sky, between underground waters and celestial thunderbolts (Rowe 1979:26). Chaquí, too, was (and is) famous for its thermal springs. In ancient times, these springs may also have been associated with the setting of Choque Illa, the evening star or lightning bolt. Chaqui's thermal link to the setting of the evening star would account for the town's political and religious importance in the southern province of Qaraqara.[51]

Venus: The Evening and Morning Star

Let us now return to the blazon. The position of Venus "in the middle," beneath the resplandor of the lightning flash and above the three amancayas, is consistent with its interpretation as the lightning bolt driven down by the celestial slinger to sow the earth with silver ore-shoots, leaving its mark on the germinating white amancaya tubers as well. But the lucero of the blazon differs in two respects from the star shown in the first quarter of the tarjeta. In the tarjeta, we see a silver star with six rays: in Old World terms, it is a Star of David, although we have seen that don Fernando also had good Andean reasons to propose a six-rayed star in his tarjeta.[52] Why, then, does the blazon change this image to a gold star with eight rays?

Stars of six and eight rays are to be found in both European and Andean traditions. The College of Heralds in Madrid probably changed the six-rayed silver star into an eight-rayed golden star because of its significance in Europe, where the latter was considered more Christian,[53] without realizing that the change might have different resonances in the Andean world. For eight-rayed stars were also part of pre-Hispanic Andean iconography (Linares Málaga 1979; Mejía Xesspe 1979; Zuidema 2011). What was their significance in the Andes?

In his article on Pukina as language and culture in Condesuyo, Mejía Xesspe (1979) writes: "Another Pukina cultural element is the use of a figure which is painted, woven or engraved, in the form of a star with eight rays, whose meaning is unknown." He further suggests that the language and iconography of Pukina in Arequipa was connected with Tiawanaku, anticipating current ascriptions of the Pukina language to this Middle Horizon religious center (Cerrón-Palomino 2000, 2010). Zuidema (2011) discusses weavings from Condesuyo in which the eight-rayed star appears regularly in a sequence of symbolic forms, possibly linked with calendrical calculations. Pachacuti Yamqui (1993:13r) distinguishes a twelve-rayed morning star (grandfather) from a six-rayed evening star (grandmother).

Recently it has been found that Pukina ritual language was used in early colonial Potosí. If the association of the eight-rayed star with Pukina persisted there in the sixteenth and seventeenth centuries, Andean observers might have linked it as morning star to the Qhapaq Iki (Capac Iqui), a Pukina name for the sun deity as first ancestor, which as we know from the report of the Franciscan Bernardino de Cárdenas in the 1620s, was used by miners transformed into shamanic jaguar priests inside the mines (Bouysse-Cassagne 2004). The eight-rayed golden star could therefore have been read as the harbinger of the Qhapaq Iki of Potosí, rephrasing in colonial Pukina ritual language the role of the morning star, Chaska Quyllur, that, like the Qaraqara themselves as part of Wayna Qhapaq's army, announced the arrival of the Inca sun to those living in darkness, while also being a transformation of the Choquilla, six-rayed lightning bolt and evening star.

The "three white lilies," blossoming after being sown by the lightning and "falling star" of

Venus, are thus a metonym for Ayra Kanchi himself: they are his title. The name White Flower reveals him as ritual patron of the white metal that grows within the earth; its association with lightning and Venus shows him to be the owner of shamanic knowledge. This quarter of the shield cued an esoteric narrative that, in the early seventeenth century, the lords of Qaraqara needed to keep alive: it underpinned their claim to the whole of the Inca "Province of Qaraqara" and to privileges appropriate to valued and powerful servants of the Inca and the king.

The Second Quarter

"And in the second quarter on the [heraldic] left, in a field of gold, a condor, which is a big black bird, and on its neck it has a white smudge, its wings extended, and made beautiful with gold" (Platt et al. 2011 [2006]:762). This quarter shows the condor-mallku with his well-known "white scarf." *Mallku*, meaning "condor," was also the title of the highest Aymara lords and a symbol of authority and of male protection and predation.[54] It would be the first word pronounced when reading the shield from the heraldic point of view.

The Third Quarter

"In the third quarter, in a red field, a tiger with its natural colors [*un tigre de su color*], rampant, made beautiful with gold" (Platt et al. 2011 [2006]:762). *Tigre* is Spanish for the jaguar (Que. = *otorongo* [*uturunku*]), which we see rampant in the tarjeta. At this time in the southern Andes, the jaguar was not a distant Amazonian creature.[55] Colonial miners transformed into jaguar priests inside the tunnels of the mines with the aid of the psychotropic *curo*, identified by Thérèse Bouysse-Cassagne (2004) as a form of wild tobacco. These jaguar miners will have paid homage to their supreme ancestor, the sun, the huaca on the peak of the silver mountain, who was worshipped in the seventeenth century under the Pukina name of Qhapaq Iki.[56] The importance of felines in 1545 is clinched by Diego Wallpa, the "dis-coverer of Potosí," who, in his deathbed statement at the end of 1572, referred to a "nest of lions of the country," or pumas, that was located in a wood of *queñoa* trees close by the huaca on the summit (Sanct Angel 1965 [1573]:360). All these felines were the *kamasqas* of *choque chinchay*, lord of the otorongos, whose feline constellation was linked to Venus as the six-rayed evening star by Pachacuti Yamqui (1993:13, 21r).[57] Such symbols were still available to don Fernando in the seventeenth century; they were probably known and discussed in Potosí itself, as the new wealth-generating "center" of the colonial Andes. The jaguar would certainly appear on the stone blazon carved on the mallku's house in nearby Pocoata or Macha.

The Last Quarter

"And in the last quarter, in a blue field, a rock and on it a strong tower of stone" (Platt et al. 2011 [2006]:762). The final quarter described in the blazon and shown on the tarjeta contains the emblem of a "tower" built on a rocky eminence, which is translated by Bertonio as *pukara*. The image is European, but the Andean resonances of pukara are numerous, and relevant examples are given in the documentation: first, the line of forts constructed as a defense against the Chiriwana, notably Oroncota (Alconini 2004; Pärssinen and Korpisaari 2003), also the palace-fortress of Tumipampa built with labor contributions from Uchatuma's men in southern Ecuador. The pukara commemorates such "towers," also referring metaphorically to the mallku as a "tower of strength" and source of protection, and perhaps to the funeral tower (or chullpa) where his mummy dwelt.

The coat of arms can now be read, from (heraldic) left to right and top to bottom:

> Mallku Resplandor-Chasca Cuyllor/Chuqui Illa-Amancaya [Ayra Kanchi]
> Pukara Uturunku
>
> which can be translated as
>
> Condor-Chief Lightning–Venus–Three White Lilies [Ayra Kanchi]
> Fortress Jaguar[58]

But there remains one element to be examined.

The Crest

"And as crest [*tiembre*], above the vizor [*celada*] of the shield, a half-indian with a red bonnet, and three gold feathers in it, as here go placed and painted" (Platt et al. 2011 [2006]:762). The half Indian on the crest appears at first sight to be female: the long hair at a time when male lords were cutting their hair (Capoche 1959 [1585]), the adornments, even the eyelashes. But the colleagues I consulted first thought the figure must be male, not only because the word *indio* is used in the blazon (rather than *india*, though this is not a conclusive argument), but because in pre-Hispanic times feather headwear was a sign of male authority.[59]

However, new evidence shows that brooches in the form of golden feathers were used by wealthy Andean women during the early colonial period. An example is the "*tipque* of gold . . . very elegant, made to resemble feathers" that don Diego Chambilla, mita captain of Pomata (a Lupaqa town near Lake Titicaca), had made for his daughter by a Potosí goldsmith in the 1620s (Medinaceli and Inch 2010:192v).[60]

If we accept that the half Indian is female, there is only one possible identity for her: doña Úrsula Ayra Kanchi Hanq'u Tutumpi herself, daughter of don Francisco Ayra and granddaughter of Inka Muruq'u; descendent of Tata Ayra Kanchi Hanq'u Tutumpi and of Uchatuma and Payku Chimpu; wife of a descendant of Ayra Kanchi's ambassador Lukalarama and his son Ariutu; and mother of don Fernando Ayra de Ariutu, the petitioner for the coat of arms. Her position on the crest and the three golden "feather brooches" she wears on her bonnet express clearly her key role in the colonial refoundation of the house of Ayra Kanchi (Figure 12.12).

Doña Úrsula's other adornments are also significant. Her silver neck-pendant with the half moon beneath is associated by Guaman Poma with the Qullasuyu in several drawings, while the

figure 12.12
Doña Úrsula Ayra Kanchi Hanq'u Tutumpi, from the Archivo General de Indias, MP-Escudos 75. Photograph courtesy of the Ministerio de Educación, Cultura y Deporte, Archivo General de Indias, Spain.

gold disks that form her "cape" are a clear reference to the golden "*roeles*" that adorned the "shirt" given to Uchutuma, her great-grandfather, by Inka Wayna Qhapaq (*una camiseta de roeles de oro*). Both were originally "male" symbols attributed to doña Úrsula in recognition of her role as the female link between Uchatuma and her son don Fernando. Naturally, Uchutuma's "shirt" and "cape" have been re-imagined by the painter in European terms, as in the case of the "tower" or pukara in the bottom right-hand quarter of the coat of arms. But this does not mean there was no memory of an Andean "iconographic event" underlying the European surface representation, as in the case of the fortresses (pukara) constructed against the Chiriwana on the Inca's southwest frontier, and at Tumipampa south of Quito.

Conclusion

It is sometimes argued that the probanzas of Amerindian lords articulate a "discourse of power" in response to colonial pressures, rather than presenting evidence for pre-Hispanic regional history and knowledge (Graña 2001). On its own, such an approach may obstruct the recovery and historical interpretation of symbols whose life began before the political pretext for the discourse. Estenssoro (2010) has shown that while coats of arms are nothing if not "discourses of power," they may also retain traces of selected "events" whose impact persists in the specific form of the blazon. In the case of the Qaraqara, we can observe a pre-Hispanic dynasty refounding itself and appealing to regional collective memories in order to prolong its jurisdiction and privileges by re-membering pre-Hispanic events and knowledge within a colonial frame.

Don Fernando's coat of arms was imagined using ancestral emblems, then sketched, probably by his own "examined painter." It was requested through the preparation of the *Probanza*, sent to Madrid by the corregidor of Potosí, granted by the king on the advice of the College of Heralds, and then returned to Potosí to be sculpted on the petitioner's house and probably on his tomb. Although Gonzalo Pizarro's encomienda eluded don Fernando, he secured official recognition for the refoundation of the house that had begun with his parents' marriage, while reiterating a persistent (if fading) claim to authority over both the old Qaraqara provinces of Macha and Chaquí. The emblems would be recognized by at least some Qaraqara as part of pre- and post-Hispanic expressions of lordly power, not simply as colonial inventions: they transmitted pre-Hispanic traditions, if through a European lens.

Let me summarize the conjunctures proposed for pre- and post-Hispanic Qaraqara ethnohistory, which show how successive generations used the Venus cycle to express historical change.

The name Qaraqara, meaning "Dawn," was clearly associated with Venus, the morning star, a link anticipated by Pachacuti's gift of the name of "Blossoming White Flower of the Dawn" to Ayra Kanchi, but consolidated by Wayna Qhapaq through his gift of the name to Mallku Uchatuma and his men. I take this to be a genuine memory-trace, not a colonial invention, that recalls how the White Charka became the Qaraqara. The new name marked a temporal change, with Venus as eight-rayed morning star now accompanying Uchatuma and his army as harbinger of the new sun of Wayna Qhapaq and complementing the six-rayed evening star driven to earth by the slinger in the sky to generate the white metal and the amancaya plant.[61] The alternation of evening and morning stars in the night sky during the 584 days of the Venus cycle probably marked an important regularity in the Qaraqara ritual calendar, which may have been celebrated close to the thermal springs of Chaquí.

The title "White Flower That Blossoms" bestowed on Ayra Kanchi by Pachacuti, and recovered in his coat of arms by don Fernando as "three white lilies," represents a spray of amancaya flowers, six-rayed kamasqas of Venus as evening star, and hence metallogenetic symbols of Tata Ayra Kanchi, mallki, and paqarisqa of his descent group. The sacred number three may have been embodied in the three sons of the founding lord—Pocota, Uchatuma, and Usuquira (see Figure 12.2)—while the three flowers may be the White Charka equivalent to the three caves of the Inca myth of Pacaritambo (Figure 12.9). The later assumption of Ayra Kanchi's title by doña Úrsula in the 1590s indicates a late sixteenth-century memory trace of a pre-Hispanic "event" in the context of the uterine refoundation of the house, later validated in the seventeenth century by her son's royal blazon. The reference in the *Probanza* to the "woven map of the lands" or "awning" (carpatira) given by Inca Pachacuti to Ayra Kanchi also seems like a genuine memory trace. Both are part of a narrative of pre-Hispanic ritual events, which renders intelligible data on the three great silver mines, such as the thermal springs and three stones of Chaquí, the shamanic cult and three stones of Porco, and the feline priests of the solar huaca of Potosí.

But the "great arch" (Corrigan and Sayer 1985) of Andean state construction in Charcas was violent as well as negotiated. We have seen that Tupaq Yupanki was suppressed from Qaraqara dynastic memories. Yet Cobo (1956 [1653]:84–85) describes how this Inca besieged twenty thousand Charcas soldiers in the fortress of Oroncota, which ended with the Aymara warriors sidling out to join an orgy with naked dancing women prepared by the Inca as bait, thus allowing the Inca's soldiers to enter and take the fortress. Tupaq Yupanki was also associated with the expansion of Inca mining in Charcas (Sarmiento) and with the imposition of a silver tribute in Porco (Betanzos). Yet he is completely absent from don Fernando's *Probanza*. His reign has disappeared from the testimonies like the planet Venus during the interstitial period between its appearances as evening and morning star.

Don Fernando's tarjeta, supported by the *Probanza* and completed in the blazon, can be seen, even through its silences, to contain a rich historical statement of his lineage identity. As Estenssoro suggests, the coat of arms is his "self-portrait" (*autoretrato*): it tells those who can read it "who he is." Elderly indians, passing the blazon engraved on their mallku's house, would recognize the symbols and reminisce by telling stories from the "old days" (*ñawpa pacha*). For a few, the blazon may have shed a glow of ancestral authority onto Mallku don Fernando. Others will have seen the pretensions of a tyrannical cacique driven by Spanish fiscal pressures to exploit his people;[62] while in Spanish eyes, a royal blazon, bolstered by alphabetic certificates issued by the king and his officers, legitimized don Fernando as an indispensable lynchpin of colonial rule (Saignes 1987).

We have argued that two periods of Qaraqara and Inca history were symbolized by two aspects of Venus. But the attempt to transfer the imperium of Wayna Qhapaq to the Spanish king through the gift of the Inca's silver mine in Porco ran into trouble as Spanish dominion evolved, particularly with the division of Qaraqara in 1570 into two colonial provinces. The refounding of the house of Ayra Kanchi in the 1590s through the marriage of doña Úrsula to don Fernando Chinchi, and the blazon of 1640, attempted to revert the trend by defending the integrity of the Inca province. The golden eight-rayed star sent from Madrid may have been read by miners and mitayos, for a while, in relation to the Pukina rite of the sun; later, it may have been seen, from a Joachimite perspective, as a harbinger not so much of the Inca sun as of the future age of the Holy Spirit.[63]

The analysis of this coat of arms has shown us a periodization "in the making," successively reconfigured in line with new political events. By comparing this pragmatic approach to time with the stunning accumulations of observations of the Venus cycle contained in the Maya Dresden Codex, we can put comparative questions to these two Amerindian groups.

One approach to time, followed by the Maya, was to accumulate observations over several centuries and to take these progressively as an astrological matrix to be invoked for rendering events on earth propitious, even to the point of incorporating correction tables to compensate for the increasing lack of fit between the real time of Venus's cycle and the tabulated estimates (Aveni 1990:195–207). Another, simpler approach, taken by the Qaraqara, was to relate the reigns of Incas Pachacuti and Wayna Qhapaq to successive lordships, linking them to surface features of the cycle: the passage of Venus from evening star, through a period of invisibility, to morning star. It will be useful, therefore, to compare the retrospective periodizations of the sixteenth- and seventeenth-century Qaraqara with the temporal systems constructed by the Maya before accumulated observations had cast a celestial web over Mesoamerican history and again after the abandonment of their Venus calendar a couple of centuries before the arrival of the Spanish.

What was the fate of South Andean ethnoperiodization in the later seventeenth century? We do not know. Perhaps the retrospective accumulation of periods ceased with the eclipse of the Province of Qaraqara, probably in the early eighteenth century. Under the republic, however, we find a new, well-known mythic periodization in the southern Andes (still influenced by Joachimism). It narrates (1) the mythic age of the moon, when "the Chullpas" (Late Intermediate federations) lived in a state of natural feracity (Old Dispensation) and (2) the age of the sun (first Inca, later Christian) under which people still live today (New Dispensation), while looking forward—after another period of warfare—to (3) the ever-postponed age of the Holy Spirit (future age of the saints) (Sendon 2010). Several versions elaborate on this schema in different ways. Yet in one respect this threefold periodization coincides with the version presented by don Fernando Ayra de Ariutu. The "event" that initiates the present age is remembered not as the invasion of the Andes by the Spanish, but as the rising of the Inca sun. The myth seems to have evolved in dialogue with the earlier dynastic narrative through a process we do not yet understand, but the emergence of the Andean solar state is still considered, from this late mythic perspective, as more memorable than the Spanish occupation itself.

Acknowledgments

The probanzas on which this essay is based are published in Platt et al. 2011 [2006]. I am grateful to Thérèse Bouysse-Cassagne, Giuseppe Ciancia, Pablo Cruz, Pablo Quisbert, Gabriela Ramos, Ivan Zambrana-Flores, Mariusz Ziółkowski, and the symposiarch for this volume, Anthony Aveni. All translations are by the author.

NOTES

1 In Potosí Spanish, the *-sqa* of "sudden discovery" is translated with the pluperfect of the verb *ser*: "¡había sido picante!" indicating a temporal pragmatics of taste.

2 For the phenomenological approach to time based on retention, intention, and protention, see Husserl 1999:186–219 on the memory, perception, and anticipation of melodic sound.

3 The creative aspect of memory is glossed as "re-membering" (reworking the themes of memory) in Platt 2012.

4 The *Probanza* of don Fernando de Ariutu is conserved in the Archivo General de Indias, Seville, Charcas 56.

5 Gabriel de Rojas and his companions were present in the Carabaya gold mines by 1534, however, before Almagro made his quick dash down the altiplano in 1535 en route to Chile.

6 Ceremonial visits were exchanged by the lords of Qaraqara and Charka, as don Pedro Ayawiri, son of the Charka lord Kuysara, remembered. In 1612, he was the eighty-year-old lord of the Charka capital, Sacaca. See the *Probanza* of don Joan de Castro y Paria, published in Platt et al. 2011 [2006]:779, doc. 18.

7 For the White and Red Charka, see "Los significados de Charcas" in Platt et al. 2011 [2006]:42–47, with Rowe 1985. The Qaraqara are still sometimes referred to as *indios charcas* in the early colonial documentation.

8 For the development of the probanza as a Hispanic documentary genre, see Macleod 1998.

9 The genealogy shows the patrilineal bias expected by the Spanish. Figure 12.2 draws on two other sources besides don Fernando's *Probanza*, while making no claim to be exhaustive, for genealogical sources notoriously only show those relationships that are

relevant to the purpose at hand. Thus, Figure 12.3a, from an earlier Macha *Probanza* of 1619, shows four sons of don Francisco Ayra omitted from the *Probanza* of don Fernando Ayra de Ariutu. The earlier *Probanza* is presented by don Alonso Uchatuma, son of don Martín Moroco (glossed "bastard" in a different hand). See Platt et al. 2011 [2006]:767–781. Don Alonso is defending his *principalazgo*, and his father is shown in Figure 12.2 as "brother" of doña Úrsula (no doubt by a different mother, hence "illegitimate": *cacical* polygyny died hard). Figure 12.3b, on the other hand, shows Tata Ayra Kanchi, "lord of 20,000 indians of the nation of the Cara[ca]ras," as father of three sons, also shown in Figure 12.2, where Pocota, the eldest son, appears as the ascendant of don Diego Ayra Kanchi, who presents this *Probanza* (also in 1619). See Platt et al. 2011 [2006]:793–804. But the dominant themes for us here are Fernando Ayra de Ariutu's paternal ascent to Ariutu and Lukularama and his maternal ascent to Uchatuma, Tata Ayra Kanchi's second son.

10 *Probanza of don Joan de Castro y Paria*, lord of Macha in 1612.

11 Two towns (*cabeceras*) called Macha and Chaquí are mentioned in Francisco Pizarro's encomienda grant of the Qaraqara to his brother Gonzalo (1540). See Platt et al. 2011 [2006]:289–301.

12 The presence of Qaraqara pottery in Cinti is attested archaeologically by Rivera (2013). For the fortresses constructed on the Chiriwana frontier before and under the Inca, see Alconini 2004; Pärssinen and Korpisaari 2003.

13 The "news" of the four provinces of Tawantinsuyu reminds us that the idea of quadripartition, if not the names of the suyus, was also present in Aymara as Pusisuyu (*pusi*, Aym. = "four"); Bertonio (1956 [1612]:278): "Pusi suu; Todo el universo, mundo."

14 *Carpa* (Sp. = "tent") is translated by Bertonio as *toldo* (also Sp. = "tent"); *carpatira* probably derives from Sp. *carpa-tierra*, literally "tent-land." The weaving may have been displayed as an awning that shaded the heads of the lords from the sun as they sat on their "stools of authority" (*duo*). The use of a Spanish name for the weaving is an example of a "tweaked" memory, suggesting a desire to communicate in Spanish a performative aspect of the Inca's "land grant." For interpretations of *carpa* with reference to land and irrigation, see Sanhueza 2008.

15 See don Fernando's questionnaire, questions two and eight, with testimony of don Juan Vissaya, cacique of Pocoata. *Anco* (*hanq'u*), Aym. = "white"; Bertonio (1956 [1612]:2:369) gives: "Thuthumpi. Flor de qualquier genero que sea. Thuthumpi uyu: Jardín de flores."

16 For Polo Ondegardo's observations of woven or painted "maps" in Cochabamba, see Del Río 2005.

17 Weavings of cumbe cloth given to regional lords by the Inca were prized out of them by early Spanish encomenderos: thirty-five pieces of cumbe cloth, given by the Inca to the Charka lords of Sacaca, were extracted from don Alonso Ayawiri by his encomendero Alonso de Montemayor, who used imprisonment, torture, and a mock execution (Platt et al. 2011 [2006]:387). Ayra Kanchi's woven map may similarly have ended up in the hands of Gonzalo Pizarro, encomendero of the Qaraqara from 1540 until 1548, or of Pedro de Hinojosa, encomendero of Macha from 1548 until his assassination in 1553.

18 For the construction of Tomebamba (Tumipampa), see document 20 (843). For the Andean confederations in the Pasto campaign, see Pärssinen 2002. Compare Guaman Poma (1980 [1612]:170 [172]) for the groups that accompanied Wayna Qhapaq to Quito, which included the *charca*.

19 See for the Hunu Mallku of the Killaka, who also took part in the gift, the *Probanza* of don Juan Colque Guarache, published by Espinoza Soriano (1981).

20 Anónimo 1934 [1539] is identified by Horacio Urteaga with Miguel de Estete, encomendero of Carangas. For the campaign of Cochabamba, the Treaty of Awkimarka, and the gift of the Porco silver mountain to Charles V, see Platt et al. (2011 [2006]:103–128). In Chuquisaca, the Pizarros rewarded Kuysara for his role in bringing about the Treaty of Awkimarka by presenting him with "a suit, a mantle of green damask with its shirt of green velvet with its golden flecks, and some lace-up boots, and a red hat in the fashion then current, and a big box of knives" ("un vestido, una manta de damasco verde con su camiseta de terciopelo verde con sus flocaduras de oro, y unos borseguíes de lazo, y un sombrero colorado que a la sazón se usaba, y un estuche grande de caja de cuchillos") (Platt et al. 2011 [2006]:844). This amounted to a lesson in the new European style of cacical self-presentation desired by the invaders.

21 For the idea of traslatio imperii in the Spanish empire, see Martínez 2014.

22 La Gasca transferred briefly the whole Province of Qaraqara to Pedro de Hinojosa, whose defection from Gonzalo Pizarro with the entire rebel fleet at Panama was a decisive contribution to the royal cause. Shortly afterward frontier groupings were distributed between Hinojosa (Pikachuri and some Kakina), Alonso de Montemayor (Kulu and other Kakina), and Pablo de Meneses (Moromoro), while Macha, Pocoata, and Chayanta (Anansaya) remained in the hands of Hinojosa. Moromoro, probably part of the pre-Hispanic Qaraqara Province of Chaquí, was assigned in 1570 to the colonial Province of Chayanta rather than Porco Province (Platt et al. 2011 [2006]:265–272; Zagalsky 2012:9).

23 See the *Título de cacique a don Francisco Ayra*, published in Platt et al. 2011 [2006]:748. For the mita captains, see Capoche 1959 [1585].

24 Don Francisco Ayra supplied Toledo with five hundred Indians, one thousand food-bearing llamas, and seven hundred pesos (Platt et al. 2011 [2006]:722).

25 In predominantly patrilineal systems of inheritance it is common for female heirs to prolong the patrilineage by transmitting their fathers' rights to their sons.

26 For the "encapsulation" of the past in the present, see Collingwood 1946; and Wingfield 2005:119–135.

27 *Ch'isi* has both temporal and spatial referents, meaning "evening" but also "threshold" or "place of transition," as in the ceremonial place of entry and exit outside South Andean rural towns known today as *ch'isiraya*.

28 The Maya Venus calendar was based on observations accumulated over more than four hundred years (see Aveni 1990:195–207). Zuidema (2011:251–275) interprets the Chuquibamba textiles from the Condesuyu as representing interlocking solar, lunar, and sidereal calendars, but the observations that might have enabled the secular accumulation of such knowledge are unknown.

29 Compare with the silver *papas* from the Inca's mine in Huantajaya (Tarapacá) in Pizarro 1986 [1571]:189–192; also Platt and Quisbert 2008.

30 Sophisticated refining technologies from the Middle Horizon have been detected in Pulacayo (Qaraqara provincia of Chaquí), probably connected to Lipez and Atacama (Lechtman et al. 2011). Betanzos (1987 [ca. 1551]:164–165) writes of Tupaq Yupanki imposing a silver tribute on the warrior-miners of Porco, probably after the siege of Oroncota (Cobo 1956 [1653]; and my conclusion).

31 Porco was studied archaeologically by Van Buren (2005; Van Buren and Presta 2010), who has argued against any pre-Inca presence at Porco, dismissing the extirpation of idolatry document and also (implicitly) Betanzos. I (2014:273) suggest a possible reason for the absence of traces of Late Intermediate or Charcas pilgrims at Porco.

32 Gisbert (2011) argues that Potosí was dedicated to the coastal deity Pachakamaq as well as to the sun.

33 Aymara lords assumed the role of mita captains following Toledo's reform of Potosí mining in 1572–1573. For details of the *capitanías*, see Capoche (1959 [1585]).

34 These mansions have not survived in Macha or Pocoata, but equivalents from the Collao are studied by Gisbert (1992). They are analogous to houses built by noble peasant hidalgos in northern Spain.

35 Compare with the evidence given by don Juan Achata, cacique principal of Sipesipe: "Some caciques discussed the matter in this witness's presence, and they said that Cumsara was a principal lord of this province . . . and they debated about it and it was ascertained, in the presence of this witness, that the said Cumsara was a very principal lord" (Platt et al. 2011 [2006]:928–929).

36 See the "Carta de unos caciques al virrey Francisco de Toledo." According to the signatory lords of the *Memorial* of Charcas (fols. 7r–v), Joan de Padilla had told Viceroy Toledo about camelid herds in Sacaca dedicated to the cult of the huacas, just as llamas were being assembled to carry provisions for Toledo's campaign against the Chiriwana. See Platt et al. 2011 [2006]:855n231).

37 Probably European halberds. Although Betanzos talks of Inca halberds, Lechtman (2008) has shown that these were tin-bronze axe-maces, widely disseminated by the Inca.

38 Contrast this approach with that of the lords of Pakasa in the Collao, whose seditious letters to the English corsair Sir Francis Drake were headed "Very Magnificent Lutheran lords." They were intercepted in 1579 by the Spanish (Platt et al. 2011 [2006]:667).

39 In 1629 don Fernando's father was referred to in old age as don Fernando Ayra Chinchi, attributing to him the name of his wife's ascendant (Ayra). At the same time, the son was referred to as don Fernando García Chinchi; he seems to have taken the name of don Fernando Ayra de Ariutu in 1632 (Platt et al. 2011 [2006]:754).

40 This undated document refers to the offending "governor" as don Fernando Ayra Chinche (@29), the name of don Fernando Ayra de Ariutu's father. The name might also refer to the son, who was also called Chinchi in 1629. Compare previous note.

41 Compare with "Amankaya, flor blanca, o colorada, como lirio o açucena" (Bertonio 1956 [1612]:2:15).

42 *Ismene narcissiflora*. Known commercially as "Peruvian daffodil" and "Sacred Lily of the Incas," this white-flowered species (http://apps.kew.org/wcsp/namedetail.do?name_id=279429) is closely related to the yellow *Ismene amancaes* (formerly known as *Hymenocallis amancaes*, http://apps.kew.org/wcsp/namedetail.do?name_id=278752). *Ismene narcissiflora* has been reported as present in Bolivia by Foster (1958) with the synonym *Hymenocallis narcissiflora*. The name shown in Figure 12.7a, *Pancratium (Ismene) Calathinum*, is also an old synonym of this species (http://apps.kew.org/wcsp/namedetail.do?name_id=310134); see Elliott (1943) and https://www.flickr.com/photos/23140499@N07/2451201939/in/photostream/. I am grateful to Bolivian conservation biologist Ivan Zambrana-Flores for the information contained in this note.

43 For the fertility of the lightning bolt Chuqui Illa see Polo Ondegardo 1585 [1559?]:ch. 11.1, fol. 13r: "Others say that in tempestuous weather some women became pregnant by Chuqui Illa."

44 For the Inca lightning cult and the close association between Inca Yupanki Pachacuti and lightning, both as Inti Illapa, the Inca's "brother" or *wawqi*, and as Chuqui Illa or Kacha (Aym. Kaqya), see Bouysse-Cassagne 1993; and Ziólkowski 1997:59–68.

45 Ziólkowski (1997:228–231) also relates the rhythm of Inca military activities to the Venus cycle. Garcilaso (1963 [1609]:72) writes that the Inca "honored [the star Venus] because they said it was the attendant of the Sun, and went closest to it, sometimes ahead of it, and at other times following after." The Maya had observed the five "diving motions" made by the lucero over its 584-day cycle (Aveni 1990). We have no evidence of such a minute observation of the Venus cycle in the Andes, but the Maya association of Venus with warfare recalls the Andean symbolism of the lightning warrior. See Staller and Stross 2013, for a comparison between Andean and Mesoamerican ideas of lightning, Venus, and warfare.

46 Thanks for this information to Giuseppe Ciancia.

47 A related meaning may be attached to the *sakaka*, "comet with flaming tail," that appears in the coat of arms of don Fernando Ayawiri, mallku of Sacaca, capital of the neighboring (Red) Charka. See Platt et al. 2011 [2006]:doc. 22; also Arze and Medinaceli ca. 1991. Two *sakakas* likened to winged dragons of the lower world are said by Pachacuti Yamqui (1993 [1613?]:21v) to have emerged from Mount Ausangate and passed over to Mount Phutina (Arequipa) and "below Guamanga": "animals with wings and ears and tails and four feet, and on their shoulders many spines like a fish, and from a distance they say they seemed all fire." In 1971, I was told in Macha that there are two kinds of lightning: one driven down from above by the celestial *arcabucero* Santiago (who has replaced in many contexts the old slinger in the sky), the other darting from the tongues of a cnthonic two-headed lizard at the mouths of caves.

48 Guaman Poma (1980 [1612]:56 [56]) sees Inca lightning beliefs as derivative, attributing them to the people of his second age, the Wari Runa: "And that is why afterward the Inka sacrificed to the lightning [rayo] and feared it deeply."

49 See Ziólkowski 1984, 1997:128ff. on the theophany, or lightning bolt of crystal mirrors, that initiated Pachacuti at the pool of Susurpuquio, transforming him into a shamanic warrior.

50 According to Albornoz (1967 [1584]:169–171), the *pacarisca*, "dawning places," of different Andean peoples were treated as huacas that could take many forms, including those of plants (*géneros de árboles y yerbas*) or mallki. In this case, the amancaya was probably the mallki of the White Charka, perhaps associated with rites for the mummy, or chullpa, of Tata Ayra Kanchi.

51 The mine of Chaquí was documented in the seventeenth century by Alonso Barba (1967 [1640]) and in 1625 by the Spanish miners Pedro de Sanabria and Juan de los Reyes. See Platt et al. 2011 [2006]:215–220; and Platt and Quisbert 2008. The recurrence of the number three in the Qaraqara sources suggests a South Andean equivalent of the Andean triple

divinity also known in Cuzco (Bouysse-Cassagne 1997). The huaca of Tata Porco was composed of three stones, together with a *mama* of pure silver; the stones can be compared with the three stones on the mountain of Chuquipalta in the Inca ceque system. These were associated with the sun, Inti Illapa (Lightning), and Pachayachachiq ("World-Teacher"; Rowe 1979; Ziólkowski 1997:63). Cruz (2009) has confirmed ethnographically the continued association of Chaquí with lightning; it appears as a third great mine, with Porco and Potosí. The entry to the Chaqui mine was triangulated by three stones, one associated with lightning, the second with the sun, and the third unidentified, perhaps equivalent to the stone of Pachayachachiq among the three stones on Chuquipalta. The three stones on the mountain of Chaquí were probably equivalent to the three stones on Porco, as well as the three on Chuquipalta.

52 Compare the early seventeenth-century paintings of the church of Curahuara de Carangas on the Bolivian altiplano, where the roof is ablaze with six-petaled flowers and six-rayed stars (with one or two eight-rayed examples); see Gisbert and Rosso 2008 and Figure 12.8. Figure 12.8b shows four six-petaled flowers with visible stamens, probably amancayas, surrounded by eight-rayed stars.

53 My thanks to Riccardo Scotti for this suggestion.

54 Bertonio (1956 [1612]) gives "Mallco; vel Mayco: Cacique, o Señor de Vasallos." The term *hunu mallku*, "lord of ten thousand tributaries," occurs in the *Probanza* of Kuysara's descendent, don Juan Ayawiri. See Platt et al. 2011 [2006]:doc. 21. See, too, the *Probanza* of Juan Colque Guarache, hunu mallku of Killaka and Asanaki, in Espinoza Soriano 1981.

55 A son of the Inca was apparently eaten by a jaguar in the valley of Sayapaya, lands of the Indians of Chayanta near San Pedro de Buena Vista (Platt et al. 2011 [2006]:628).

56 See Bouysse-Cassagne 2004. Cerrón-Palomino (2011) argues that *Capac* and *Yupanqui* are Pukina words. The presence of Pukina toponyms in north Potosí (the colonial province of Chayanta; Mendoza and Patzi 1997) may indicate a Pukina substrate beneath the predominantly Aymara language spoken there by the Qaraqara and Charka in the sixteenth century.

57 The Andean feline constellation *choque chinchay*, "an animal painted all over with all colors," also associated with Venus, was probably invoked by jaguar priests at the Inca's gold mines at Carabaya, by jaguar miners at Oruro and Potosí, and no doubt in many other mines. Cf. Polo Ondegardo 1585 [1559?]:ch. 1.1, fol. 7r: "Others who live in the mountains adore another star called Chuqui chinchay, which they say is a Tiger [jaguar], which is in charge of the Tigers [jaguars], Bears [hukumaris] and Lions [pumas]."

58 I thank Jan Szeminski for suggesting to me that coats of arms might be read as honorific texts.

59 On the use of feathers as Inca signs of authority, see Bouysse-Cassagne 1998.

60 This amplifies the definition given by Gonzalez Holguin (1989 [1609]:343), who translates *Ttipque* as "alfiler o topo pequeño con que prenden la manta de encima."

61 *Karakara* is translated by Bertonio (1956 [1611]) as "la cresta" and *karakara jaque* as "la cumbre de las peñas" (literally, "people of the high peaks")—that is, the high mountaintops whence the morning star and sunrise can best be viewed.

62 For don Fernando's tyrannical government *within* his colonial repartimiento, see Platt et al. 2011 [2006]:756, app. to doc. 16.

63 Franciscans, particularly active followers of Joachim di Fiore, were already present in Macha between 1548 and 1553 (Platt et al. 2011 [2006]:771). For the postponement of the age of the Holy Spirit, see Platt 1993.

REFERENCES CITED

Acosta, José de

1987 [1590] *Historia natural y moral de las Indias.* Historia 16, Madrid.

Albornoz, Cristobal de

1967 [1584] Instrucción para descubrir todas las guacas del Piru y sus camayos y haziendas. Transcribed and edited by Pierre Duviols. *Journal de la Société des Américanistes* 56(1):7–40.

Alconini, Sonia

2004 The Southeastern Inka Frontier against the Chiriwanos: Structure and Dynamic of the Inka Imperial Borderlands. *Latin American Antiquity* 15(4):389–418.

Alonso Barba, Álvaro

1967 [1640] *Arte de los metales.* Edited by Armando Alba. Editorial "Potosí," Potosí.

Anonymous

1934 [1539] Relación del sitio del Cuzco y principios de las guerras civiles del Perú. In *Colección de libros y documentos referentes a la historia del Perú,* 2a ser., 10. Librería e Imprenta Gil, Lima.

Arze, Silvia, and Ximena Medinaceli

ca. 1991 *Imagenes y presagios: El escudo de los Ayavire, Mallkus de Charcas.* HISBOL, La Paz.

Assadourian, Carlos Sempat

1982 *El sistema de la economía colonial.* Instituto de Estudios Peruanos, Lima.

1989 Base técnica y relaciones de producción en la minería de Potosí. In *Ciencia, vida y espacio en Iberoamérica*, edited by José Luis Peset, vol. 2, pp. 185–205. Consejo Superior de Investigaciones Científicas, Madrid.

Aveni, Anthony F.

1990 *Empires of Time: Calendars, Clocks, and Culture.* Tauris, London.

Benino, Nicolás de

1966 [1573] Relación muy particular del cerro y minas de Potosí y de su calidad y labores, dirigida a Don Francisco de Toledo, virrey del Perú. In *Relaciones geográficas de Indias*, vol. 1, edited by Marcos Jiménez de la Espada, pp. 362–371. Atlas, Madrid.

Bertonio, Ludovico

1956 [1612] *Vocabulario de la lengua aymara.* Universidad Mayor de San Andrés, La Paz.

Betanzos, Juan de

1987 [ca. 1551] *Suma y narración de los incas.* Transcribed and edited by María del Carmen Martín Rubio. Atlas, Madrid.

Bouysse-Cassagne, Thérèse

1993 "El rayo que no cesa": D'un dieu pukina à un dieu inca. In *Religions des Andes et langues indigènes,* edited by Pierre Duviols, pp. 165–180. Université de Provence, Aix-en-Provence.

1997 Le palanquin de l'Inca: Petite enquête d'ethnohistoire à propos d'un objet absent. *Techniques et culture* 29:69–111.

1998 "Si votre plumage . . .": Signes d'identité, signes de pouvoir chez les incas. In *Des Indes Occidentales a l'Amérique Latine,* edited by Alain Musset and Thomas Calvo. ENS Éditions. École Normal Supérieure de Lyon, Lyon.

2004 El sol de adentro: Wakas y santos en las minas de Charcas y en cl lago Titicaca (siglos XV–XVII). *Boletín de arqueología PUCP* 8:59–97.

2005 Las minas del centro-sur andino, los cultos pre-hispánicos y los cultos cristianos. *Boletín del Institut Français d'Études Andines* 34(3):443–462.

2011 Apuntes para la historia de los Puquina-hablantes. *Boletín de arqueología PUCP* 14:283–307.

Burger, Richard L., Craig Morris, and Ramiro Matos Mendieta (editors)

2008 *Variations in the Expression of Inca Power.* Dumbarton Oaks Research Library and Collection, Washington, D.C.

Capoche, Luis de

1959 [1585] Relación general de la villa imperial de Potosí. In *Relaciones histórico-literarias de la América meridional*, pp. 70–198. Ediciones Atlas, Madrid.

Cerrón-Palomino, Rodolfo

2000 El origen centro-andino del aymara. *Boletín de arqueología PUCP* 4:131–142.

2010 Contactos y desplazamientos lingüísticos en los Andes centro-sureños: El Puquina, el Aimara y el Quechua. *Boletín de arqueología PUCP* 14:119–130.

2011 El legado onomástico puquina: A propósito de "Capac" y "Yupanqui." *Estudios atacameños* 41.

Cobo, Bernabé

1956 [1653] *Historia del Nuevo Mundo*. Biblioteca de Autores Españoles, Madrid.

Collingwood, Robin J.

1946 *The Idea of History*. Oxford University Press, Oxford.

Condarco, Carola, Edgar Huarachi Mamani, and Mile J. Vargas Rosquellas

2002 *Tras las huellas del tambo real de paria*. Fundación PIEB, La Paz.

Corrigan, Philip, and Derek Sayer

1985 *The Great Arch: English State Formation as Cultural Revolution*. Basil Blackwell, Oxford.

Cruz, Pablo

2009 Huacas olvidadas y cerros santos: Apuntes metodológicos sobre la cartografía sagrada en los Andes del sur de Bolivia. *Estudios atacameños* 38:55–74.

Del Río, María de las Mercedes

2005 *Etnicidad, territorialidad y colonialismo en los Andes: Tradición y cambio entre los soras de los siglos XVI y XVII*. Instituto Francés de Estudios Andinos (IFEA), Lima, and Instituto de Estudios Bolivianos (IEB), La Paz.

Duviols, Pierre

1978 Camaquen, upani: Un concept animiste des anciens peruviens. In *Amerikanistiche Studien I*, edited by Roswith Hartmann and Udo Oberem, pp. 132–144. Haus Völker und Kulturen, Anthropos-Institut, St. Augustin.

Elliott, Frances Helen

1943 *Hymenocallis Calathina, the Peruvian Daffodil*. Cornell University, Ithaca.

Espinoza Soriano, Waldemar

1981 El reino aymara de Quillaca-Asanaque: Siglos XV–XVI. *Revista del Museo Nacional* 45:175–274.

Estenssoro Fuchs, Juan Carlos

2010 Autoretrato del conquistador como vencido o la invención del Perú: La aparición del Inca y de sus atributos políticos en las representaciones plásticas, 1526–1548. *Colonial Latin American Historical Review* 19(1):151–205.

Foster, Robert Crichton

1958 *A Catalogue of the Ferns and Flowering Plants of Bolivia*. Gray Herbarium of Harvard University, Cambridge, Mass.

Garcilaso de la Vega, El Inca

1963 [1609] *Comentarios reales de los incas*. Ediciones Atlas, Madrid.

Gisbert, Teresa

1992 Los curacas del Collao y la conformación de la cultura mestiza andina. In *500 años de mestizaje en los Andes*, edited by Hiroyasu Tomoeda and Luis Millones, pp. 75–119. SENRI Ethnological Studies 33. Museo Nacional de Etnología, Osaka.

2011 La sacralidad del cerro de Potosí. *Historia y cultura* 35:11–18.

Gisbert, Teresa, and Carlos Rosso (editors)

2008 *La iglesia de Curahuara de Carangas*. Universidad Católica de Bolivia San Pablo, Museo Nacional de Etnografía y Folklore, Plural Editores, La Paz.

González Holguín, Diego

1952 [1608] *Vocabulario de la lengua general de todo el Perú llamada lengua quichua o del inca*. Instituto de Historia, Universidad Nacional Mayor de San Marcos, Lima.

1989 [1608] *Vocabulario de la lengua general de todo el Perú llamada quichua o del Inca*. Facsimile ed. Universidad Mayor de San Marcos, Lima.

Graña, Mario Julio

2001 La verdad asediada: Discursos de y para el poder; Escritura, institucionalización y élites indígenas surandinas, Charcas, siglo XVI. *Andes* 12:1–14.

Guaman Poma de Ayala, Felipe
1980 [1612] *Nueva corónica y buen gobierno.* Transcription and edition by John V. Murra, Rolena Adorno, and George Urioste. Siglo XXI, Mexico City.

Halbwachs, Maurice
1992 *On Collective Memory.* University of Chicago, Chicago.

Husserl, Edmund
1999 A Phenomenology of the Consciousness of Internal Time. In *The Essential Husserl: Basic Writings in Transcendental Phenomenology,* edited by Donn Welton, pp. 186–219. Indiana University Press, Bloomington.

Lechtman, Heather
2008 The Inka, and Andean Metallurgical Tradition. In *Variations in the Expression of Inka Power,* edited by Richard L. Burger, Craig Morris, and Ramiro Matos Mendieta, pp. 313–356. Dumbarton Oaks Research Library and Collection, Washington, D.C.

Lechtman, Heather, Pablo Cruz, Andrew Macfarlane, and Sidney Carter
2011 Procesamiento de metales durante el Horizonte Medio en el Altiplano Surandino (Escaramayu, Pulacayo, Potosí, Bolivia). *Boletín del Museo Chileno de Arte Precolombino* 15(2):9–27.

Linares Málaga, Eloy
1979 La estrella de ocho puntos en la arqueología del area meridional andina. In *Amerikanistische Studien/Estudios Americanistas II,* edited by Roswith Hartmann and Udo Oberem, pp. 16–23. Haus Völker und Kulturen, Anthropos-Institut, St. Augustin.

Macleod, Murdo
1998 Self-Promotion: The "Relaciones de méritos y servicios" and Their Historical and Political Interpretation. *Colonial Latin American Historical Review* 7(1):25–42.

Malpass, Michael A., and Sonia Alconini (editors)
2010 *Distant Provinces in the Inka Empire: Toward a Deeper Understanding of Inka Imperialism.* University of Iowa Press, Iowa City.

Martínez, María Elena
2014 Indigenous Genealogies: Lineage, History, and the Colonial Pact in Central Mexico and Peru. In *Indigenous Intellectuals: Knowledge, Power, and Colonial Culture in Mexico and the Andes,* edited by Gabriela Ramos and Yanna Yannakakis. Duke University Press, Durham and London, in press.

Medinaceli, Ximena, and Marcela Inch
2010 *Pleitos y riqueza: Los caciques andinos en Potosí del siglo XVII. Transcripción y estudios del expediente de don Diego Chambilla contra los bienes de su administrador.* Archivo y Biblioteca Nacionales de Bolivia, Fundación Cultural del Banco Central de Bolivia, and Instituto de Estudios Bolivianos de la Facultad de Humanidades, Sucre.

Mejía Xesspe, Toribio
1979 Cultura Pukina. In *Amerikanistische Studien/Estudios Americanistas II,* edited by Roswith Hartmann and Udo Oberem, pp. 40–49. Haus Völker und Kulturen, St. Augustin.

Mendoza, Fernando, and Felix Patzi
1997 *Atlas de los ayllus del norte de Potosí: Territorio de los antiguos Charka.* Comisión Europea, Potosí.

Pachacuti Yamqui Salcamaygua, Joan de Santacruz
1993 [1613?] *Relación de antigüedades deste reino del Pirú.* Estudio etnohistórico y etnolingüistico de Pierre Duviols y César Itier. Travaux de l'Institut Français d'Études Andines 74 and Centro "Bartolomé de las Casas," Cuzco.

Pärssinen, Martti
2002 Confederaciones interprovinciales y grandes señores interétnicos en el Tawantinsuyu. *Boletín de arqueología PUCP* 6:23–41.
2010 Paria, the South Inka Capital Rediscovered. *Chungará revista chilena de antropología* 42(1):235–246.

Pärssinen, Martti, and Antti Korpisaari (editors)
2003 *Western Amazonia–Amazonia Occidental: Multidisciplinary Studies on Ancient Expansionistic Movements, Fortifications and Sedentary Life.* Renvall Institute Publications 14. University of Helsinki, Helsinki.

Pizarro, Pedro

1986 [1571] *Relación del descubrimiento y conquista de los reinos del Perú.* Edited by Guillermo Lohmann Villena. Pontificia Universidad Católica del Perú (PUCP), Lima.

Platt, Tristan

1993 Simon Bolivar, the Sun of Justice and the Amerindian Virgin: Andean Conceptions of the Patria in Nineteenth-Century Bolivia. *Journal of Latin American Studies* 25(1):159–185.

2012 Between Routine and Rupture: The Archive as Field Event. In *The Sage Companion to Social Anthropology,* edited by Richard Fardon and John Gledhill, vol. 2, pp. 21–38. Sage, London.

2014 Power and Propitiation in the Andes. In *Inca Sacred Space: Landscape, Site, and Symbol in the Andes,* edited by Frank Meddens, Katie Willis, Colin McEwan, and Nicholas Branch, pp. 269–276. Archetype, London.

Platt, Tristan, Thérèse Bouysse-Cassagne, and Olivia Harris

2011 [2006] *Qaraqara-Charka: Mallku, Inka y rey en la provincia de Charcas (siglos XV–XVII). Historia antropológica de una confederación aymara.* Instituto Francés de Estudios Andinos (IFEA), Lima, and Plural editores, La Paz.

Platt, Tristan, and Pablo Quisbert

2008 Tras las huellas del silencio: Potosí, los incas y el virrey Toledo. In *Minería y metalurgia en los Andes del sur,* edited by Pablo Cruz and Jean Vacher, pp. 231–277. Instituto Francés de Estudios Andinos (IFEA), Lima, and Institut de Recherches pour le Développement (IRD), Sucre.

Polo Ondegardo, Juan

1585 [1559?] Los errores y supersticiones de los indios, sacadas del tratado y averiguación que hizo el Licenciado Polo, fols. 7r–16r of *Instrucción contra las ceremonias y ritos que usan los indios,* in *Confessionario para los curas de indios, . . .* Antonio Ricardo, Lima. Electronic document, https://archive.org/details/confessionariopaoocath, accessed March 27, 2014.

Rivera, Claudia

2013 Dinámicas regionales prehispánicas entre los siglos XIV–XVI: Las provincias de Pilaya y Paspaya (Cinti). In *Qaraqara-Charka: Reflexiones a tres años después,* edited by Ana María Presta. Plural editores, La Paz.

Rowe, John

1979 An Account of the Shrines of Ancient Cuzco. *Ñawpa pacha* 17:1–80.

1985 Probanza de los incas nietos de conquistadores. *Histórica* 9(2):193–245.

Saignes, Thierry

1987 De la borrachera al retrato: Los caciques andinos entre dos legitimidades (Charcas). *Revista andina* 5(1):139–170.

Sanct Angel, Rodrigo de La Fuente

1965 [1573] Relación del cerro de Potosí y su descubrimiento. In *Relaciones geográficas de Indias,* edited by Marcos Jiménez de Espada, vol. 1, pp. 357–361. Atlas, Madrid.

Sanhueza, Cecilia

2008 Territorios, prácticas rituales y demarcación del espacio en Tarapacá en el siglo XVI. *Boletín del Museo Chileno de Arte Precolombino* 13(2):57–75.

Sendon, Pablo

2010 Los límites de la humanidad: El mito de los Ch'ullpa en Marcapata (Quispicanchi), Perú. *Journal de la Société des Américanistes* 96(2):133–179.

Staller, John, and Brian Stross

2013 *Lightning in the Andes and Mesoamerica: Pre-Columbian, Colonial, and Contemporary Perspectives.* Oxford University Press, Oxford.

Taylor, Gerald

1974 *Camay, camac* et *camasca* dans le manuscrit quechua de Huarochirí. *Journal de la Société des Américanistes* 63:231–244.

Urton, Gary

1981 *At the Crossroads of the Earth and the Sky: An Andean Cosmology.* University of Texas Press, Austin.

Van Buren, Mary

2005 Huayrachinas and Tocochimbos: Traditional Smelting Technology of the Southern Andes. *Latin American Antiquity* 16(1):3–25.

Van Buren, Mary, and Ana María Presta

2010 The Organization of Inka Silver Production in Porco, Bolivia. In *Distant Provinces in the Inka Empire*, edited by Michael A. Malpass and Sonia Alconini, pp. 173–192. University of Iowa Press, Iowa City.

Wingfield, Chris

2005 Historical Time versus the Imagination of Antiquity: Critical Perspectives from the Kalahari. In *The Qualities of Time: Anthropological Approaches*, edited by Wendy James and David Mills, pp. 119–135. Berg, Oxford.

Zagalsky, Paula

2012 Nuevas preguntas sobre una antigua federación aymara: Algunos aportes en torno a la Federación Qharaqhara (Charcas, siglo XVI). *Prohal monográfico* 2(2), segunda sección: *Surandino monográfico.*

Ziólkowski, Mariusz

1984 La piedra del cielo: Algunos aspectos de la educación e iniciación religiosa de los príncipes incas. *Ethnologia polona* IX:219–234.

1997 *La guerra de los wawqis: Los objetivos y los mecanismos de la rivalidad dentro de la élite inca, siglos XV–XVI.* Abya-Yala, Quito.

Zuidema, R. Tom

1974–1976 La imagen del sol y la huaca de Susurpuquio en el sistema astronómico de los incas en el Cuzco. *Journal de la Société des Américanistes* 63:199–230.

2011 Chuquibamba Textiles and Their Interacting Systems of Notation: The Case of Multiple Exact Calendars. In *Their Way of Writing: Scripts, Signs, and Pictographies in Pre-Columbian America*, edited by Elizabeth Hill Boone and Gary Urton, pp. 251–275. Dumbarton Oaks Research Library and Collection, Washington, D.C.

Zuidema, R. Tom, and Gary Urton

1976 La constelación de la llama en los Andes peruanos. *Allpanchis phuturinqa* 9(9):59–119.

13

Of Calendars and Computers

Comparing Mesoamerica and Bali

JOHN MONAGHAN

In his definition of Mesoamerica, Paul Kirchoff identifies eight broad classes of traits that set its civilization apart from other Native American civilizations. One of these concerns the marking of time: "The year of 18 months of 20 days, plus 5 additional days; [the] combination of 20 signs and 13 numerals to form a period of 260 days; [the] combination of the two previous periods to form a cycle of 52 years; festivals at the end of certain periods; good and bad omen days; persons named according to the day of their birth" (Kirchoff 1952 [1943]:26). In the Americas, he concludes, this kind of calendric system is typically and exclusively Mesoamerican. Some of the earliest decipherable texts inscribed by Mesoamerican people were dates, and the 260-day calendar continues to be kept by native people despite official suspicion and the homogenizing forces of globalization.

When the Spanish friars were compiling their dictionaries for the languages of Mesoamerica, one of the words they usually asked for was the indigenous term for calendar. The answer they received was usually something like the "count of the days" or "paper on which is written the sequence of days." At first glance these definitions are not too far off from the Spaniards' own definition of *calendar.* Sebastián de Cobarruvias (1977 [1611]:269), in his *Tesoro de la lengua casellana o española*, defined a calendar as "tables for the observations of the days of the month." But calling these instruments calendars also leaves a lot unsaid, since, as Kirchoff suggests, they have broad cultural significance and multiple uses and meanings. For example, the word *day* could also mean "name," since one's name was made up in part by the day of the year on which one was born. It could also mean "fate" since everything from one's personality to one's social roles would be heavily influenced by the day on which one was born. One Mixtec man defined *day* as "our limit" because the day of one's birth also established the day of one's ending. Beings that come into the world at the same time are linked by the day of their birth. Hence, the co-essence phenomenon, by which people acquire parts of their selves that are not physically connected to their bodies.

Elsewhere, I (Monaghan 2002) have suggested that in the light of these features, one could conceive of the Mesoamerican 260-day calendar as forming a kind of charter or constitution. It is generally true that a shared temporal order is the basis of group formation and sociality (Hallowell 1955; Schutz 1951). It defines group boundaries, and control of the calendar is tied to power and authority (Bourdieu 1977; Gell 1992; Hassig 2001:71). Failure to adhere to temporal norms is a sign of mental illness on an individual level and on a group level can constitute a form of resistance to a central power. In Mesoamerica, we see that the calendar, in particular the 260-day calendar, goes further by formalizing two fundamental values that we associate with a charter or constitution. On the one hand, the Mesoamerican calendars define the nature of persons who compose the group. They do this in a clearly relational as opposed to individual way. On the other hand, the calendars define the principles that organize society. The Mesoamerican ritual calendar does so by putting forward the various social roles that make up a community. It is true that there were procedures to allow one to avoid undesirable roles, but nonetheless the 260-day calendar lays out the panorama of social types and establishes a normative distribution of power since the roles it defines are often ordered hierarchically. The calendar, thus, defines the basic building blocks of a comprehensive social and political ideology (Monaghan 2002).

This idea that the Mesoamerican ritual calendar is a kind of charter or constitution is very much in keeping with a focus on Mesoamerican time reckoning, which has largely been on what we might call "high-end" uses—making amazingly complex temporal calculations, calling forth important rituals, curing, divining, observing astronomical phenomena, prophesying, and the like. A particular focus has been on how these activities and their temporal dimensions constitute a distinctive Mesoamerican worldview, which Miguel León Portilla (1988 [1968]:112) has famously labeled for the Maya a "chronovision."

Yet even something as esoteric as divination has a practical side. Alfred Gell (1992) has pointed out that one can look at divination as something that helps one to assess opportunity costs, or the benefits one must forgo in order to make something or perform some service. In modern economics, the idea of opportunity cost expresses the relationship between scarcity and choice and is one of the keys to analyzing market behavior. The opportunity costs of an activity are an obstacle to action, and these must be considered, evaluated, and resolved before one can move forward. People sometimes get stuck at this point since it is hard to know what the best course is. I saw firsthand the role divination can play in helping people assess opportunity costs in the 1980s, when I was working in the Maya community of Sacapulas, Guatemala. One of my neighbors was a blood diviner. What he would do was roll up his pant leg and ask a question. If a vein on his calf throbbed, the answer was affirmative. One day a couple came who were pondering a trip to El Quiche to sell the celebrated black salt Sacapultecos made in the beds along the Río Chixoy. They asked if it would be a productive trip. The diviner put the question to his leg, and the vein failed to throb. The next day, as I was walking on a path along the river, I saw the couple working in their salt bed. After witnessing the divination and negative response in my neighbor's house the day before, I thought that the couple would have just stayed at home, perhaps with the door shut and the lights out. But after seeing them at work I understood that what they were trying to determine was not just whether they should go to El Quiche to sell salt, but whether it was better to sell salt or make salt. In other words, they were trying to figure out which activity had the greater opportunity cost, something that for them was difficult to evaluate beforehand; thus, they decided that divination might give them an answer. Since it turned out to be an inauspicious day to travel, this information substantially lowered the opportunity cost of staying home and making salt. Over and above this interpretation, divination allowed them to avoid what is perhaps the worst outcome: that of being stymied and not making any choice at all.

Setting to one side for the moment the practical aspects of divination, I would like to call attention to one of the ways the Mesoamerican calendar has been represented. Inspired by a deep Western tradition that represents time as a wheel (Aveni 2012:4–7, 11–16), scholars have commonly depicted the Mesoamerican solar and ritual calendars as a set of intermeshing gears. In anthropology, this tradition appears to begin with Sylvanus Morley (1956 [1946]:234), when he produced a drawing of the relationship between the thirteen numbers of the ritual calendar, the twenty named days, and the 365-day solar calendar as two cogs (Figure 13.1). J. Eric S. Thompson (1966 [1954]:ix) followed Morley but put the information on three cogwheels instead of two (the actual drawing in Thompson's book appears to have been executed by Avis Tulloch, staff artist of the Department of Archaeology of the Carnegie Institution) (Figure 13.2). Although Thompson (1966 [1954]:175) did not want people to think this model was native ("A Maya would not approve of this illustration, for to him it is not a matter of a complex machine, but of series of gods who take it in turn to rule the world"), it has proven to be very useful and has been reproduced in increasingly vivid, and one might say "industrial," versions, although the basic principle remains unchanged from Morley's and Thompson's original illustrations (e.g., Culbert 1993; Sabloff 1990:35).

What makes the gear imagery so effective is that it clearly represents time as being made up of a number of cycles. It also aptly illustrates the way these cycles intermesh in a regular and predictable way, so that in combination they produce other kinds of cycles. Aveni (2002:284) has called our attention to the fact that this kind of phenomenon is generally associated with state-level societies. What I will do here, in relation to my discussion of the uses of divination to assess opportunity costs, is focus on the practical advantage that might be

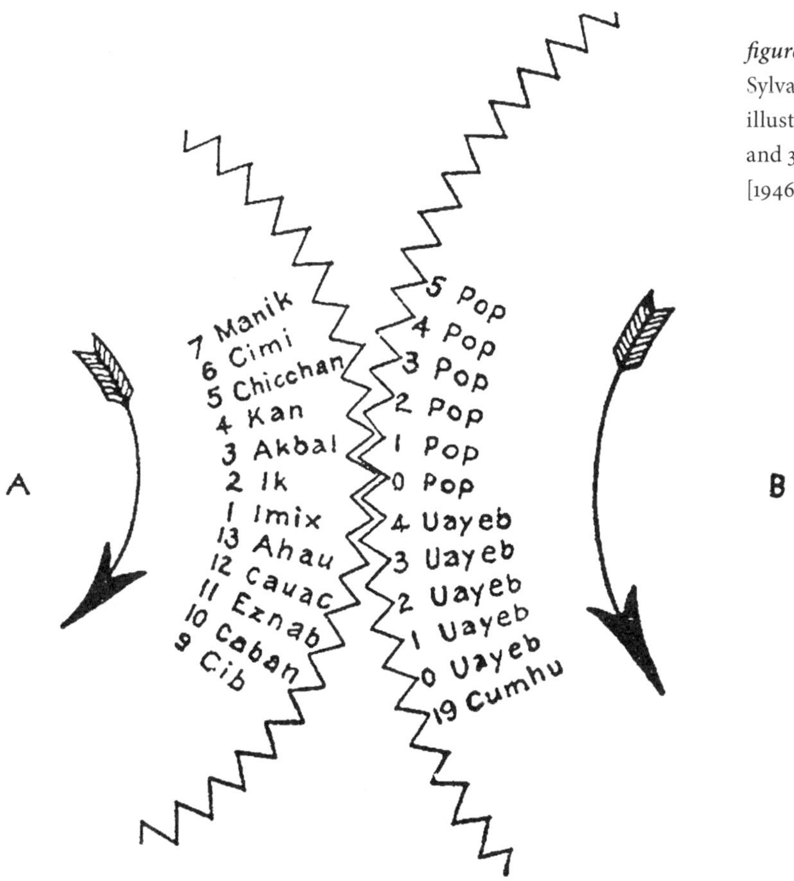

figure 13.1
Sylvanus Griswold Morley's cogwheel illustration of the Maya 260-day and 365-day calendars (Morley 1956 [1946]:230).

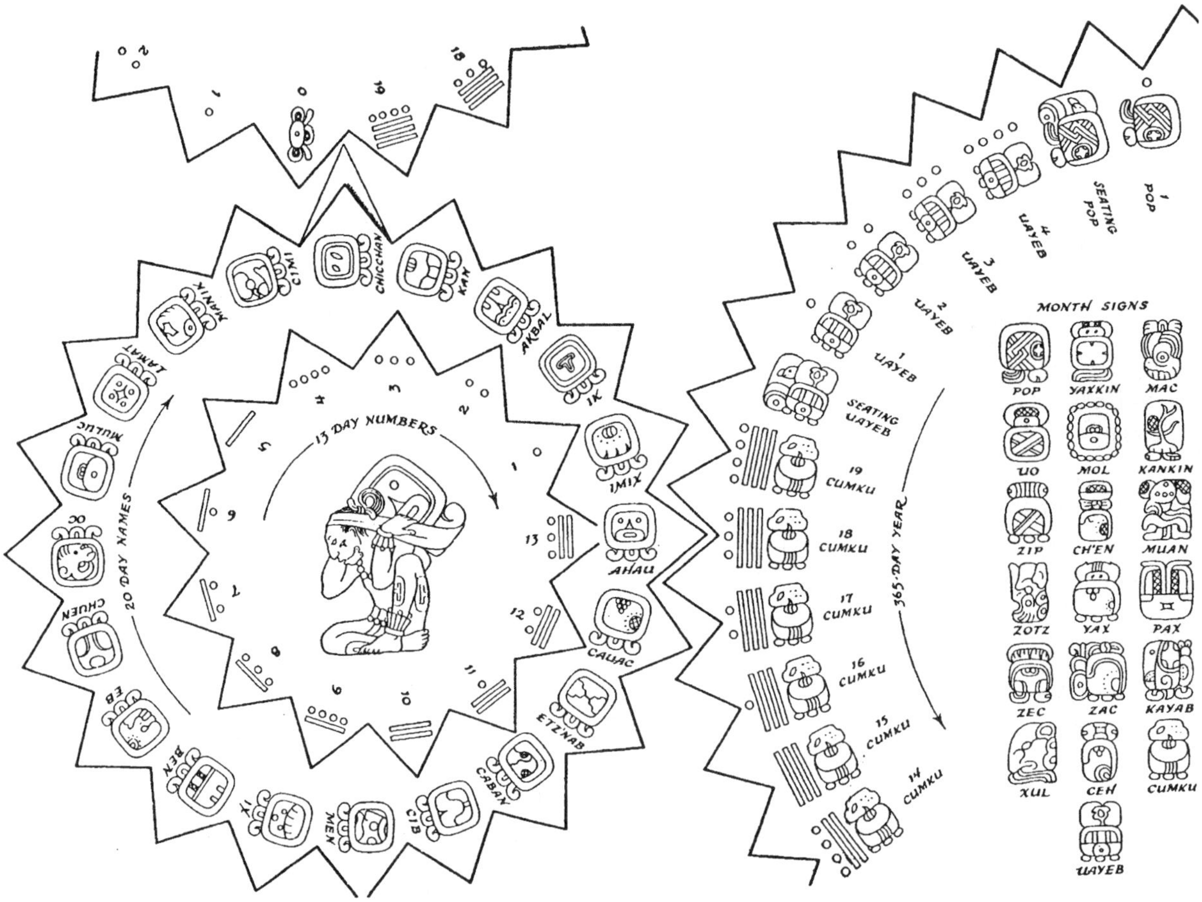

figure 13.2
The three-cogwheel illustration of the Maya calendars (Thompson 1966 [1954]:175).

derived from having a system that highlights a multiplicity of concurrent cycles. I now turn to Southeast Asia, whose peoples have developed many extremely complex calendric systems, perhaps the best known being the one found on the island of Bali.

The Balinese Calendars

Miguel de Covarrubias produced two great travelogues in his lifetime. While Latin Americanists know him best for his monumental *Mexico South*, published in 1946, it was largely prefigured by an earlier work, *Island of Bali*, published in 1937. Covarrubias journeyed through the islands of Southeast Asia with his wife, Rose, in 1930 on their honeymoon and was so taken with Bali that he returned in 1933 on a Carnegie fellowship to carry out ethnographic investigations. Like *Mexico South*, *Island of Bali* is filled with his vivid drawings; he illustrates everything from domestic animals and agricultural implements to dance positions and Balinese gods, as well as providing the photographs he and Rose took.

The 210-day Balinese ritual calendar, or Pawukon, was of particular interest to Covarrubias (1986 [1937]:314). It is composed of ten different kinds of weekly cycles: a week of one day, a week of two days, a week of three days, and so forth all the way up to a week of ten days. Each day of each week has a unique name. The day of the one-day week is Luang. The days of the two-day week are Menga and Pept (Covarrubias 1986 [1937]:314–315). The seven-day week is special because not only are its seven days named, but each of its thirty weeks is named, too (Covarrubias 1986 [1937]:314). The names of the weeks recall the defeat of a powerful ruler, named Watu Gunung, "Stone Mountain." After fathering twenty-seven offspring with his aunt and his mother, Stone Mountain sought the hand of the goddess Sri in marriage. When the gods refused him, he declared war. The gods defeated him and executed each of Stone Mountain's children at seven-day intervals. Then the gods named the thirty seven-day weeks after each of the slain children, as well as Stone Mountain, his mother, and his aunt, to commemorate their victory (Covarrubias 1986 [1937]:314–315).

Theoretically, a day can have ten different names, as well as a name derived from the solar-lunar calendar (Covarrubias 1986 [1937]:283). But the days of the one-, two-, and ten-day weeks do not always occur in series. Complicated calculations are made involving five-, seven-, and ten-day weeks to determine the exact nature of a particular day. As a result of these calculations, it sometimes happens that the one-day week does not occur at all. Furthermore, the 210-day year is not divisible by 4, 8, or 9. The dilemma is solved by repeating certain days in each of these weeks to get to 210. As one can see, the sequence of days of the Pawukon are not fixed by any astronomical or other cycles (although it is said that originally the 210-day round was related to the growing period of rice), and one must have a great deal of training and computational skill to operate the calendar at a high level.

Physical representations of the Pawukon are called *tika* or *tuku* (Figure 13.3). In the literature, *tika* is sometimes exclusively used for the calendar, although properly it refers only to the charts composed on paper or palm leaves or carved onto wood boards by calendar specialists (in other Southeast Asian traditions, such as that of the Batak, calendars and divinatory manuals are sometimes composed on screenfolds, just like Mesoamerican codices). The traditional tika does not have writing on it but has symbols that stand for the various names of the day. The basis of the tika is the special seven-day week. The vertical axis consists of seven rows, each corresponding to one of the days of the seven-day week, with the days listed at the left, Redite at the top and Saneschara at the bottom. Across the horizontal axis are thirty vertical columns, each representing one of thirty seven-day weeks with the week names beginning on the left with Sinta (the name of Stone Mountain's mother) and ending at the far right with Watu Gunung, "Stone Mountain." Each cell contains the symbols of significant days from the other weeks, which in turn contain the information needed by a knowledgeable user to determine the days of other cycles that are not listed and symbols for the important conjunctions of different weeks (Lansing 1991:69).

The principal uses of the Pawukon are to determine lucky and unlucky days, when markets are scheduled, personal anniversaries, what kind of offerings to make, and the days of religious ceremonies. So, for example, whenever a certain two days of the three- and five-day weeks correspond, temple ceremonies must be performed. Temple priests are specialists in the use of the calendar (Covarrubias 1986 [1937]:284). Parents, for example, might consult them to find out what day of the eight-day week a child was born on, since this information provides a clue to his or her identity in a previous life. If the child was born on Sri, the first day in the eight-day week, he or she is probably a reincarnation of a woman from the mother's side. If the child was born on Indra, the second day of the week, he or she is probably a reincarnation of a man from the father's side. Sometimes a child is born owing debts to beings in the spirit world. Until these are paid, the child will suffer. A priest will deal with this situation by determining the child's birthday in both the five- and seven-day weeks. Each one of thirty-five possible combinations is related to a specific debt-offering. Today, the Balinese also make use of Western-style calendars, but some are published with each day of the week accompanied by the corresponding days of the Pawukon. Clifford Geertz (1966) observed that the Balinese calendar is not so much focused on the sequence of days, but what kind of day each is. Another student of the Balinese calendar, J. Stephen Lansing (1997:8), says the Pawukon

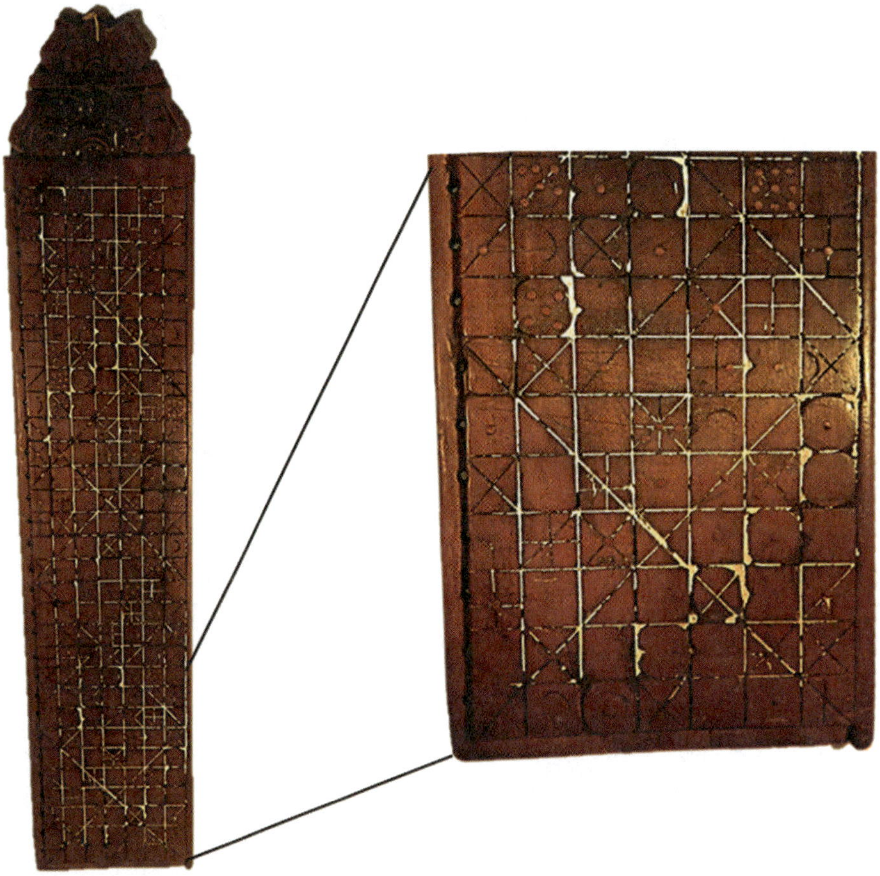

figure 13.3
Wooden tika based on the seven-day week of the Pawukon. (Photograph by John Monaghan, from the collection of J. Stephen Lansing.)

"suggests that time is dense." By this, I understand him to mean that the user of the Pawukon becomes aware, to a degree not possible in its absence, of the multiplicity of cycles of activities that concur at any instant and the complex social interdependencies that undergird them.

Lansing's insight into the way the Pawukon conveys a sense of density to time is in part due to the way he conducted his fieldwork on Bali. In contrast to most of the traditional ethnographic work on the island, which took as its unit of analysis the local community, Lansing conducted research along an entire watershed. This methodology allowed him to observe uses of the calendar that are not apparent at the village level. Lansing studied Balinese wet-rice farmers on an island that is made up of a cluster of volcanoes running roughly east to west and forming a central spine. The highest of the volcanoes are at the eastern end, and their craters contain large lakes of freshwater. The Balinese channel this water, through tunnels carved from living rock, canals, pipes, and dams, to their fields on the slopes of the volcanoes. These fields are terraced and diked and look like they have been sculpted out of the red soil (Figure 13.4). This landscape makes the island seem as if it is covered by gigantic staircases descending down the mountainsides, twisting through ravines, and then spreading out over the plains nearer the coast (Covarrubias 1986 [1937]:71). When they are all filled with water, Covarrubias writes, they look "like mosaics of mirrors that reflect the clouds." After they have been planted with seedlings from the nursery, they appear "chartreuse on a ground of brown ooze . . . [thickening eventually] into a

figure 13.4
Balinese rice terraces. (Photograph courtesy of Peter Just.)

tender yellow-green carpet which turns to a rich gold ochre as the grain ripens" (Covarrubias 1986 [1937]:71–72). Finally, after the harvest has been gathered and the water drained, they are only dry cracked mud.

Wet-rice farming is incredibly productive—I have seen figures of six tons per hectare—and Bali was considered one of the granaries of its part of the world. This kind of agriculture requires a great deal of water and intensive labor, but farmers traditionally produced enough rice for the island to be a wealthy exporter of the grain. One of the strategies for maintaining a high level of productivity is to sow the rice crop in a given area all at the same time, so that all the fields contain rice at the same stage of maturity. While farmers do not have to worry too much about weeds, since the crop is underwater for much of the time, they do have to worry about a variety of pests that infest the rice at particular stages in its growth. By keeping all the fields on the same schedule they reduce the window the pests have for feeding on the rice plants. But if the fields in an area have been planted at different times and the crops are at different stages, the pests have the opportunity to increase in number as they migrate from one field to the next and severely reduce the harvest (Lansing 1991:112–125).

The fundamental social unit responsible for rice cultivation is the *subak*, which Covarrubias (1986 [1937]:72, 412) defines variously as an "agricultural co-operative society," "village rice-growers association," and "water board." As these descriptions suggest, the subak oversees matters relating to irrigation and agriculture. Disputes are negotiated within and between subaks, and the subak arranges for the distribution of water to its members, oversees the repair of local dams and canals, and hosts rice festivals. The subak leaders are elected, and they preside over meetings of farmers that are held at least once a month at a small temple in the middle of the subak rice fields. These temples are located at the junction of the major canal that feeds the fields of the subak. At the meetings, the leaders discuss finances and negotiations with other subaks and make future plans (Covarrubias 1986 [1937]:72–73; Lansing 1991:40–49, 77–82).

In addition to the small subak temples, as one moves up the drainage one finds increasingly important temples (called Ulun Ski and Maceti) at each of the major canal junctures, until one reaches the Temple of the Crater Lake at the center of the island. This temple is the most important (a temple's importance is signaled by the number of roofs it has—the Temple of the Crater Lake has eleven roofs), and Crater Lake is felt to be the source of the approximately eighty rivers that flow down to the sea in Bali. A goddess resides in the Crater Lake, and the Temple of the Crater Lake is the residence of the Jero Gde, who is believed to be the human incarnation of the goddess. Traditionally, before planting, subaks would send a delegation, along with the local priest, to the main water sources in Bali to bring back a small container of water that was placed on the subak temple and was believed to embody the gods. The fields were sprinkled with the holy water, and the rest was poured into the main canal of the subak (Covarrubias 1986 [1937]:74; Lansing 1991:59–67, 73–94).

Those familiar with irrigation systems know that people in a watershed have to work out a system for allocating water, what might be called the "coordination problem." If people upstream take the water whenever they like, then those downstream will not be able to organize their irrigation activities very effectively. If the people upstream use all the water, then those downstream will be unable to plant irrigated crops. So people downstream are dependent on arrangements with people upstream, which is particularly true in Bali, where tens of thousands of farmers depend on the water flowing out of just a few sources.

While all the farmers in the same subak or in adjacent subaks should be on the same cycle so the pest population will be held down, the cycle does not have to be the same everywhere. In one region, the subaks can be planting their rice, while in another they are harvesting. It is important that the agricultural cycles and crops of the different regions on a watershed not match because otherwise there would come a time when everyone would need water at the same time and there would not be enough to go around.

According to Lansing, the key to the solution of the coordination problem in Bali is the temple priests. Temple priests along the watershed are in charge of overseeing a number of agricultural activities. They resolve disputes and decide whether and how much new cropland can be opened up. They also decide when the different subaks will get water, and, in fact, they plan the distribution of water across the entire watershed. In order to calculate and allocate these distributions, they employ the Pawukon calendar. Along the watershed studied by Lansing were 172 subaks with their local temples, twenty-eight lower-level or Ulun Ski temples (which might be only used a few days out of the year and do not have resident priests), and fourteen of the larger Maceti temples. The important managerial role played by the temple priests was missed by earlier fieldworkers, who focused on social institutions within localities and not between localities, and also because of the bias that spiritual priests and a "ritual" calendar like the Pawukon would not be involved in practical matters.

As we have seen, an important feature of the Pawukon is that it defines time as composed of a number of interlocking cycles of one to ten days. What the priests do is match cyclical agricultural events, such as periods of crop growth and maturity, to cycles in the Pawukon. In this way, they can plan the allocation of water. As Lansing describes it:

> The principal practical use of the tika . . . is to synchronize concurrent production cycles, which may be of different lengths. Consider the

> complexity of timing water use and planting cycles for a productive system like the fifteen subaks of the Masceti temple Pamos Apuh. A particular block of terraces may be flooded on a certain date for a specified number of weeks depending on the crop. Later, the flow may be reduced when the ground beneath the plow pan becomes saturated. Perhaps another block of terraces is scheduled to grow vegetables for a 105-day cycle and should receive irrigation water every third day of the five-day week during this 105-day interval. By using the tika, multiple concurrent cycles can be specified with ease and precision, with the assurance that they will mesh neatly together, synchronizing the labors of thousands of farmers. . . . In the water temples, the uses of the tika extend beyond the synchronization of productive schedules, to structure the hierarchy of productive relationships. The personal growing cycles of individual farmers are aggregated into cycles of the subak, weir [diversionary dam], Ulun Swi and Masceti temples, and ultimately into the annual cycle of the Temple of the Crater Lake. . . . This wheels-within-wheels view of time assigns different ranks to different water temples, based on their role in the productive system. The highest rank belongs to major water temples, which control productive cycles for whole sections of rivers and blocks of terraces. Lesser temples control smaller cycles, involving smaller congregations of farmers. (Figure 13.4; Lansing 1991:70)

From the point of view of individual farmers, all they need to know is what day or days they can expect to receive water so they can plan on preparing the canals, shutoff points, and fields. But the priests have to see the bigger picture, since their task is to ensure that all the farmers and subaks are taken care of.

The computational power of the Pawukon makes the bigger picture possible because it allows priests to model regional, subregional, and local cycles of water use and agricultural activity. It is interesting that the watersheds managed by a series of temples crosscut the old kingdom boundaries. The idea was that this activity was too important and holy to be subject to the political whims of rulers.

Mesoamerica and Bali

To briefly summarize the main points, the Balinese maintain a ritual calendar of 210 days as well as a traditional lunar-solar calendar and the solar calendar introduced by the Dutch. The ritual calendar is not tied to either the lunar-solar calendar or to any other larger astronomical cycles. But this lack of correspondence is one of the things that makes the Pawukon of practical use. Because the Pawukon can define many cycles of varying lengths, and because it can track them simultaneously, it is used to model a variety of social and natural cycles, specifying with precision when these cycles intersect. Applied to wet-rice cultivation, it becomes a kind of technology that is crucial to Balinese environmental adaptations. In essence, the Pawukon, when employed in this manner, is an analog computer.

What insights might be gained about the Mesoamerican calendar from the Balinese example? As is well known, the ancient people of Mesoamerica maintained a solar calendar along with a 260-day "ritual" calendar. The ritual calendar, like the Pawukon, ran alongside the solar calendar and the two intersected, such as at the end of fifty-two years, when their end dates coincided, but the ritual calendar was not in itself determined by the shifting of the sun along the horizon or any other kind of natural cycle (although some feel it may have been originally related to the human period of gestation). Both the Mesoamerican ritual calendar and the solar calendar define various subcycles, including cycles of 365 days, 260 days, 100 days, 80 days, 65 days, 20 days, 13 days, 12 days, 7 days, 6 days, 5 days, 4 days, and 1 day.

It is well known that cyclical activities were regulated by the calendars. So Tenochtitlan held markets on the year-bearer days, while Tlateloco held markets on the third, eighth, thirteenth, and eighteenth days of the month, using the standard five-day week and twenty-day month of the solar year. Ross Hassig (2001) points out that scheduling markets based on fundamental units of time creates efficiencies that would not otherwise exist and is essential to the coordination necessary for a periodic market system to flourish. We also know of thirteen-day markets, twenty-day markets (which may have been higher- or lower-order markets than those based on the five-day week), and markets held on the occasions of specific feasts.

Early Spanish observers made some general comments on the way calendars were used by Mesoamerican people to regulate and coordinate activities. In his overview of the Aztec calendars, for example, Diego Durán (1971:319, 387) writes, "I shall now continue and shall explain the calendar according to whose system the natives patterned their lives, governed themselves, and counted the course of time. . . . Regarding all of this there was careful computation, so that things had to take place on such and such a day or season and so that everything might be in its right time," while Diego de Landa (1975:106) states that the Maya of the Yucatan "composed a sort of calendar by which they arranged not only their feast days but also their numerical system, contracts and affairs, as we do." Just as in Bali there was a priesthood in Mesoamerica that cultivated an expertise in calendrical calculations. Moreover, as in Bali, where the feasts scheduled by temple priests marked the arrival of water at important irrigation junctions, so, too, in Mesoamerica the priests scheduled feasts that were celebrated in tandem with important economic activities, such as markets, the arrival of labor drafts, and the payment of taxes (Hassig 2001:122–123).

Were there activities in Mesoamerica that required complex scheduling and coordination of the order of irrigation in Bali? We know of some complex irrigation systems in Mesoamerica, so the Mesoamerican calendars could have conceivably been used in the way the Balinese priests use the Pawukon to model the distribution of water along watersheds. Unfortunately, we do not have the accounts of accomplished ethnographers like Miguel de Covarrubias, Clifford Geertz, or J. Stephen Lansing for sixteenth-century Mexico and Guatemala. It is true that Spaniards like Diego Durán and Alonso de Zorita were brilliant observers of New World cultures and societies, and Spanish observers did document how the calendar worked. In explaining the Mesoamerican calendars, however, they transformed them into something Europeans could understand, the full extent of which is only recently becoming known (e.g., Aveni 2012; Hassig 2001). Moreover, the Spanish were looking at the ritual calendar in a way that was sure to downplay its practical or computational uses. Recall that in the sixteenth century astrology was a form of legitimate knowledge. The influence of the planets on human life was considered an established fact, and some kinds of divination could be used to predict the future. Those familiar with the sources will know that when they described the 260-day ritual calendar, Spanish descriptions focused heavily on its divinatory uses. So the best that can be done is to present some indirect, but suggestive, evidence.

Tribute

A general activity in Mesoamerica that generated complex scheduling problems was tribute collection. While in Bali the coordination problem centered on how to get enough of a single scarce substance to numerous users, in Mesoamerica the coordination problem with regard to tribute was how to manage the contributions of numerous producers who produced different things so that the right amount of each product would end up where and when it was needed.

It is obvious that the collection of tribute was a major concern of political leaders and not surprising that elites were highly involved in the management of tribute collection, as illustrated by Zorita's observation that

> The ruler, taking into account the number and bent of the people in each town, indicated the amount of land it must cultivate and what should be grown on this land. As I have said, the most common form of tribute was produce grown

locally. To the merchants and artisans they also indicated what they should give taking into account the production of each province or the things they did in it. They did the same with the people who had to give domestic service and those who had to provide the needs of war. (Zorita 1963 [ca. 1569]:193)

He then goes on to say, "The lords looked over everything and saw that the commoners paid their tribute, that they cultivated their fields and worked at their crafts" (Zorita 1963 [ca. 1569]:193).

Scholars have observed that tribute payments were set on cycles marked by the solar year. But we also know that tribute could be scheduled using other kinds of cycles. S. L. Cline (1993:81) observes: "In Quahchichinollan, generally speaking, tribute was delivered quarterly, although there were various cycles of delivery specified in numbers of days. The most common interval was 80 days." Tribute set at eighty-day intervals would not, of course, be on a 360- or 365-day cycle.

Given that "there were various cycles of delivery," the question then becomes, were managers interested in coordinating cycles of tribute payment? The Aztecs, probably not too long before the Spanish arrived, broke with earlier patterns of staggered tribute payments by imposing on all their subjects the same tribute-paying schedule of four times a year. These events coincided with special festivals, and the spectacle of thousands of tribute payers arriving in Tenochtitlan and carrying their payments was meant "to emphasize the hegemonic dominance of the empire" (Hassig 2001:82–83). But Zorita (1963 [ca. 1569]:193), who reported on groups outside central Mexico, observed: "Accordingly, not all men paid tribute [every] 20 or 80 days, for their turns came at different times. Thus there were tribute payers all year round, and the houses of the lord lacked for nothing. It was the same with tribute of fruits, fish, game, pottery, and other things for the table and service of the lords; each tribute payer gave his tribute in two of three or more payments, as was convenient."

It is clear from this statement not only that it was the norm to place people on staggered cycles of tribute payment, but that the end result—that "the houses of the lords lacked for nothing"—could only be achieved through close attention to the coordination of the payments. Moreover, Zorita specifically mentions a number of goods—fruit, fish, game, and labor—that are by nature a lot like water for the Balinese. What I mean is that they are not easy to store, that having too much does no one any good, and that not having enough might create problems. All of which suggests that a fine-tuned system for tribute payments was in place and was designed to meet, but not to exceed or fall short of, consumption demands. Finally, Zorita and most other sources who mention tribute payments tie them to calendrical intervals, suggesting that the calendar was essential to the way Mesoamerican people solved the coordination problem.

In the beginning of this essay, I contrasted the high-end uses of the calendar with more mundane, practical ones. But the two were probably so intertwined that any separation would be artificial. Indeed, the complex divinatory calculations that we know Mesoamerican people made and continue to make would have been good training for the calculations required for the coordination of tribute payments or other kinds of production cycles. Barbara Tedlock (1983) gives the example of a set of offerings made in Momostenango by priests and novices who are training to be new calendar priests. The cycle begins with a sixty-five-day period called "washing for the work service" at the end of which the novices will be able to make offerings independently at important shrines. In the sixty-five-day period, shrines are visited and offerings made at a thirteen-day interval, then on a seven-day interval, followed by a six-day interval, and continuing in this sequence until the sum of the intervals equals sixty-five. Anthony Aveni (2011:191) finds this kind of intervallic sequencing in Maya almanacs, and he describes the pattern of cyclical calculations focused on the use of commensurate numbers, which means that long cyclic intervals would be seen as composed of multiples of smaller ones or in the Momostecan example, $65 = 13 + 7 + 6 + 13 + 7 + 6 + 13$. Balinese temple priests also employ commensurate numbers when they use the Pawukon to synchronize different productive cycles within larger cycles of water usage. Interestingly, the Momostecan word for work service in the sixty-five-day "washing for the work service" is *patan*, the old word for tribute (Tedlock 1983:60, 62).

One of the things that the Balinese solution to their coordination problem allows them to do is avoid the time and expense of constructing large reservoirs. The subaks do not need to do this work because they can be confident that they will receive the water they need at the right time, thanks to the calculations of the temple priests. More generally, this means that sophisticated scheduling—achieved with the aid of a temporal model of the production process—can eliminate the need to stockpile large amounts of materials, reducing the associated costs of storage. A curious thing about Mesoamerica is the paucity of large storage facilities. It is a fact that the kinds of warehouses found in the Andes have not been discovered in the archaeological record of the Mesoamerican region. This is not to say that Mesoamerican people did not have storage facilities, since they are mentioned in ethnohistorical sources. But if planners, with the use of the calendar, were able to model cycles of tribute so that they could ensure an even supply or a supply at times when it was most needed, as Zorita says, then the need for an extensive system of warehouses would be obviated. This idea is also behind the Japanese Kanban or "Just in Time" model of inventory management, which begins with the premise that inventory, which had traditionally been seen as adding or storing value, is simply a waste and that stockpiles instead represent inefficiencies in the productive process. So it may be that extensive centralized storage facilities in Mesoamerica were not necessary due to a similar level of sophisticated management of tribute collection. The one place where we do have solid evidence for the existence of warehouses is in Tenochtitlan, where, as I have mentioned, traditional patterns of staggered tribute payments were replaced by a system that created a quarterly flood of tribute brought by the masses of subordinated people arriving to pay homage to the Aztecs.

It is important to keep in mind that even though descriptions of tribute payment sometimes leave the impression that once set up, payments would have functioned automatically with little need for change, we have descriptions that indicate planners needed to deal with contingencies on a regular basis. For example, due to natural calamities or other disasters groups could not always pay and might even be in need of tribute paid by others (e.g., Zorita 1963 [ca. 1569]:194). Likewise, tribute such as labor service might be reassigned to different projects as needed. Military operations would also have required planners to reconfigure both the kinds of goods collected through tribute and their places of delivery.

Conclusion

The simple idea of this essay is that, like the Pawukon, the Mesoamerican calendars have features that allow them to function as analog computers. Specifically, the intermeshing of cycles defined by the Mesoamerican calendars would have given planners a sophisticated tool to both cope with contingencies in a flexible and efficient way as well as to model and plan a complex coordination that ensured supply met consumption demand. The Balinese temple priests traditionally recalculated water deliveries on a fairly regular basis, as crop patterns changed, new lands were brought into cultivation, and other contingencies arose. But was there an instrument like the tika that would have aided Mesoamerican planners in making their calculations? It is worth pointing out that if we did not know anything about the Balinese time reckoning but did have totally decontextualized examples of the tika, we would likely not recognize it as a calendar. Fortunately, our knowledge of the way Mesoamericans reckoned time, although not as complete as it is for Bali, enables us to recognize many of the instruments they used, even if we do not always understand them. One such device may be the pecked-cross circles found in central and southern Mesoamerica. Typically, two or three circles of dots centered on a cross of dots (many are "pecked" into rock) add up 260 or a number referring to some other important calendrical cycle.

These are associated with astronomical alignments, and the numbers of dots are often calendrically significant (Aveni 2005). Some of these pecked crosses mark sequences that intersect more than thirty times, which could model the correspondence of distinct cycles, just like the tika does.

To return to the drawings of the Mesoamerican calendars as a set of intermeshing gears, Aveni (2012:73–89) has determined that in Pre-Columbian written sources time-keeping devices are represented instead as the sides of a square or rectangle or as rectilinear diagrams that were designed to be read as spirals. Furthermore, Aveni's work suggests that in Mesoamerica the calculations that defined distinct temporal cycles would have been based on the directional quadripartition of time and the quincunx rather than the simultaneously occurring weeks of the Balinese system. Nonetheless, I would not reject the use of the gear imagery for our own understanding of the uses of the calendar as a computational tool, especially if we add cog wheels for other cycles beyond the solar year, thirteen numbers, and twenty day names. And this metaphor may not have been completely foreign to Mesoamerican people: the circular pecked crosses are one example (Aveni 2012:81–82). Either way the result was, according to Durán (1971:413), that "the natives thought of the year as a series of many months and days, composed, like a bouquet, of many branches and leaves." In this image, time is a complex interlacing of temporal cycles that give it, as noted by Lansing for the Balinese, a kind of density.

REFERENCES CITED

Aveni, Anthony F.

2002 *Empires of Time: Calendars, Clocks, and Cultures.* University Press of Colorado, Boulder.

2005 Observations on the Pecked Designs and Other Figures Carved on the South Platform of the Temple of the Sun at Teotihuacan. *Journal for the History of Astronomy* 36:31–47.

2011 Maya Numerology. *Cambridge Archaeological Journal* 21(2):187–216.

2012 *Circling the Square: How the Conquest Altered the Shape of Time in Mesoamerica.* American Philosophical Society, Philadelphia.

Bourdieu, Pierre

1977 *Outline of a Theory of Practice.* Cambridge University Press, Cambridge.

Cline, S. L.

1993 *The Book of Tributes: Early Sixteenth-Century Nahuatl Censuses from Morelos.* UCLA Latin American Studies 81. UCLA Latin American Center Publications, Los Angeles.

Covarrubias, Miguel de

1986 [1937] *Island of Bali.* KPI, London.

Cobarruvias, Sebastián de

1977 [1611] *Tesoro de la lengua casellana o española.* Ediciones Turner, Madrid.

Culbert, T. Patrick

1993 *Maya Civilization.* Smithsonian Institution Press, Washington, D.C.

Durán, Diego

1971 *Book of the Gods, Rites, and the Ancient Calendar.* Translated by Fernando Horcasitas and Doris Heyden. University of Oklahoma Press, Norman.

Geertz, Clifford

1966 *Person, Time, and Conduct in Bali: An Essay in Cultural Analysis.* Southeast Asia Program, Cultural Report Series. Yale University, New Haven, Conn.

Gell, Alfred

1992 *The Anthropology of Time: Cultural Constructions of Temporal Maps and Images.* Berg, Oxford.

Hallowell, Irving

1955 *Culture and Experience.* University of Pennsylvania Press, Philadelphia.

Hassig, Ross

2001 *Time, History, and Belief in Aztec and Colonial Mexico.* University of Texas Press, Austin.

Kirchoff, Paul

1952 [1943] Mesoamerica: Its Geographic Limits, Ethnic Composition, and Cultural Characteristics. In *Heritage of Conquest*, edited by Saul Tax, pp. 17–30. Macmillan, New York.

Landa, Diego de

1975 *The Maya: Diego de Landa's Account of the Affairs of Yucatán.* Translated and edited by A. R. Pagden. J. Philip O'Hara, Chicago.

Lansing, J. Stephen

1991 *Priests and Programmers: Technologies of Power in the Engineered Landscape of Bali.* Princeton University Press, Princeton, N.J.

1997 A Thousand Years in Bali. Transcript of a talk given at the University of Michigan, Ann Arbor.

León Portilla, Miguel

1988 [1968] *Time and Reality in the Thought of the Maya.* University of Oklahoma Press, Norman.

Monaghan, John

2002 Los calendarios mesoamericanas como constituciones. In *Procesos de cambio y conceptualizacíon del tiempo: Memoria de la Primera Mesa Redonda de Monte Albán*, edited by Nelly Robles Garcia, pp. 247–254. Instituto Nacional de Antropología e Historia, Mexico City.

Morley, Sylvanus Griswold

1956 [1946] *The Ancient Maya.* Stanford University Press, Stanford, Calif.

Sabloff, Jeremy A.

1990 *The New Archaeology and the Ancient Maya.* Scientific American Library, New York.

Schutz, Alfred

1951 Making Music Together. A Study in Social Relationships. *Social Research* 18(1):76–97.

Tedlock, Barbara

1983 Quichean Time Philosophy. In *Calendars in Mesoamerica and Peru: Native American Computations of Time*, edited by Anthony F. Aveni and Gordon Brotherston, pp. 59–72. BAR International Series 174. B.A.R., Oxford.

Thompson, J. Eric S.

1966 [1954] *The Rise and Fall of Maya Civilization.* University of Oklahoma Press, Norman.

Zorita, Alonso de

1963 [ca. 1569] *Life and Labor in Ancient Mexico: The Brief and Summary Relation of the Lords of New Spain.* Translated by Benajmin Keen. Rutgers University Press, New Brunswick, N.J.

14

An Overview of the Measure and Meaning of Time in Mesoamerica and the Andes

ANTHONY F. AVENI

> It is, perhaps, as irrational to expect a satisfactory penetration of the mystic and emotional aura of the Maya philosophy of time by a creature of twentieth-century Western culture as it is to hope for a balanced, sympathetic, and understanding study of the ecstasy of St. Francis from the pen of a militant atheist of our materialistic age. Our outlooks are too far from those of the Maya and, on top of that terrible handicap, there are so many aspects of the problem which are imperfectly known or completely unknown to us. The atheist student of St. Francis has at his disposal incomparably richer sources than we can ever dream of having.
>
> —J. Eric S. Thompson, "Foreword" to Miguel León Portilla, *Aztec Thought and Culture*

Beginning and Endings

SIR J. ERIC S. THOMPSON'S SOBERING STATEMENT reminds us all that understanding the vagaries of time among indigenous cultures—how time was and is remembered and recorded, how people experienced and cognized time, and what time means or meant to them—is a difficult task. That task is made even more onerous when we realize that even with a wealth of data bearing on time's measure and meaning in the West, we are still confronted by huge gaps in ways of conceiving our own temporality. Clearly, we in the West are different, and Richard Landes's paper, which initiated our symposium, articulated some of the differences. Only in the West, for example, is time perceived as an entity that lies outside of human existence. Indigenous America knew no clocks, nor scalar maps for that matter, but in stark contrast, the measure of time in the West came with a technology unto itself. In the New World, on the other hand, time's currency was the day. Never was it apportioned into minuscule divisions; at most, segments of days might be indicated by pointing to where the sun was situated in the sky. Nor was indigenous time keyed to an apocalyptic endpoint.

Often precisely timed, to judge by the evidence in the liturgical record, as Landes's essay also demonstrates, the reading of prophecy was designed to appeal to the downtrodden, those who were promised a salvatory eternity that would take place "on earth, as it is in heaven." This eternity would commence at the end of the series of nonrepeatable events that we call history, arranged along a linear continuum and terminating in the apocalypse. The unstable, liminal period of interrogation of Antichrist separates the events of this historical world from eternal salvation.

Time's precise measure in the West is rooted not in the realm of science, nor even in that of business and commerce developed in medieval times and which bequeathed us precise mechanical clocks, but in the demands of the religion that preceded it. Specifically, people needed to arrange for the precise times to pray in unison, that the deity might hear their collective voices. Are the roots of Mesoamerican timekeeping so different? The Maya record implies that religion lay at the root of the precise timing of the rituals of reciprocation via sacrifice to the gods that we read about in the codices. Maya monumental inscriptions suggest that accurate day keeping was concerned with certifying dynastic connections with the ancestor gods. Might the same be true in the Andes? The prescriptive offerings to the huacas in and around the Inca capital of Cuzco, along with the precise designation of pilgrimage routes, served both religious and sociopolitical needs.

Setting the Western time clock, that is, establishing time's zero point, began with the sequential ordering of the estimated lengths of the lives of characters in the Bible. Quite like the Maya long count, a true number-based chronology was achieved during the Roman empire, when dates were reckoned *ab urbe condita* or "from the founding of the city." The serial numbering of the years, as we know them in the calendar the Romans handed down to us, did not actually begin until the sixth century, during the reign of Dionysius Exiguus in the Eastern (Christian) Holy Roman Empire. The sequence of years reckoned in the BC/AD format, now termed BCE/CE, was not initiated until more than a thousand years later, in 1627 CE, by the Jesuit astronomer Dionysius Petavius (Denis Pétau). The starting point of the Julian era, 4713 BCE, was fabricated in the sixteenth century by the French philologist Joseph Scaliger, who rolled back three major time cycles, two of them purely astronomical and one sociopolitical, to a fixed point of coincidence. These were the period that includes all possible matching combinations of the days of the week and the first day of the year (twenty-eight years); the period over which a given phase of the moon is returned to the same date of the seasonal year, the so-called Metonic cycle (nineteen years); and the decidedly unnatural fifteen-year cycle of *indiction*, a period during which certain governmental acts took place. First instituted by Constantine the Great in the fourth century CE, the indiction cycle was geared to the imposition of taxes, specifically those paid to troops who were discharged from the Roman army after fifteen years, the maximum length of service permitted by the emperor. This practice led to the keeping of monetary accounts on a fifteen-year cycle.

I have emphasized the Western analog of setting the zero point of deep time as a caveat to those who investigate how such systems were set up in the New World. For example, take the origin of the zero point of the Maya calendar (August 11, 3114 BCE). Like that of its Julian counterpart, its choice may well harbor elusive political vagaries such as those that affected the arrangement of the Christian calendar. At least we know of no astronomical phenomena that might have served to set the Maya zero point, save for the possibility that its seasonal component, August 11, is close to one of the two annual passages of the sun across the zenith, or overhead point, in the general area where the long count was initiated. More likely, the starting point was back-calculated sometime during *bak'tun* seven, the earliest recorded bak'tun, to commemorate the birth of a mythical figure associated with dynastic rulership (cf. Aveni 2009:82–83). We know nothing of the mythic events that might have marked the present or "eral" creation, to use the term employed by Victoria and Harvey Bricker. They tell us that it is most likely that an indivisible pre-eral "eon," or, to apply Alfredo López Austin's terminology, anecumene, going back to minus infinity preceded the present era. The Maya say nothing of a future age, nor do they proffer cataclysms separating major long count overturnings.

For the Aztecs, beginnings and endings of deep time seem to have been characterized quite differently. The celebrated Aztec Calendar Stone, which, as William Barnes shows, is one of a class of dedicatory "sun stones" rife with dynastic metaphors, clearly depicts five world ages called "suns." Various chroniclers have indicated that the one in which we live had a definite starting point; for example, the chronicler Ixtlilxochitl writes: "In the year 5097 after the creation of the world, which was [a year named] 1 *Tecpatl* [1 Flint Knife] . . . all of the Toltec sages . . . assembled in Huehuetlapallan, capital city of their empire, where they dealt with many different things, including the events and calamities that had occurred, the movements of the heavens since the creation of the world, and many other things . . . among which was the adding of a leap year [*bisiesto*] in order to adjust the solar year with the equinox" (Aveni et al. 1988:309). Incidentally, the erroneous statement alluding to indigenous leap years continues to be propagated through the literature.

Regarding deep time in the Andean world, Juan Ossio challenges the prevailing notion that the ages of the world depicted in chronicler Guaman Poma's *El primer nueva corónica y buen gobierno* were derived entirely from European or, in part, copied from Aztec sources. He finds both indigenous elements and indigenous logic embedded in Guaman Poma's letter. Here, I would note especially the way the Christian millennium-based Augustinian chronology conflates with the seminal number five (see Urton 1997) in Ossio's account of Andean world ages. It is also worth reporting that symposium discussant Oswaldo Chinchilla suggested that it might be appropriate for scholars to invert the oft-anticipated question about Western influence on indigenous concepts by asking, to what degree did indigenous concepts of long cycles and ages influence what we read in the chronicles?

The Sense of Time

One result of our symposium was to dispel the oft-quoted perspective mentioned by Stella Nair: If Mesoamerica is all about time, the Andean world is all about space. Mesoamericanists at the symposium were introduced to *pacha*, the Andean word for space-time, variously referred to by presenters in its temporal aspects as a moment or an eternity, even potentiality. But in a symposium commentary, Andeanist Catherine Allen explained that more than space-time, pacha is material, and it contains a moral dimension as well. We might think of it as a world-moment that exists as a complex of interrelated animate beings, a category that would include stones as well as humans. This complex concept, in which the person emerges as partial and relative rather than discrete and individual, reveals the density of time in Andean thought. This sort of personhood also resonates in Tristan Platt's essay as well as Linda Brown's essay on Mesoamerica.

Jalh Dulanto's paper dealt with pacha in the early seventeenth-century Huarochirí Manuscript, in which the voyages of ancestor Cuni Raya Viracocha and his hero descendants are traced. These travel tales segment space-time in the vertical ecology that surrounds the people whose myth/history the manuscript purports to narrate. The segmented qualities of the pacha that apply to the up (highlands) and down (coast) were further reflected in divisions between day/night, wet/dry, past/present, and center/periphery. Once an undifferentiated unity, all things are brought back together in the thrice-told journey of the founding Huarochirí lineage. Here is a creation story that possesses a similar narrative structure to that found in the Quiché Maya Popol Vuh. Thus, as in parts IV and V of the Popul Vuh, the Huarochirí peregrinations take the heroic descendants on a counterclockwise route that incorporates the prominent peripheral villages and mountain peaks visible from Huarochirí.

Comparing these narratives from two continents, one senses the profound effect exerted by the ecosphere on creation mythology. Like that implied in the Huarochirí narrative, the courses of timed contemporary Maya rituals often involve routes that pass from one side of quadripartite space to the next in a counterclockwise fashion, usually from east to north to west to south, ending at the center. This spiral shape of ritual time is also indicated in Late Postclassic Maya codices. Moreover, a similar counterclockwise progression is described in detail in Bernardino de Sahagún's Aztec chronicle (Aveni 1991). It involves rites related to the sacrifice of children at mountain stations arranged about the periphery of the Valley of Mexico, beginning in the north during the month of Atlcahualo in the Aztec calendar.

Stella Nair's study of Inca pacha, expressed architecturally in her "memory palaces," reminds one, if only superficially, of European historian Frances Yates's (1966) study of Renaissance architecture, *The Art of Memory*, especially in the way she analyzes how one utilizes the built environment as a memory device, for example, by associating a corner, an arch, a space between columns, and so forth with an event. One places the mind in various loci that then become identified with related entities to create a mnemonic scheme. A significant difference is that in Nair's account one must think of the relationship between things and people as interactive and reciprocal. As Catherine Allen pointed out, like the *nawales* in Linda Brown's essay, stones testify to the past, and they may even guide one there. Nair demonstrates the way time is articulated via movement, an activity of profound importance in Andean cultures, for example, in the sinuous pilgrimage routes such as that associated with the ritual of Qolluri R'iti (Gow 1974) and the labyrinthine ambulations that likely took place among the geoglyphs on the Pampa de Nasca (Aveni and Silverman 1991; Reinhard 1988; Silverman 1990). One can only own a landscape by moving over it in a prescribed manner.

Drawing on philosophical works on the phenomenology of temporal consciousness, Markus Eberl's "To Put in Order" offers us a theoretical background. Regarding our experience of time, Eberl reads Maya iconography to give us the sense that Maya time is part of a process of putting things in order (*tz'ak*) based on a selection of precepts. He argues that the subjectivization of time is an inescapable feature of the process of perception. Thanks to our ticking clocks and digital watches, we envision a temporal now as a knife-edge moment between past and present. In the Maya world, in contrast, one might best think of it as an extended musical monotone or, as Platt suggests in the opening of his essay on refounding the Andean past, the experience of tasting. Time occupies a succession of moments. Thus, one recalls the initial tone or the first bite throughout its audible, or gustatory, constancy. In Eberl's essay, as well as those of Ossio and López Austin, one acquires a sense of time units as physical persons who enter and leave the normal world.

Eberl's epigraphic discussion of the expression of conflated "now moments"—for example, the glyphic bringing together of green-yellow that appears within the physical frame or cartouche that likens the Maya ruler to the growth stages of the maize plant—reminded me of certain works of Aztec sculpture (cf. Aveni 1992:3–4). For example, tucked away in one corner of the Aztec Hall of Mexico City's National Museum of Anthropology is a jadeite representation of a calabash, or gourd, about a foot across and carved out of a single chunk of stone. One cannot imagine a real-life squash that ever could have looked like this one. The artist rendered the mature, ready-to-eat Aztec staple with its attached flower opened to full bloom. But even a weekend tiller of soil knows that when the flower is fully open and developed, the squash barely will have sprouted. On the other hand, if the squash has fully matured and is ready to pick, its flower will have long since withered and dropped off.

I have seen sculpted representations of maize, that other basic Aztec commodity, executed in similar guise. One image, carved in granite, is that of a serpent, which symbolizes the earth's fertility. On its back are a number of fully matured maize cobs, their husks pulled back and neatly braided. A flowery tassel in full bloom is attached to each cob. Like that of the squash, the maize sculpture depicts two stages of growth that can never happen at the same time. And, like the glyphs Eberl discusses, each of the aspects of the calabash and maize sculptures seems totally faithful to what one sees in a garden at the beginning and at the end of the growing season: the correct number of petals on the squash blossom, the perfectly shaped ribs on the mature squash, the corn tassel that hangs down and blows in the wind just the way it does in the field, and the precisely articulated kernels of maize.

Clearly, these artists were expressing knowledge of the world about them in a way that is unfamiliar to us. In both cases, they conflated different stages of plant metamorphosis to represent a single coherent image. This sort of image, a combination of realities drawn from different time frames and brought together by the human imagination into a composite whole, seems to have held great significance in Mesoamerican temporal thought.

Similar to Nair's "memory places," where the past is experienced via architectural features and space, Brown's study explores the conjunction of materiality and memory via sacred objects and dream spaces. For Brown, the collective memory of time past arrives in the present via the nawales, who both communicate and reciprocate with their human collectors. Unlike the antiques we collect, which for most of us hold nostalgic remembrances of a past gone forever, the bits and pieces of artifacts Brown engages instead are animate. They are embodied by ancestral beings who empower the present. As Victoria Bricker commented during the discussion period, time, at least in the Maya worldview, is an ordering process reflected grammatically by the idea that all actions are either ongoing or completed. There is no present tense. In like manner, Platt shows us how past political events are reconfigured to construct a now event. Like Eberl's example of the ruler of Palenque's "portal," the imagery on the colonial Aymara coat of arms he analyzes collects memories.

Twilights are the interfaces between light and dark. And if you experience time either in a jungle or a forest, you will find that they are the noisiest, most active portions of the diurnal cycle. These are the times when animals scurry about and screech, when flowers open or close rapidly. Especially in tropical latitudes, twilit nows are fleeting. These liminal phases of disorganized time delimit the stark contrast between two radically contrasting optical sensory experiences. Several of the presentations transported us to time's liminal regions, where things and events undergo rapid change. At another level, liminality also characterizes periods of social instability (accompanied by floods, eclipses, and so on), for example, events that punctuate the journeys of Cuni Raya and his hero descendants about the landscape of Huarochirí. Likewise, the liminal time between two structural periods becomes the "Pachacuti" in Ossio's essay.

The liminal zone affecting Platt's Aymara coat of arms relates to the cosmic process of metallogenesis thought to have been initiated by the appearance of Venus. That this planet, unlike all others, usually lies low in the sky in the realm of the clouds during both twilights makes it the ideal causal agent. As Platt points out, the changing aspects of Venus from evening star, when one witnesses it throwing down its light to earth, to morning star, when it announces the arrival of the sun, symbolic of the Inca empire, are transformed into agents that identify the lineages of those who bore the seal on their coats of arms. Through their worship of the sun, these people became the ultimate processors of time.

López Austin's paradigmatic representations of the Mesoamerican space-time of creatures (ecumene) and supernaturals (anecumene) occupy the liminal realm between the two principal aspects of deep time. The zone that separates them is characterized by the earth solidifying, the mountains ceasing to move around, and the gods no longer engaging in transformation. In the Nahuatl myth of creation, the extended now of twilight happens when Nanahuatzin, the "Pimpled One," the underdog or "ugly duckling" of Mexica supernaturals, dives into the New Fire at Teotihuacan, the place where time began, as the Aztecs termed it, and emerges as the Sun of Movement, the initiator of the age in which we live. There is, as well, an extended liminal moment of the bringing together of the contents of Eberl's "portal." Interestingly, Barnes's analysis centers on the 13 Reed entry date, written at the top of the Aztec Sun Stone, which, he aptly demonstrates, happens to occupy the liminal zone between halves of the fifty-two-year calendar round, and the like-in-kind rites associated with change that take place on other 13 Reed dates.

Registering Time

Another important theme that emerged at the symposium was concerned with the study of calendars or how we remember and record time. As in the West, deep or "pre-eral" time, reminiscent of López Austin's anecumene, played a decided role in the complex calculations that appear in the Maya monumental inscriptions and codices.

Like its counterpart in the West, eral time is ordered arithmetically in lengthy chainlike statements. Mesoamerican time consists of the creation of large cycles built up out of smaller ones that are commensurate in the ratio of whole numbers. For example, the 584-day Venus cycle, the 260-day ritual count, and the 365-day year all divide, without remainder, into the 37,960-day length of the Venus Table in the Dresden Codex, which Maya astronomers employed to chart the course of that planet. Not to be lost in the oft-complex calendrical manipulations is the mandate sought by the Maya to divine the venerean omens written in the table in accord with ritual dictates, in this instance imposed by the centrality of the 260-day sacred round that guided the process of precisely fitting the cycles together. The role of pure number in the inscriptions that appear on the wall of a room at the recently excavated Classic Maya site of Xultun, consisting of numbers that are commensurate not only with calendrical periods but also with each other, emerges as starkly Pythagorean (Aveni et al. 2013; Saturno et al. 2012).

To judge from their complexity, one might be led to characterize the whole of Maya calendrics as purely esoteric, especially when one examines the extraordinary long cycles that appear, for example, on Stela 1 of Cobá or in the Serpent Numbers in the Dresden Codex, discussed by Victoria and Harvey Bricker. Barnes characterizes central Mexican calendrics as a divining device to foster the continuity of rulership, its prescriptive ritual details "shouldered" (to utilize Eberl's term) by the very units that compose the calendar. The continuity of dynastic succession is also the driving force in the creation of carved monumental texts.

My essay on astronomy and calendar in the *ceque* system of Cuzco deals with the extent to which precise knowledge of time (and space) is evident in Andean culture. Here, the text is a vast mnemonic scheme, a map of precise space-time that overlays the Inca capital itself. The huacas that make up the radially organized ceques included temples (natural or man-made), intricately carved rock formations, bends in rivers, fields, springs or other natural wells called *puquios*, hills, and even impermanent objects such as trees. Here were places where servants and attendants came to offer the appropriate sacrifices at the proper times in a hierarchical system that represented the various social classes of Cuzco.

Most important in the context of the symposium is that the ceque system is a pacha, that is, an interactive medium comprised of both space and time. I follow R. Tom Zuidema's (1977) interpretation of the ceque system, in which the 328 huacas constitute a precise calendar, the year beginning with the heliacal rise or first appearance of the Pleiades star group following its thirty-seven-day period of invisibility (328 + 37 = 365 days). This event occurs during the lunar cycle leading up to the June solstice, when the Inti Raymi festival was, and still is, celebrated.

The calendar of Bali, as John Monaghan illustrates, is no less complex and precise than that of the Maya or the Inca. Monaghan reminds us that there are cultures other than the West against which we Mesoamericanists and Andeanists might reflect the time systems we study. Religious or dynastic concerns did not always drive the increasing complexity of the registration of time. Benjamin Franklin wrote: "Time is money." Balinese time is "econotime," and it is centered around the stages of development of the rice crop. Monaghan also deals with the moral dimension of time in Mesoamerica. Time exists as a part of social personhood, especially in regard to the importance of one's birthday in determining one's character and destiny.

Heeding my own warning expressed at the outset, I realize, in closing, that it can be all too easy to force connections between Andean and Mesoamerican worldviews and ideology when it comes to so elusive a set of concepts incorporated in various definitions of time. As López Austin reminded us during the discussion, the big question—"what is the cause?—looms at the top. Was there a common origin? Was there contact? Or are the parallels we draw regarding the sensation and registration of time across the two continents the result of universal neurological processes? In this regard, John Justeson, also a participant in the discussion period, noted that the contrast between registering time and experiencing it happens simultaneously—that is, we experience time by registering and unconsciously enacting it whenever we relate events, states, or objects to one another in time. This process of relativizing temporality, he noted, is not about what precedes what, but how near things are with respect to some situationally relevant standard. Languages reflect these parallel ways in which spatial and temporal information is processed in that, to a great extent, locations, distances, and travel in time and space make use of the same locational vocabulary and grammar. So, following the course set by so many previous Pre-Columbian symposia hosted by Dumbarton Oaks, we tread the turf on which we compare cultures cautiously but with continued determination.

REFERENCES CITED

Aveni, Anthony F.

1991 Mapping the Ritual Landscape: Debt Payment to Tlaloc during the Month of Atlcahuálo. In *To Change Place: Aztec Ceremonial Landscapes*, edited by David Carrasco, pp. 58–73. University Press of Colorado, Niwot.

1992 *Conversing with the Planets: How Science and Myth Invented the Cosmos*. Times, New York.

2009 *The End of Time: The Maya Mystery of 2012*. University Press of Colorado, Boulder.

Aveni, Anthony F., Edward Calnek, and Horst Hartung

1988 Myth, Environment, and the Orientation of the Templo Mayor of Tenochtitlan. *American Antiquity* 53(2):287–309.

Aveni, Anthony F., William Saturno, and David Stuart

2013 Astronomical Implications of Maya Hieroglyphic Notations at Xultun. *Journal for the History of Astronomy* 44:1–15.

Aveni, Anthony F., and Helaine Silverman

1991 Between the Lines: Reading the Nazca Markings as Ritual Writ Large. *The Sciences* 31(4):36–42.

Gow, David

1974 Taytacha Qoyllur Ríti. *Allpanchis phuturinqa* 7:49–100.

Reinhard, Johan

1988 *The Nazca Lines*. Editorial Los Piños, Lima.

Saturno, William, David Stuart, Anthony F. Aveni, and Franco Rossi

2012 Ancient Maya Astronomical Tables from Xultun. *Science* 236(6082):714–717.

Silverman, Helaine

1990 The Early Nasca Pilgrimage Center of Cahuachi and the Nazca Lines: Anthropological and Archaeological Perspectives. In *The Lines of Nazca*, edited by Anthony F. Aveni, pp. 207–244. American Philosophical Society, Philadelphia.

Thompson, J. Eric S.

1963 "Foreword" to *Aztec Thought and Culture* by Miguel León Portilla. University of Oklahoma Press, Norman.

Urton, Gary

1997 *The Social Life of Numbers: A Quechua Ontology of Numbers and Philosophy of Arithmetic*. University of Texas Press, Austin.

Yates, Frances

1966 *The Art of Memory*. University of Chicago Press, Chicago.

Zuidema, R. Tom

1977 The Inca Calendar. In *Native American Astronomy*, edited by Anthony F. Aveni, pp. 219–259. University of Texas Press, Austin.

CONTRIBUTORS

Anthony F. Aveni is the Russell Colgate Distinguished University Professor of Astronomy, Anthropology, and Native American Studies at Colgate University. He has more than three hundred research publications to his credit, including three cover articles in *Science* magazine and key works in *American Scientist, The Sciences, American Antiquity, Latin American Antiquity,* and the *Journal of Archaeological Research*. Two of his short pieces have been cited as "notable essays" in the volumes *Best American Essays of 2002* and *Best American Science Writing of 2002*. Aveni's books include *Empires of Time*, on the history of timekeeping; *Conversing with the Planets*, a work that weaves together cosmology, mythology, and the anthropology of ancient cultures by showing how various cultures discovered harmony between their beliefs and their study of the sky; and *The End of Time: The Maya Mystery of 2012. Skywatchers,* his revised and updated version of *Skywatchers of Ancient Mexico*, has served as a basic textbook in cultural astronomy.

William Landon Barnes is an assistant professor of art history at the University of St. Thomas in Saint Paul, Minnesota. His 2009 dissertation, "Icons of Empire: The Art and History of Aztec Royal Presentation," focused on depictions of Aztec (Tenochca Mexica) rulers. His articles have focused on the art and iconography of Pre-Columbian and early colonial central Mexico and include "Partitioning the Parturient: An Exploration of the Aztec Fetishized Female Body" (1997), "Secularizing for Survival: The Changing Depictions of Aztec Rulers in Early Colonial Texts" (2004), and "The Teocalli of Sacred Warfare: An Example of Late Imperial Calendrical Rhetoric in the Court of Moteuczoma II" (forthcoming). He was a past junior fellow in Pre-Columbian Studies at Dumbarton Oaks and is currently working on projects that examine the use of calendrical discourse in central Mexican sculptures and manuscripts of the Late Postclassic and early colonial periods, and the original impetus for the creation of what could be called an Aztec imperial art style.

Harvey M. Bricker, an archaeologist, is Emeritus Professor of Anthropology at Tulane University and Courtesy Professor of Anthropology and Research Associate at the Florida Museum of Natural History at the University of Florida. His early research focused on Palaeolithic sites in southwestern France. Since the early 1980s, he has worked with Victoria R. Bricker in research on Maya archaeoastronomy. Their coauthored book, *Astronomy in the Maya Codices*, was published in 2011 by the American Philosophical Society. Bricker is a fellow of the American Association for the Advancement of Science, and in 1987 he was named "Chevalier dans l'Ordre des Palmes Académiques" by the government of France.

Victoria R. Bricker is Emeritus Professor of Anthropology at Tulane University and Courtesy Professor of Anthropology and Research Associate at the Florida Museum of Natural History at the

University of Florida. Her research interests include Mesoamerican ethnology, ethnohistory, folklore, linguistics, epigraphy, and archaeoastronomy. She is the author of *Ritual Humor in Highland Chiapas*; *The Indian Christ, the Indian King: The Historical Substrate of Maya Myth and Ritual*; and *A Grammar of Mayan Hieroglyphs,* and she is the coauthor, with Harvey M. Bricker, of *Astronomy in the Maya Codices*. She was elected to membership in the National Academy of Sciences in 1991 and to the American Philosophical Society in 2002.

Linda A. Brown is an anthropological archaeologist who works in the Maya area and professorial lecturer in anthropology at The George Washington University. Her research involves both archaeologically focused ethnographic studies as well as archaeological excavations, and she is the codirector of the Say Kah Archaeological Project in Belize. Her research interests include the archaeology of ritual and religion, household archaeology, and the uses of archaeological artifacts and sites by contemporary descendant communities. She is the coeditor, with William Walker, of a special issue of the *Journal of Archaeological Method and Theory* focused on archaeology and animism, and she has published various articles on research at Maya sacred sites.

Jalh Dulanto is an associate professor of archaeology at the Pontificia Universidad Católica del Perú. His research focuses on the archaeology and ethnohistory of death and mortalities and time and temporalities in the pre-Hispanic Andes. He is the author of several articles on Andean archaeology and ethnohistory. He has been a visiting professor at the École des Hautes Études en Sciences Sociales—Paris, Colgate University, and several other universities in the United States. Since 2010, he has been the director of the Paracas Archaeological Program in the Paracas Peninsula on the South Coast of Peru.

Markus Eberl is assistant professor of anthropology at Vanderbilt University. He is an archaeologist and epigrapher who works in the Classic Maya lowlands. He has conducted archaeological fieldwork in Germany, Mexico, and Guatemala, and he is currently director of the Tamarindito Archaeological Project in Guatemala's tropical lowlands. He has also been working with Maya glyphic inscriptions. His publications include *Muerte, entierro y ascensión: Ritos funerarios entre los antiguos mayas* (2005) and *Community and Difference: Change in Late Classic Maya Villages of the Petexbatun Region* (2014).

Richard Landes is professor of history at Boston University. His early work focused on cultural and religious changes in the period around 1000 CE. From 1996 to 2003, he directed the Center for Millennial Studies at Boston University. Recent publications include *Heaven on Earth: The Varieties of the Millennial Experience* (2011) and *The Paranoid Apocalypse: A Hundred-Year Retrospective on the Protocols of the Elders of Zion* (2011). He is currently working on two books: *While God Tarried: Disappointed Millennialism from Jesus to the Peace of God (33–1033)* and *A Medievalist's Guide to the 21st Century*.

Alfredo López Austin is an emeritus researcher at the Instituto de Investigaciones Antropológicas (Universidad Nacional Autónoma de México). He specializes in history and thought of Pre-Columbian Mexico. He has authored *The Human Body and Ideology* (1988); *The Myths of the Opossum* (1993); *The Rabbit on the Face of the Moon* (1996); *Tamoanchan, Tlalocan* (1997); *Mexico's Indigenous Past* (with Leonardo López Luján, 2001); *Dioses del norte, dioses del sur* (with Luis Millones, 2008); and *Monte Sagrado—Templo Mayor* (with Leonardo López Luján, 2009), among others.

John Monaghan received his PhD in anthropology from the University of Pennsylvania in 1986 and has held appointments at a number of universities. He is currently the head of the Anthropology Department at the University of Illinois at Chicago. Monaghan carried out ethnographic fieldwork early in his career among the Maya of highland Guatemala and later came to focus on Mixtec speakers of western Oaxaca, eastern Guerrero, and southern Puebla. He has published books and articles on ethnological and ethnographic topics that combine field observations and archival research. He is currently working on a project that examines how indigenous people responded to the state-imposed liberal and later agrarian reforms in nineteenth- and twentieth-century Mexico.

Stella Nair is an associate professor in the Department of Art History and the Cotsen Institute of Archaeology at the University of California, Los Angeles. Trained as an architect and architectural historian at the University of California, Berkeley, Nair has conducted fieldwork in Bolivia, Mexico, Peru, and the U.S. Midwest, with ongoing projects in the south-central Andes. In support of her research, Nair has received grants and fellowships from the American Philosophical Association, the Center for the Study of the Visual Arts (National Gallery of Art), Dumbarton Oaks, the Fulbright Institute, the Getty Foundation, and the John Carter Brown Library. Recently, Nair and Jean-Pierre Protzen published the monograph *The Stones of Tiahuanaco: A Study of Architecture and Construction* (2013).

Juan M. Ossio A. (PhD, Oxford, Anthropology) is *catedrático principal* at the Pontificia Universidad Católica del Perú. He has authored or coauthored ten books and more than one hundred articles in national and international journals, and he has held appointments at L'École des hautes études in Paris, the University of Virginia, the University of Chicago, and Harvard University. He also served as the first minister of culture for Peru.

Tristan Platt is a retired full professor in anthropology and history at the University of St. Andrews, Scotland. Educated at Oxford and the London School of Economics, he carried out extended fieldwork in Bolivia (Macha), which was followed by many ethnographic and ethnohistorical publications. Author of *Estado boliviano y ayllu andino* (1982), he has written on Andean nineteenth-century economic and anthropological history, held a Guggenheim Fellowship, taught at the École des hautes études en sciences sociales (Paris), and published *Qaraqara-Charka: Mallku, inka y rey en la provincia de Charcas (siglos XV–XVII)* (with Thérèse Bouysse-Cassagne and Olivia Harris, 2006; 2nd ed., 2011). His latest research project concerns quicksilver flows in the nineteenth century.

INDEX

DUMBARTON OAKS PRE-COLUMBIAN SYMPOSIA AND COLLOQUIA

PUBLISHED BY DUMBARTON OAKS RESEARCH LIBRARY AND COLLECTION, WASHINGTON, D.C.

The Dumbarton Oaks Pre-Columbian Symposia and Colloquia series volumes are based on papers presented at scholarly meetings sponsored by the Pre-Columbian Studies program at Dumbarton Oaks. Inaugurated in 1967, these meetings provide a forum for the presentation of advanced research and the exchange of ideas on the art and archaeology of the ancient Americas.

Further information on Dumbarton Oaks Pre-Columbian series and publications can be found at www.doaks.org/publications.

Dumbarton Oaks Conference on the Olmec, edited by Elizabeth P. Benson, 1968

Dumbarton Oaks Conference on Chavín, edited by Elizabeth P. Benson, 1971

The Cult of the Feline, edited by Elizabeth P. Benson, 1972

Mesoamerican Writing Systems, edited by Elizabeth P. Benson, 1973

Death and the Afterlife in Pre-Columbian America, edited by Elizabeth P. Benson, 1975

The Sea in the Pre-Columbian World, edited by Elizabeth P. Benson, 1977

The Junius B. Bird Pre-Columbian Textile Conference, edited by Ann Pollard Rowe, Elizabeth P. Benson, and Anne-Louise Schaffer, 1979

Pre-Columbian Metallurgy of South America, edited by Elizabeth P. Benson, 1979

Mesoamerican Sites and World-Views, edited by Elizabeth P. Benson, 1981

The Art and Iconography of Late Post-Classic Central Mexico, edited by Elizabeth Hill Boone, 1982

Falsifications and Misreconstructions of Pre-Columbian Art, edited by Elizabeth Hill Boone, 1982

Highland-Lowland Interaction in Mesoamerica: Interdisciplinary Approaches, edited by Arthur G. Miller, 1983

Ritual Human Sacrifice in Mesoamerica, edited by Elizabeth Hill Boone, 1984

Painted Architecture and Polychrome Monumental Sculpture in Mesoamerica, edited by Elizabeth Hill Boone, 1985

Early Ceremonial Architecture in the Andes, edited by Christopher B. Donnan, 1985

The Aztec Templo Mayor, edited by Elizabeth Hill Boone, 1986

The Southeast Classic Maya Zone, edited by Elizabeth Hill Boone and Gordon R. Willey, 1988

The Northern Dynasties: Kingship and Statecraft in Chimor, edited by Michael E. Moseley and Alana Cordy-Collins, 1990

Wealth and Hierarchy in the Intermediate Area, edited by Frederick W. Lange, 1992

Art, Ideology, and the City of Teotihuacan, edited by Janet Catherine Berlo, 1992

Latin American Horizons, edited by Don Stephen Rice, 1993

Lowland Maya Civilization in the Eighth Century AD, edited by Jeremy A. Sabloff and John S. Henderson, 1993

Collecting the Pre-Columbian Past, edited by Elizabeth Hill Boone, 1993

Tombs for the Living: Andean Mortuary Practices, edited by Tom D. Dillehay, 1995

Native Traditions in the Postconquest World, edited by Elizabeth Hill Boone and Tom Cummins, 1998

Function and Meaning in Classic Maya Architecture, edited by Stephen D. Houston, 1998

Social Patterns in Pre-Classic Mesoamerica, edited by David C. Grove and Rosemary A. Joyce, 1999

Gender in Pre-Hispanic America, edited by Cecelia F. Klein, 2001

Archaeology of Formative Ecuador, edited by J. Scott Raymond and Richard L. Burger, 2003

Gold and Power in Ancient Costa Rica, Panama, and Colombia, edited by Jeffrey Quilter and John W. Hoopes, 2003

Palaces of the Ancient New World, edited by Susan Toby Evans and Joanne Pillsbury, 2004

A Pre-Columbian World, edited by Jeffrey Quilter and Mary Ellen Miller, 2006

Twin Tollans: Chichén Itzá, Tula, and the Epiclassic to Early Postclassic Mesoamerican World, edited by Jeff Karl Kowalski and Cynthia Kristan-Graham, 2007

Variations in the Expression of Inka Power, edited by Richard L. Burger, Craig Morris, and Ramiro Matos Mendieta, 2007

El Niño, Catastrophism, and Culture Change in Ancient America, edited by Daniel H. Sandweiss and Jeffrey Quilter, 2008

Classic Period Cultural Currents in Southern and Central Veracruz, edited by Philip J. Arnold III and Christopher A. Pool, 2008

The Art of Urbanism: How Mesoamerican Kingdoms Represented Themselves in Architecture and Imagery, edited by William L. Fash and Leonardo López Luján, 2009

New Perspectives on Moche Political Organization, edited by Jeffrey Quilter and Luis Jaime Castillo B., 2010

Astronomers, Scribes, and Priests: Intellectual Interchange between the Northern Maya Lowlands and Highland Mexico in the Late Postclassic Period, edited by Gabrielle Vail and Christine Hernández, 2010

The Place of Stone Monuments: Context, Use, and Meaning in Mesoamerica's Preclassic Transition, edited by Julia Guernsey, John E. Clark, and Barbara Arroyo, 2010

Their Way of Writing: Scripts, Signs, and Pictographies in Pre-Columbian America, edited by Elizabeth Hill Boone and Gary Urton, 2011

Past Presented: Archaeological Illustration and the Ancient Americas, edited by Joanne Pillsbury, 2012

Merchants, Markets, and Exchange in the Pre-Columbian World, edited by Kenneth G. Hirth and Joanne Pillsbury, 2013

Embattled Bodies, Embattled Places: War in Pre-Columbian Mesoamerica and the Andes, edited by Andrew K. Scherer and John W. Verano, 2014

The Measure and Meaning of Time in Mesoamerica and the Andes, edited by Anthony F. Aveni, 2015